SDGs and Textiles

Series Editor

Hafeezullah Memon, International Institute of Silk, Zhejiang Sci-Tech University, Hangzhou, China

The book series "SDGs and Textiles" addresses the strategies to achieve sustainable development goals (SDGs) in the present, past, and future. It presents books about the present and future policies of textile ministries of different countries, and books related to sustainability education around different parts of the world in the textile sector. Moreover, it would welcome the conference proceeding related to SDGs and Textiles. The series would cover books comparing the sustainability and SDGs of different institutions and countries. The individual book volumes in the series are thematic. The goal of each book is to give readers a comprehensive overview of a different area of sustainability in the textile sector. As a collection, the series provides valuable resources to a broad audience in academia, the research community, industry, and anyone looking to expand their knowledge of SDGs and Textiles.

Textiles and life are together – life cannot be separated from textiles as it is the most important need for human beings after food. In 2015, the United Nations General Assembly proposed 17 interlinked global goals to be achieved by 2030. Since then, academia and industry have paid much attention to achieving these goals. Textile found its close relation with almost all of these 17 goals.

SDG 1 - No Poverty: Poverty would never be overcome by a charity only; it is essential to develop people's skills to have a better and wealthy life. Thus, the textile can be considered an excellent discipline to achieve this goal by creating jobs and small and medium businesses.

SDG 2 - Zero Hunger: Through the effective utilization of advanced application of Agrotech Textiles, it is possible to have higher crop yields and save crops from rough weather, unexpected rains, floods, insects, etc.; thus, geotextiles play an essential in achieving this goal of sustainable development.

SDG 3 - Good Health & well-being: There has been much health consciousness after Covid19, and medical textiles assist in getting good health and well-being.

SDG 4 - Learning & Education: Textile or fashion has remained a significant discipline for societies for ages, and there has always remained much to explore in this field. Textile-related universities may play a vital role by offering free access to their education resources, training and spreading information among the locals.

SDG 5 - Gender Equality: The textile sector is one of the industrial sectors that accepted gender equality long ago; in particular, the garment sector has more females than males. Thus, the textile sector has been doing gender equality. Moreover, there has been a recent trend for Gender Neutral Clothing, which need worth studying and may further assist gender equality.

SDG 6 - Clean Water & Sanitation: Textiles could be achieved through filtration, and of course, textile is one of the critical materials for filtration.

SDG 7 - Affordable & Clean Energy: With the recent advancement in material science and engineering, the textile sector has come on the front for, not only by using this clean energy during textile production but also by assisting the production of this clean energy, either in the form of wind turbines blades made of textile composites or by energy harvesting from T-Shirts, etc.

SDG 8 - Decent Work: Recently, there has been much attention that the textile workers are not paid well, labor rights are not cared about, etc.

SDG 9 - Industry and innovation: Textile Industry always follows innovation; the textile companies that do not chase innovation cannot survive in the market.

SDG 10 - Reduced Inequalities: Getting better life and well-being would help reduce inequalities in the textile industry.

SDG 11 - Sustainable Cities: Sustainable Textile Cities through Buildtech and transport textiles.

SDG 12 - Consumption and Production: Textile and garment consumption and production all come under.

SDG 13 - Climate Action: Oekotech or Ecotech Textile, waste management of textiles are upfront to achieve this goal of sustainable development.

SDG 14 - Life Below Water: Mitigating microfiber waste in rivers and oceans may come under the context of it. There has been much attention on this subject after passing the bill at the parliament level of the UK.

SDG 15 - Life on Land: Geotech or Geotextiles studies life on land.

SDG 16 - Peace, Justice, and Strong Institutions: Protective textiles are doing their best to achieve peace, justice, and strong institutions.

SDG 17 - Partnerships for the Goals: The application of textiles to achieve sustainable development goals is only an example. In all textiles sectors, combined efforts of all the goals are essential to achieve true sustainability.

Mohsin Khurshid
Muhammad Aamer Mehmood
Muhammad Kashif
Editors

Microbial Biotechnology in Textile Innovation

Springer

Editors
Mohsin Khurshid
Institute of Microbiology
Government College University Faisalabad
Faisalabad, Pakistan

Muhammad Aamer Mehmood
Department of Bioinformatics and Biotechnology
Government College University Faisalabad
Faisalabad, Pakistan

Muhammad Kashif
Department of Chemistry
Emerson University Multan
Multan, Pakistan

ISSN 2948-1236 ISSN 2948-1244 (electronic)
SDGs and Textiles
ISBN 978-981-95-0346-9 ISBN 978-981-95-0347-6 (eBook)
https://doi.org/10.1007/978-981-95-0347-6

This Springer imprint is published by the registered company Springer Nature Singapore Pte Ltd.
The registered company address is: 152 Beach Road, #21-01/04 Gateway East, Singapore 189721, Singapore

To my sons,
Muhammad Saad Khurshid and Muhammad Anas Khurshid:
May you grow in a world where knowledge lights your path.
(Mohsin Khurshid)

Foreword

The textile industry stands at a pivotal juncture where innovation must intersect with sustainability. The accelerating demand for eco-friendly, efficient, and high-performance materials calls for bold scientific solutions—and microbial biotechnology offers just that. This book, *Microbial Biotechnology for Textile Innovation*, is a timely and comprehensive compilation that brings together state-of-the-art advances and practical applications of microbes in transforming the textile sector.

From microbial enzymes that redefine bioprocessing to bio-based pigments, biosurfactants, biopolymers, and biodegradable textiles, the chapters present a compelling narrative of how nature's smallest organisms can address some of the industry's biggest challenges. The inclusion of emerging areas such as green nanomaterials, biofilms, and textile waste management adds depth and foresight to the volume.

As someone deeply invested in environmental biotechnology, I find this book to be both visionary and practical. It not only advances scientific understanding but also opens pathways for industry transformation. I commend the editors and contributors for their scholarly contributions and commitment to sustainable innovation.

I am confident this book will serve as an essential resource for researchers, students, and industry professionals striving to create a more sustainable future through microbial solutions.

Nicholls State University
Thibodaux, LA, USA

Ramaraj Boopathy

Preface

The textile industry stands at a pivotal moment in its history, compelled to balance innovation with environmental stewardship. Traditional manufacturing methods, which rely heavily on intensive chemical processes and non-renewable materials, have demonstrated their detrimental impact on ecosystems and human health. The demand for sustainable alternatives has never been more urgent.

Microbial Biotechnology in Textile Innovation emerges from this critical need. It is founded on the conviction that microorganisms with their remarkable diversity of biological processes and natural compatibility with the environment hold immense promise in shaping a sustainable future for the textile sector.

This book presents cutting-edge research and practical advancements, bridging the gap between laboratory discoveries and industrial applications. As you progress through the chapters, you will explore foundational principles of microbial biotechnology applied specifically to textiles, as well as advanced developments such as enzymatic bioprocessing, microbial pigments as natural dyes, and the integration of green nanomaterials. This book delves into how microbial enzymes revolutionize textile preparation and finishing, and how microbial pigments derived from bacteria and fungi offer vibrant, eco-friendly alternatives to synthetic dyes. The exploration also encompasses biosurfactants for textile cleaning and finishing, as well as biopolymers and bio-composites sourced from microbes and their practical applications. Furthermore, this book highlights the formation of microbial biofilm on textile surfaces, the latest progress in antimicrobial textiles, and the dynamic field of biodegradable textile development. Critical challenges, including textile waste management through microbial interventions, regulatory frameworks, and ethical considerations, are also discussed. Each topic is presented not merely as scientific advancement but as a step toward building circular systems and fostering deeper environmental responsibility.

This book invites readers including students, researchers, and industry professionals to engage with its content both critically and creatively. The aim is to provide insights and tools that support meaningful contributions to sustainable innovation. The path toward a more resilient and responsible textile industry is a shared journey, and we are pleased to contribute to this essential discourse.

Faisalabad, Pakistan — Mohsin Khurshid
Faisalabad, Pakistan — Muhammad Aamer Mehmood
Multan, Pakistan — Muhammad Kashif

Acknowledgments

We extend our sincere gratitude to the Springer Nature Group for their support in publishing this work, Microbial Biotechnology in Textile Innovation, and for recognizing its relevance to advancing sustainable industrial practices. Their platform has enabled a timely contribution to this critical field of study.

This edited volume stands as a testament to collaborative scholarship. We warmly thank all contributing authors for sharing their expertise, insights, and research through their chapters. Their dedication and intellectual rigor were indispensable in shaping this book, and we deeply appreciate their commitment to its success. We are profoundly grateful to our families and to Government College University Faisalabad for their unwavering support, encouragement, and patience throughout this endeavor. Their understanding and flexibility provided the foundation needed to complete this project.

Finally, we extend special thanks to the Springer Nature publishing team, particularly Vidyaa Shri Krishna Kumar, Nobuko Hirota, Bharath Kumar Dhamodharan, Maggie Guo, and Taeko Sato, alongside the editorial and production staff. Their expertise, guidance, and professionalism were invaluable in bringing this book to fruition.

Mohsin Khurshid
Muhammad Aamer Mehmood
Muhammad Kashif

Contents

Editors and Contributors

About the Editors

Mohsin Khurshid holds the position of Associate Professor at the Institute of Microbiology, Government College University Faisalabad, Pakistan. His academic journey also includes roles as a faculty member at King Saud University in Riyadh, Saudi Arabia, and as a visiting researcher at the Institute of Antibiotics, Fudan University, Shanghai, China. His research is primarily dedicated to three key areas: Microbial biotechnology, antimicrobial resistance, and gut microbiota. Over the years, Dr. Khurshid has made significant contributions to the field, boasting a portfolio of more than 160 published articles in esteemed scientific journals, collectively carrying an impressive impact factor of approximately 450. Additionally, he has authored over 28 book chapters featured in publications by renowned publishers, such as Elsevier and Springer. Dr. Khurshid has been serving as a referee for numerous international journals, including but not limited to *Briefings in Bioinformatics*, *Frontiers in Microbiology*, *Cancer Letters*, *Expert Review of Anti-infective Therapy*, *Microbial Ecology*, and *Lancet*. Furthermore, he holds esteemed positions as an Editorial Board Member for various high-impact journals affiliated with Frontiers Group, Hindawi, and BMC group. Beyond his academic achievements, Dr. Khurshid is recognized as an esteemed speaker, regularly invited to share his insights and expertise at conferences and symposiums.

Muhammad Aamer Mehmood is a Professor and Group Leader of the BioEcoTech Research Cluster at the Department of Bioinformatics and Biotechnology, Government College University Faisalabad, Pakistan. He is also a Founding Executive Board Member and the Secretary-General of the Asian Federation of Biotechnology, Pakistan Chapter. His research team is dedicated to optimizing processes for the sustainable utilization of bio-resources, including wastewater recycling, resource recovery, and the biotransformation of biomass into bioproducts, such as biofertilizers, industrial enzymes, biopolymers, and functional foods. They achieve this by using microalgae, cyanobacteria, and fungi as microbial cell factories within a multiproduct biorefinery framework.

Muhammad Kashif is currently serving as an Assistant Professor of Chemistry and researcher at Emerson University Multan, Pakistan. He obtained his MSc in Chemistry from Bahauddin Zakariya University, Multan, followed by an MPhil in Chemistry with distinction from The Islamia University of Bahawalpur, Pakistan. He subsequently completed his PhD in Pharmaceutical Drug Designing with distinction (*Mención Honorífica*) under the supervision of Prof. Dr. Gildardo Rivera Sánchez at the Laboratory of Pharmaceutical Biotechnology, CBG, IPN, Mexico. One of his notable achievements is a patent registered in Mexico, which stems directly from his PhD work. His research interests encompass drug design, environmental chemistry, biochemistry, enzyme inhibition, medicinal chemistry, molecular docking, and molecular dynamics. Dr. Kashif has received several accolades, including the Best Student Award, Best Thesis Award, and second position for Best Drug Design at the European Federation for Medicinal Chemistry and Chemical Biology (EFMC). Dr. Kashif has authored or co-authored approximately 25 research articles in peer-reviewed journals. His ongoing research is dedicated to advancing medicinal and pharmaceutical chemistry, with a particular focus on computational drug design and the biochemical study of therapeutically important enzymes.

Contributors

Khalil Ahmad Department of Chemistry, Emerson University Multan, Multan, Pakistan

Muhammad Sajjad Ahmad Department of Biotechnology, University of Sargodha, Sargodha, Pakistan

Waqas Ahmad Department of Biotechnology, University of Sargodha, Sargodha, Pakistan

Saeed Ahmed Department of Biological Sciences, National University of Medical Sciences, Rawalpindi, Pakistan

Wajid Ali Department of Biotechnology, University of Sargodha, Sargodha, Pakistan

Muhammad Ashfaq Institute of Chemistry, The Islamia University of Bahawalpur, Bahawalpur, Punjab, Pakistan

Usman Ali Ashfaq Department of Bioinformatics and Biotechnology, Government College University Faisalabad, Faisalabad, Pakistan

Zia Ashraf College of Allied Health Professionals, Government College University Faisalabad, Faisalabad, Pakistan

Sehar Aslam Department of Bioinformatics and Biotechnology, Government College University Faisalabad, Faisalabad, Pakistan

Maryam Ayyaz Atta-ur-Rahman School of Applied Biosciences, National University of Sciences and Technology, Islamabad, Pakistan

Fatima Azeem Centre for Applied Molecular Biology (CAMB), University of the Punjab, Lahore, Pakistan

Heitor B. S. Bento Department of Bioprocess Engineering and Biotechnology, School of Pharmaceutical Sciences, São Paulo State University (UNESP), Araraquara, Brazil

Farah Deeba Department of Biochemistry and Biotechnology, The Women University Multan, Multan, Pakistan

Apridah Cameliawati Djohan Research Center for Applied Microbiology, National Research and Innovation Agency (BRIN), Cibinong Bogor, Indonesia

Hasan Ejaz Department of Clinical Laboratory Sciences, College of Applied Medical Sciences, Jouf University, Sakaka, Saudi Arabia

Marcelo Franco Department of Exact Science, State University of Santa Cruz, Ilheus, Brazil

Saba Gulnaz University Institute of Diet and Nutritional Sciences (UIDNS), The University of Lahore, Lahore, Pakistan

Ali Haider Department of Pharmaceutical and Natural Sciences, Institute of Pharmacy named after A.P. Nelyubin, First Moscow State Medical University named after I.M. Sechenov (Sechenov University), Moscow, Russia

University Institute of Medical Laboratory Technology (UIMLT), The University of Lahore, Lahore, **Pakistan**

Nazim Hussain Centre for Applied Molecular Biology (CAMB), University of the Punjab, Lahore, Pakistan

Azalfah Ibrar Institute of Microbiology, Government College University Faisalabad, Faisalabad, Pakistan

Muhammad Irfan Department of Biotechnology, University of Sargodha, Sargodha, Pakistan

Muhammad Islam Department of Biotechnology, University of Sargodha, Sargodha, Pakistan

Muhammad Israr Department of Biological Sciences, Abasyn University, Islamabad Campus, Islamabad, Pakistan

Muhammad Kashif Department of Chemistry, Emerson University Multan, Multan, Pakistan

Muhammad Arif Khan Department of Chemistry, Emerson University Multan, Multan, Pakistan

Mohsin Khurshid Institute of Microbiology, Government College University Faisalabad, Faisalabad, Pakistan

Ika Rahmatul Layly Research Center for Applied Microbiology, National Research and Innovation Agency (BRIN), Cibinong Bogor, Indonesia

Saher Mahmood Department of Microbiology and Molecular Genetics, The Women University Multan, Multan, Pakistan

Tazeen Malik Centre for Applied Molecular Biology (CAMB), University of the Punjab, Lahore, Pakistan

Manal Department of Biological Sciences, National University of Medical Sciences, Rawalpindi, Pakistan

Ghulam Muhayyudin University Institute of Diet and Nutritional Sciences (UIDNS), The University of Lahore, Lahore, Pakistan

Department of Public Health, I.M. Sechenov First Moscow State Medical University, Moscow, Russia

Fatima Mujahid Institute of Microbiology, Government College University Faisalabad, Faisalabad, Pakistan

Faiqa Munir Department of Biological Sciences, National University of Medical Sciences, Rawalpindi, Pakistan

Samman Munir Department of Bioinformatics and Biotechnology, Government College University Faisalabad, Faisalabad, Pakistan

Ghulam Mustafa Department of Biotechnology, University of Sargodha, Sargodha, Pakistan

Riffat Naseem Department of Biological Sciences, Abasyn University, Islamabad Campus, Islamabad, Pakistan

Khadeeja Nasir Faculty of Allied Health Sciences, Superior University, Lahore, Pakistan

Zeeshan Nawaz Institute of Microbiology, Government College University Faisalabad, Faisalabad, Pakistan

Kinza Niaz Department of Biotechnology, University of Sargodha, Sargodha, Pakistan

Fatima Qasim Department of Biotechnology, University of Sargodha, Sargodha, Pakistan

Muhammad Qasim Department of Bioinformatics and Biotechnology, Government College University Faisalabad, Faisalabad, Pakistan

Laiba Rasheed University Institute of Radiological Sciences and Medical Imaging Technology (UIRSMIT), Lahore, Pakistan

Muhammad Khuram Razzaq Faculty of Agriculture and Veterinary Sciences, Superior University, Lahore, Pakistan

Abdur Rehman Department of Biotechnology, University of Sargodha, Sargodha, Pakistan

Waseem Safdar Department of Chemical and Biomolecular Engineering, Whiting School of Engineering, Johns Hopkins University, Baltimore, MD, USA

Faiza Saleem Department of Biological Sciences, National University of Medical Sciences, Rawalpindi, Pakistan

Muhammad Shafiq Institute of Clinical Pharmacy, Department of Pharmacology, Shantou University Medical College, Shantou, Guangdong, China

Musrat Shaheen Institute of Microbiology, Government College University Faisalabad, Faisalabad, Pakistan

Sabeeka Shahwar Institute of Microbiology, Government College University Faisalabad, Faisalabad, Pakistan

Faluk Shair Department of Chemistry, Emerson University Multan, Multan, Pakistan

Aleena Sumrin Centre for Applied Molecular Biology (CAMB), University of the Punjab, Lahore, Pakistan

Ahsen Taqveem School of Allied Health Sciences, Faculty of Sciences, Multan University of Science and Technology, Multan, Pakistan

Swebba Waheed Institute of Microbiology, Government College University Faisalabad, Faisalabad, Pakistan

Memona Yasmin Knowledge Research and Innovation Centre (KRIC), Lahore, Punjab, Pakistan

Muhammad Asif Zahoor Institute of Microbiology, Government College University Faisalabad, Faisalabad, Pakistan

Maryam Zain Department of Biochemistry and Biotechnology, The Women University, Multan, Multan, Pakistan

Chapter 1
Microbial Biotechnology in Textile Industry: An Introduction

Ali Haider, Muhammad Israr, Riffat Naseem, Laiba Rasheed, Saba Gulnaz, and Ghulam Muhayyudin

Abstract The integration of microbial biotechnology into diverse industrial processes particularly in the textile industry has revolutionized traditional manufacturing practices by offering sustainable, cost-effective, and eco-friendly alternatives. Microbial enzymes, biopigments, and bio-based materials are increasingly replacing hazardous chemicals in critical textile processes such as fiber treatment, dyeing, and finishing, thereby significantly reducing energy consumption, waste generation, and environmental pollution. Furthermore, advances in genetic engineering and synthetic biology have expanded the capabilities and efficiency of microbial systems for textile applications, enabling innovations such as the development of antimicrobial fabrics, biofunctionalized textiles, and sustainable bioremediation strategies for wastewater treatment. Despite these advancements, challenges persist,

A. Haider (✉)
Department of Pharmaceutical and Natural Sciences, Institute of Pharmacy named after A.P. Nelyubin, First Moscow State Medical University named after I.M. Sechenov (Sechenov University), Moscow, Russia

University Institute of Medical Laboratory Technology (UIMLT), The University of Lahore, Lahore, Pakistan

M. Israr · R. Naseem
Department of Biological Sciences, Abasyn University, Islamabad Campus, Islamabad, Pakistan

L. Rasheed
University Institute of Radiological Sciences and Medical Imaging Technology (UIRSMIT), Lahore, Pakistan

S. Gulnaz
University Institute of Diet and Nutritional Sciences (UIDNS), The University of Lahore, Lahore, Pakistan

G. Muhayyudin
University Institute of Diet and Nutritional Sciences (UIDNS), The University of Lahore, Lahore, Pakistan

Department of Public Health, I.M. Sechenov First Moscow State Medical University, Moscow, Russia

M. Khurshid et al. (eds.), *Microbial Biotechnology in Textile Innovation*, SDGs and Textiles, https://doi.org/10.1007/978-981-95-0347-6_1

including scalability issues, regulatory hurdles, and integration challenges with existing industrial infrastructure. Nevertheless, microbial biotechnology continues to pave the way for transformative innovations, including circular economy models, biodegradable fabric production, and smart textiles embedded with microbial sensors. This chapter delves into the intersection of microbial biotechnology and the textile sector, examining its historical evolution, pivotal microbial species and consortia, groundbreaking applications across the textile value chain, and emerging trends poised to redefine sustainable textile innovation. By bridging scientific research and industrial practice, this exploration underscores the transformative potential of microbial solutions in fostering a greener, more resilient textile industry.

Keywords Microbial biotechnology · Textile industry · Enzymatic processing · Antimicrobial textiles

1.1 Introduction to Microbial Biotechnology

Microbial biotechnology is a multidisciplinary branch of science that focuses on harnessing the ability of microorganisms to solve pressing challenges in health, agriculture, industry, and environmental sustainability. By integrating expertise from microbiology, genetic engineering, systems biology, and bioprocess engineering, researchers collaborate to leverage microbial capabilities for developing novel, eco-efficient solutions (Timmis and Hallsworth 2024). Microorganisms, owing to their near-ubiquitous presence and remarkable metabolic diversity, represent an unparalleled reservoir of biochemical and genetic potential. These characteristics enable them to perform diverse functional roles, such as depolymerizing complex organic matter, synthesizing valuable products (e.g., pharmaceuticals, biofuels, and biodegradable plastics), and remediating polluted ecosystems.

As global emphasis on sustainability intensifies, microbial biotechnology has emerged as a critical enabler of technologies that reduce environmental footprints while enhancing productivity across industrial sectors (Ahmed 2022). In particular, the textile industry has begun to capitalize on microbial innovations such as enzymatic fiber processing and bio-based dye synthesis to replace resource-intensive, chemically driven practices. This shift aligns with broader efforts to transition toward circular bioeconomies, where waste minimization and renewable resource utilization are prioritized. By bridging fundamental microbial research with applied industrial needs, microbial biotechnology continues to redefine traditional manufacturing paradigms, offering scalable, environmentally sound alternatives to conventional methods.

1.2 Key Components of Microbial Biotechnology

1.2.1 Microbial Diversity and Functions

Microbial biotechnology leverages the metabolic versatility of bacteria, fungi, yeast, and viruses to address diverse challenges. These organisms exhibit remarkable adaptability, enabling their deployment across a wide spectrum of biotechnological applications. For instance, model microorganisms such as *Escherichia coli* and *Saccharomyces cerevisiae* are extensively utilized in industrial fermentation processes, while extremophiles organisms thriving in extreme environments are harnessed for niche applications like bio-mining, bioremediation of toxic substrates, or synthesis of stress-resistant enzymes (De Sousa et al. 2018; Navale et al. 2021). This functional diversity underscores the potential of microbial systems to replace conventional, resource-intensive industrial practices.

1.2.2 Advances in Genetic Engineering

The advent of CRISPR-Cas systems has revolutionized microbial genome editing, enabling precise modifications to enhance productivity, optimize metabolic pathways, or confer novel functionalities such as toxin resistance or substrate specificity (Parsaeimehr et al. 2022). Building on these tools, synthetic biology provides frameworks for designing entirely artificial genetic circuits and biosynthetic pathways, facilitating the synthesis of complex biomolecules ranging from biodegradable polymers to high-value pharmaceuticals with unprecedented efficiency (Lawson 2021). These advancements bridge the gap between laboratory research and scalable industrial applications.

1.2.3 Synthetic Microbial Communities

Traditional reliance on single-strain microbial systems is increasingly being supplanted by engineered microbial consortia, which mimic the complexity of natural microbial communities while being tailored for specific tasks. These multispecies systems excel in applications requiring coordinated metabolic activity, such as nutrient cycling in bioreactors, degradation of mixed pollutants in wastewater, or cooperative synthesis of biofuels (Mccarty and Ledesma-Amaro 2019). By distributing metabolic workloads among specialized strains, consortia enhance process stability and efficiency in output.

1.2.4 Bioprocess Optimization

Innovations in High-Throughput Screening (HTS) and automated cultivation systems have accelerated the selection of microbial strains with desirable phenotypes, such as thermotolerance or high enzyme yield. These technologies drastically reduce development timelines and improve scalability, enabling rapid transition from laboratory prototypes to industrial-scale production (Kehe et al. 2019). Coupled with advanced bioreactor designs and real-time monitoring, such tools ensure consistent product quality and resource efficiency in large-scale biomanufacturing.

1.3 Applications for Microbial Biotechnology

1.3.1 Industrial Bioprocessing

Microbial cell factories are widely employed in the synthesis of enzymes, organic acids, antibiotics, and biofuels. These processes are inherently less energy-intensive and generate fewer waste byproducts compared to conventional chemical synthesis methods. Notable examples include the production of insulin through genetically engineered *E. coli* and the fermentation of bioethanol using *S. cerevisiae*, both of which exemplify how microbial systems offer sustainable alternatives to traditional industrial practices (Nikel and Mattanovich 2021; Mariyes et al. 2024). By leveraging microbial metabolism, industries can achieve higher efficiency while minimizing environmental impact.

1.3.2 Environmental Applications

Microbial biotechnology plays a pivotal role in addressing environmental challenges through processes such as bioremediation, waste valorization, and detoxification of pollutants. For instance, extremophiles microorganisms thriving in extreme environments are especially utilized to degrade hazardous chemicals in contaminated soils, while engineered microbial consortia break down plastic waste into non-toxic compounds (Nikel and Mattanovich 2021). These applications not only mitigate ecological damage but also contribute to the development of circular economy models by converting waste into reusable resources.

1.3.3 Healthcare Innovations

Microbes are indispensable in modern healthcare, underpinning the production of life-saving pharmaceuticals such as antibiotics (e.g., penicillin derived from *Penicillium notatum*), vaccines, and therapeutic proteins. Advances in genetic engineering have enabled microbial systems to synthesize complex biologics, including monoclonal antibodies and growth hormones, with high precision and scalability (De Sousa et al. 2018; Kapoor et al. 2020). For example, recombinant yeast strains are employed to produce hepatitis B vaccines, while *E. coli* serves as a workhorse for insulin and interferon production. These innovations highlight the synergy between microbial biotechnology and medical advancement.

1.3.4 Agricultural Benefits

In agriculture, microbial biotechnology reduces reliance on chemical inputs by enhancing soil fertility and crop resilience through biofertilizers and biopesticides. Symbiotic microorganisms such as *Rhizobium* spp. fix atmospheric nitrogen, enriching soil nutrients and promoting plant growth, while *Bacillus thuringiensis* produces insecticidal proteins that protect crops without harming non-target species (Singh 2019). Additionally, mycorrhizal fungi improve water and nutrient uptake in plants, further reducing the need for synthetic fertilizers. These microbial solutions not only boost agricultural productivity but also align with sustainable farming practices.

1.4 Emerging Trends and Innovations

1.4.1 Omics Technologies

Advances in genomic, proteomic, and metabolomic studies are revolutionizing our understanding of microbial metabolism, functional diversity, and interspecies interactions. By mapping microbial genomes, profiling protein expression, and analyzing metabolic networks, researchers gain actionable insights into strain optimization and consortium design. These data-driven approaches enable the engineering of high-efficiency microbial strains and synergistic consortia tailored for specific industrial or environmental applications (De Sousa et al. 2018). For instance, metagenomic sequencing has identified novel extremozymes from unculturable microbes, expanding the toolkit for sustainable biomanufacturing.

1.4.2 Integration with AI and Automation

The convergence of machine learning, robotics, and microbial biotechnology is transforming strain development and bioprocess optimization. Machine learning algorithms analyze vast datasets to predict microbial behavior under varying physicochemical conditions, enabling rapid identification of optimal growth parameters or metabolic pathways. Concurrently, robotic platforms automate high-throughput screening, accelerating the selection of strains with desired traits such as thermostability or substrate specificity (Kehe et al. 2019). This integration reduces trial-and-error experimentation, shortening the R&D cycle and enhancing scalability for industrial deployment.

1.4.3 Climate Change Mitigation

Microbial systems are increasingly harnessed to address pressing environmental challenges, including carbon sequestration, plastic degradation, and renewable energy production. Engineered cyanobacteria, for example, utilize photosynthetic pathways to convert atmospheric CO_2 and sunlight directly into biofuels, offering a carbon-neutral alternative to fossil fuels (Kehe et al. 2019; Agarwal et al. 2022). Similarly, synthetic microbial consortia are being deployed to degrade recalcitrant plastics like polyethylene terephthalate (PET) into biodegradable monomers. These innovations not only mitigate greenhouse gas emissions but also align with global sustainability goals by transforming waste into valuable resources.

1.5 Challenges and Future Directions

While microbial biotechnology holds immense promises for sustainable innovation, several critical challenges must be addressed to unlock its full potential.

Transitioning laboratory-scale microbial processes to industrial applications remains a major hurdle. Scaling up requires substantial optimization of growth conditions, metabolic pathway balancing, and significant infrastructure investments to maintain efficiency and cost-effectiveness. For instance, fermentative biofuel production often faces challenges in achieving consistent yields at large volumes due to microbial sensitivity to industrial bioreactor conditions (Lawson 2021).

The deployment of genetically modified organisms (GMOs) in biotechnology is constrained by stringent regulatory frameworks designed to address biosafety and ethical concerns. Regulatory approval processes, particularly in regions like the European Union, can be protracted and complex, delaying the commercialization of microbial innovations such as engineered probiotics or bioremediation consortia (Woźniak et al. 2020).

Designing and maintaining stable synthetic microbial consortia for long-term applications is fraught with technical challenges. Unpredictable interspecies interactions, competition for resources, and sensitivity to environmental fluctuations (e.g., pH, temperature) can destabilize consortia, reducing their efficacy in applications like wastewater treatment or soil remediation (Mccarty and Ledesma-Amaro 2019).

To overcome these barriers, future research will likely focus on integrating cutting-edge tools such as synthetic biology, multi-omics technologies, and AI-driven predictive modeling. For example, machine learning algorithms could optimize microbial consortia by simulating interactions under dynamic conditions, while CRISPR-based genome editing might enhance strain robustness for industrial settings. These advancements aim to expand the scope of microbial biotechnology to address global challenges, including food security (e.g., microbial biofertilizers for drought-resistant crops) and climate change mitigation (e.g., carbon-capturing cyanobacteria). By bridging gaps between innovation, regulation, and scalability, microbial biotechnology can transition from experimental promise to real-world impact.

1.6 Overview of the Textile Industry

The textile industry, one of the oldest and most economically vital sectors globally, plays a pivotal role in job creation and development across nations. It spans the production of natural fibers like cotton, wool, and silk to synthetic alternatives such as polyester and nylon, serving diverse applications from fashion to technical textiles. Despite its economic significance, the industry faces mounting criticism for its resource-intensive practices and environmental footprint. Textile manufacturing involves multiple stages: fiber preparation, spinning, weaving or knitting, dyeing and finishing, and garment production. Conventional methods rely heavily on water, energy, and hazardous chemicals, contributing to pollution and resource depletion. For instance, natural fiber processing such as cleaning cotton or wool consumes vast volumes of water and detergents, often generating contaminated runoff. Synthetic fibers like polyester, derived from petroleum through energy-intensive polymerization, further exacerbate carbon emissions and fossil fuel dependence (Sharif et al. 2020). Subsequent stages, such as weaving and knitting, demand significant electrical energy to power machinery, while dyeing which is the most environmentally taxing phase uses toxic azo dyes, heavy metals, and formaldehyde-based fixatives, releasing carcinogenic wastewater that threatens aquatic ecosystems (Shukla et al. 2021). Finishing processes, including waterproofing or softening, introduce persistent pollutants like perfluorinated compounds. Regulatory pressures and sustainability imperatives are now driving innovations such as bio-based dyes and closed-loop water systems, though scalability and cost barriers persist. The structure of industry encompassing spinning, weaving, processing, and apparel

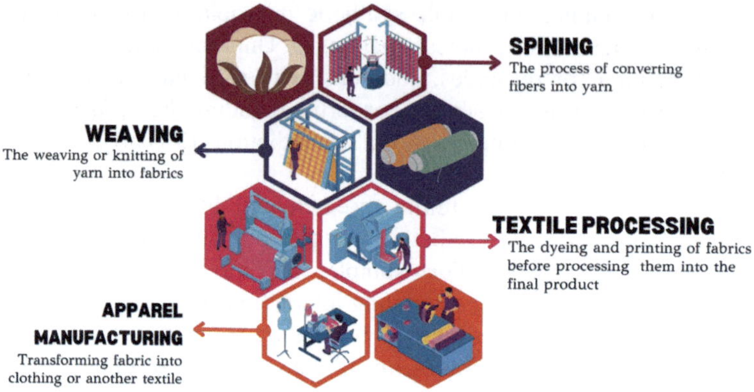

Fig. 1.1 The flow chart of the textile manufacturing process (Jasińska et al. 2024)

manufacturing (Fig. 1.1) (Jasińska et al. 2024), underscores both its complexity and the urgent need for systemic reforms to align with global sustainability goals.

1.7 Challenges in Conventional Textile Manufacturing

As a cornerstone of global industry, the textile sector continues to face pressing challenges tied to environmental degradation, unsustainable resource consumption, and inadequate waste management. Despite its economic importance, the reliance of industry on water-intensive processes, toxic chemicals, and non-renewable materials undermines ecological sustainability. For instance, the textile industry accounts for approximately 20% of global freshwater pollution and 10% of carbon emissions, with synthetic fiber production and dyeing processes being major contributors (Mazotto et al. 2021; Ahmed 2022).

1.7.1 Water Consumption and Pollution

Textile manufacturing is among the largest consumers of freshwater globally, with dyeing and finishing alone requiring up to 200 liters of water per kilogram of fabric (Mazotto et al. 2021). The chemicals used in these stages including heavy metals (e.g., lead, cadmium), formaldehyde, and perfluorinated compounds (PFCs) which leach into water bodies, disrupting aquatic ecosystems and posing risks to human health (Ahmed 2022). Untreated effluent containing dyes, salts, and organic pollutants elevates the chemical oxygen demand (COD) and biochemical oxygen demand (BOD) of receiving waters, depleting dissolved oxygen and creating dead zones in aquatic habitats (Kumar et al. 2021).

1.7.2 Energy Intensity and Greenhouse Gas Emissions

Conventional textile production, particularly synthetic fiber manufacturing, is highly energy intensive. For example, polyester that are derived from fossil fuels can generates ~5.5 kg of CO_2 per kilogram of fabric during production (Mazotto et al. 2021). The energy demand for spinning, weaving, and finishing processes further exacerbates the carbon footprint of industry and relying predominantly on non-renewable energy sources.

1.7.3 Solid Waste and Microplastic Pollution

The industry generates substantial solid waste, including fabric scraps, defective garments, and post-consumer textiles, with 92 million tons of textile waste landfilled annually due to fast fashion trends (Yalcin-Enis et al. 2019; Ahmed 2022). Synthetic textiles like polyester and nylon also shed microplastic fibers during washing, contaminating waterways and entering the food chain, where they pose long-term ecological and health risks (Mazotto et al. 2021).

1.7.4 Recycling Limitations

While recycling offers a pathway to mitigate waste, current methods face significant barriers. Mechanical recycling, which shreds and reprocesses textiles into fibers, often degrades material quality. The open-loop system downcycles fabrics into lower-value products like insulation or industrial wipes, which ultimately end up in landfills (Wang and Salmon 2022) (Fig. 1.2). In contrast, closed-loop recycling aims to reprocess textiles into equivalent products but struggles to preserve fiber integrity, functionality, and aesthetics (Tang 2023). Both approaches remain constrained by

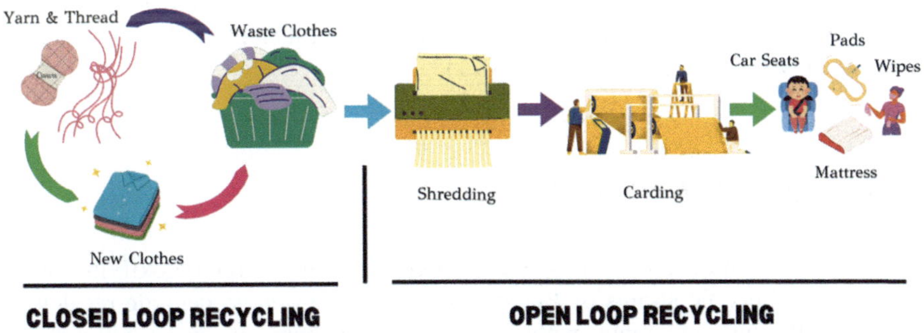

Fig. 1.2 Closed-loop and open-loop mechanical recycling of textiles

technical and economic challenges, limiting their scalability and environmental benefits.

1.8 Microbial Biotechnology and Innovative Solutions

To address the environmental and operational challenges of conventional textile manufacturing, microbial biotechnology offers groundbreaking innovations that prioritize sustainability, efficiency, and circularity. By leveraging the metabolic versatility of microorganisms, these solutions reduce reliance on hazardous chemicals, minimize waste, and lower energy consumption, positioning microbial systems as transformative tools for the textile industry.

1.8.1 Enzymatic Processes

Microbial enzymes such as cellulases, pectinases, and laccases are replacing toxic chemicals in fiber preparation, dyeing, and finishing. For example, cellulases are used in bio-polishing cotton to remove surface fibrils, while laccases decolorize textile effluents by breaking down synthetic dyes. These enzymes operate under mild temperature and pH conditions, reducing energy demands and generating biodegradable byproducts (Shukla et al. 2021). Their specificity enhances process efficiency, enabling targeted modifications without damaging fibers.

1.8.2 Biodegradable Dyes

Microbial pigments derived from fungi (e.g., *Aspergillus* spp.) and bacteria (e.g., *Streptomyces* spp.) provide sustainable alternatives to synthetic dyes. For instance, bacterial indigo produced by engineered *E. coli* eliminates the need for harmful reducing agents used in conventional indigo dyeing. These bio-based dyes require fewer chemical additives and lower temperatures, significantly reducing water and energy consumption (Mazotto et al. 2021).

1.8.3 Effluent Treatment

Microbial bioremediation offers a cost-effective solution for detoxifying textile wastewater. Specific strains of *Pseudomonas* and *Shewanella* degrade recalcitrant azo dyes through enzymatic pathways, converting them into non-toxic aromatic amines. Additionally, bioaugmentation with metal-resistant microbes like *Bacillus subtilis* sequesters heavy metals (e.g., chromium, lead) via biosorption, reducing their bioavailability in aquatic ecosystems (Kumar et al. 2021).

1.8.4 Biofabrication

Advances in microbial biotechnology enable the synthesis of novel biomaterials with superior properties. Bacterial cellulose, produced by *Komagataeibacter xylinus*, forms lightweight, biodegradable fabrics with high tensile strength, ideal for medical textiles. Similarly, recombinant spider silk proteins, expressed in *E. coli* or yeast, mimic natural elasticity and durability of silk while being fully compostable (Mazotto et al. 2021).

1.8.5 Recycling Innovations

Microbial systems are revolutionizing textile waste management. Enzymatic hydrolysis using fungal cellulases converts cotton waste into fermentable sugars for bioethanol production, while bacterial polyesterases (e.g., PETase from *Ideonella sakaiensis*) depolymerize polyester fibers into reusable monomers (Shukla et al. 2021). These processes align with circular economy principles by transforming waste into valuable feedstocks.

Overall, the textile industry faces urgent sustainability challenges linked to water pollution, waste generation, and carbon emissions. Microbial biotechnology addresses these issues through innovations such as enzyme-driven processes, bio-based dyes, and advanced recycling systems. By replacing hazardous chemicals with biodegradable alternatives, detoxifying effluents, and valorizing waste, microbial solutions not only reduce environmental impacts but also enhance resource efficiency. These advancements underscore the potential of biotechnology to redefine textile manufacturing, fostering a transition toward a circular, low-carbon economy.

1.9 Intersection of Microbial Biotechnology and Textiles

Microbial biotechnology has emerged as a transformative force in the textile industry, addressing long-standing challenges while unlocking novel opportunities for sustainable innovation. This interdisciplinary synergy leverages microbial enzymes, biopigments, and bioactive compounds to replace or enhance conventional textile processes, fostering eco-efficient production, cost-effective bioprocessing, and environmentally benign products. By integrating microbiology with textile engineering, the sector is reimagining traditional practices through bio-based solutions that prioritize circularity and resource efficiency. This section explores the historical evolution of microbial applications in textiles, key microbial groups driving innovation, and the revolutionary potential of this convergence.

Recent advancements in biohybrid materials systems integrating living microorganisms or biomolecules with textile matrices have expanded the frontiers of textile functionality. These materials hold promises for applications in biomedical engineering, environmental biosensing, architecture, and pollutant remediation. For instance, textile-microbial hybrids embed bacteria or fungi within fibrous scaffolds, enabling dynamic interactions such as pollutant degradation or responsive structural adaptations. A groundbreaking interdisciplinary study merging artistic design and scientific inquiry demonstrated how textile fibers can act as microbial highways, facilitating bacterial locomotion via aqueous layers surrounding threads (Sherry et al. 2023). Building on the discovery of bacterial movement along fungal mycelial networks (termed "fungal highways"), researchers investigated microbial translocation across diverse fiber types.

Using *Pseudomonas putida* as a model motile bacterium, confocal microscopy revealed that bacterial cells traverse liquid films coating synthetic fibers (polyester, nylon) and plant-based linen, but not proteinaceous fibers like silk or wool. This suggests that fiber hydrophobicity, surface chemistry, and water retention capacity critically influence microbial mobility (Fig. 1.3). Remarkably, bacterial translocation persisted even in the presence of crude oil which is a complex pollutant that highlights the potential applications of microbes in oil spill bioremediation.

In a parallel innovation, fungal mycelium (*Pleurotus ostreatus*) was cultivated within knitted textile structures to create living composites. These biohybrids

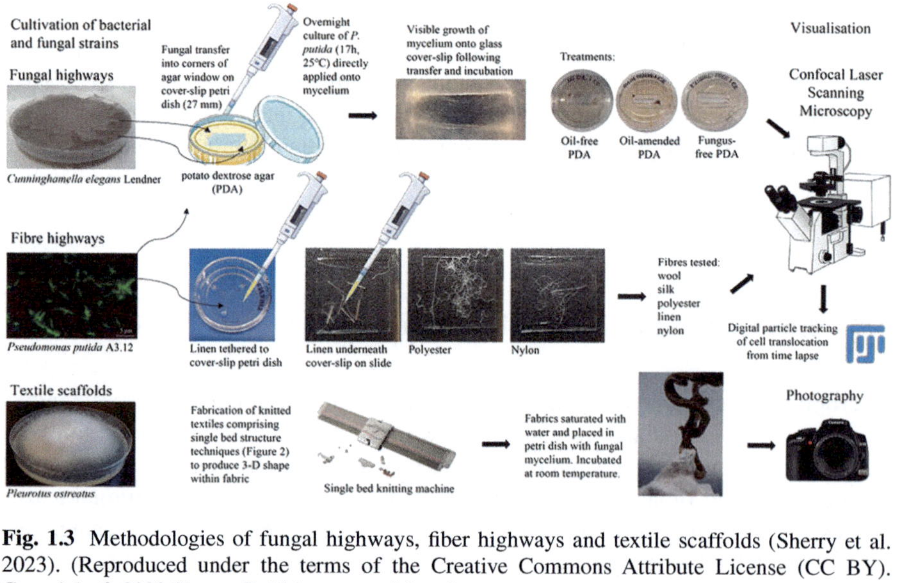

Fig. 1.3 Methodologies of fungal highways, fiber highways and textile scaffolds (Sherry et al. 2023). (Reproduced under the terms of the Creative Commons Attribute License (CC BY). Copyright © 2023 Sherry, Dell'Agnese and Scott)

retained the ability to dynamically alter their form in response to environmental stimuli, exemplifying the potential of textiles as scaffolds for functional microbial communities. The prototype *Ebb&Flow*, crafted from locally sourced UK wool, demonstrated dual functionality: (1) absorption of hydrocarbon pollutants by wool fibers and (2) guided microbial movement along fiber pathways to degrade contaminants (Sherry et al. 2023).

The intersection of microbial biotechnology and textiles is redefining the ecological and functional paradigms of industry. From enzymatic fiber processing to biohybrid materials capable of environmental remediation, microbial systems offer scalable solutions to reduce chemical use, waste, and carbon footprints. By harnessing microbial diversity and interdisciplinary collaboration, this synergy paves the way for smart textiles, biodegradable fabrics, and circular production models, positioning biotechnology as a cornerstone of sustainable textile innovation.

1.10 Key Microbial Groups in Textile Applications

The integration of microbial biotechnology into the textile industry relies on harnessing the metabolic versatility of specific microbial groups, each contributing uniquely to sustainable processes. Bacteria are highly adaptable microorganisms and are indispensable due to their ability to produce enzymes and bioactive compounds. For instance, species such as *Bacillus subtilis* and *Bacillus licheniformis* secrete amylases, enzymes critical for the desizing process in cotton production. Desizing involves removing starch-based coatings from raw cotton fibers to prepare them for dyeing, a step traditionally reliant on harsh chemicals. By replacing these chemicals with bacterial enzymes, the process becomes eco-friendly, reducing water pollution and energy consumption (Ahmed 2022). Similarly, *Pseudomonas aeruginosa* synthesizes rhamnolipids which are a class of biosurfactants that impart oil-resistant properties to fabrics. These biodegradable surfactants are used in fabric finishing to enhance stain resistance, offering a sustainable alternative to synthetic coatings (Ahmed 2022). Additionally, bacteria like *Shewanella* spp. and genetically modified *E. coli* play pivotal roles in wastewater treatment. These organisms utilize enzymes such as azoreductases and oxidoreductases to break down toxic azo dyes which are quite common in textile effluents. This bioremediation capability significantly reduces the environmental toxicity of dye-laden wastewater, addressing one of the most persistent challenges of the industry (Ismail et al. 2019; Sharif et al. 2020).

Fungi, particularly filamentous species, are equally transformative due to their robust enzymatic systems and pigment-producing capabilities. *Aspergillus niger* and *Trichoderma* spp. are prominent examples, secreting cellulases and pectinases that revolutionize fabric finishing. These enzymes hydrolyze microfibrils on cotton surfaces during bio-polishing, a process that replaces mechanical abrasion. By enzymatically smoothing fabric surfaces, bio-polishing reduces pilling (the formation of small fabric balls) and enhances softness, all while operating under mild conditions that conserve energy and water (Gmoser et al. 2017; Sharif et al. 2020). Beyond enzymes, fungi like *Monascus purpureus* and *Penicillium* spp. are harnessed for bio-based pigment production. These fungi synthesize vibrant red, yellow, and orange pigments through natural fermentation, offering biodegradable alternatives to synthetic dyes. Unlike conventional dyes, fungal pigments do not require toxic mordants (fixing agents) and degrade harmlessly in the environment, drastically reducing water pollution and health risks for workers (Mazotto et al. 2021).

Yeasts, particularly *S. cerevisiae* and *Pichia pastoris*, are prized for their genetic tractability and capacity to produce high-value compounds. *S. cerevisiae* has been genetically engineered to synthesize carotenoids, such as lycopene and β-carotene, which serve as natural dyes. These pigments provide vivid colors without the ecological drawbacks of synthetic dyes, such as heavy metal contamination (Branduardi and Porro 2012). Meanwhile, *Pichia pastoris* is engineered to express antimicrobial peptides and enzymes like laccases. These compounds are integrated into fabric finishes to inhibit pathogen growth, enhancing the functionality of medical textiles and reducing reliance on chemical antimicrobial agents (Branduardi and Porro 2012).

The collaborative potential of microbial consortia represents a frontier in textile biotechnology. These consortia mimic natural ecosystems, where diverse microorganisms synergistically perform complex tasks. For example, a consortium combining *Pseudomonas Bacillus* and *Aspergillus* can simultaneously degrade textile dyes, metabolize organic pollutants, and immobilize heavy metals in wastewater. *Pseudomonas* degrades azo dyes via enzymatic pathways, *Bacillus* neutralizes toxic metals through biosorption, and *Aspergillus* breaks down organic sludge, collectively enhancing bioremediation efficiency (Eltarahony et al. 2021; Ahmed 2022). Such consortia exemplify the power of microbial teamwork, enabling holistic solutions to multifaceted challenges like effluent treatment.

From bacterial enzymes that replace toxic chemicals to fungal pigments that eliminate synthetic dyes, microbial groups are redefining textile manufacturing. Their applications span the entire production chain—fiber preparation, dyeing, finishing, and waste management—demonstrating their versatility. As genetic engineering and synthetic biology advance, the potential to tailor microbial strains for specialized tasks (e.g., hyper-efficient dye degradation or self-healing textiles) grows exponentially. By bridging biology and industry, these microorganisms not only mitigate environmental harm but also pave the way for innovations such as biodegradable smart textiles and closed-loop production systems. Their integration underscores a paradigm shift toward a circular, sustainable textile economy, where microbial biotechnology is no longer an alternative but a necessity.

1.11 Microbial Biotechnology in Textile Applications

The integration of microbial biotechnology into the textile industry has catalyzed transformative advancements in enzymatic processing, bio-pigmentation, biofinishing, antimicrobial textiles, and wastewater bioremediation. These innovations not only reduce environmental impacts but also enhance the quality, functionality, and sustainability of textile products. Below, we explore these applications in detail, emphasizing their technical mechanisms, benefits, and future potential.

1.11.1 Enzymatic Processing

Microbial enzymes have revolutionized textile manufacturing by replacing toxic chemicals in processes such as desizing, bio-polishing, and wastewater treatment. Enzymes like amylases from *B. subtilis* remove starch-based coatings from cotton fibers during desizing, eliminating the need for corrosive alkalis. Cellulases from fungi like *Trichoderma reesei* hydrolyze microfibrils on fabric surfaces, enhancing softness and reducing pilling through bio-polishing. Laccases, oxidative enzymes from fungi such as *Trametes versicolor*, degrade synthetic dyes in effluents and enable chlorine-free bleaching. These enzymes operate under mild conditions, reducing water and energy use by up to 50% while minimizing environmental harm. Their specificity preserves fiber integrity, extends garment lifespan, and aligns with sustainability goals (Vermelho et al. 2012; Shukla et al. 2021).

1.11.2 Bio-Pigmentation

Microbial pigments offer sustainable alternatives to synthetic dyes, which are often derived from petroleum and linked to water pollution. *Chromobacterium violaceum* produces violacein, a vibrant purple pigment, while *Monascus purpureus* synthesizes red and yellow hues through fermentation. Carotenoids from *Rhodotorula* spp. and melanins from *Aspergillus niger* provide UV-protective and antimicrobial properties, making them ideal for medical and sportswear textiles. These bio-pigments are biodegradable and often produced using agro-industrial waste, reducing reliance on toxic mordants. By integrating microbial dyes, the textile industry can drastically cut chemical runoff and embrace circular production models (Mazotto et al. 2021; Ahmed 2022).

1.11.3 Biofinishing

Biofinishing with microbial enzymes enhances fabric aesthetics and functionality. Pectinases remove pectin impurities to improve dye uniformity, while laccases create distressed denim effects without the environmental toll of traditional stonewashing. These processes reduce energy use by 30% and water consumption by 40%, exemplifying how microbial solutions enhance efficiency. For instance, laccase-treated denim achieves superior colorfastness with minimal resource input, aligning with eco-conscious fashion trends (Ahmed 2022).

1.11.4 Antimicrobial Textiles

Microbial biotechnology enables the development of textiles with built-in pathogen resistance. Biosurfactants like rhamnolipids from *P. aeruginosa* inhibit bacterial adhesion, while biogenic silver nanoparticles from *Bacillus cereus* provide durable antimicrobial activity. These innovations are critical in healthcare, where antimicrobial PPE reduces hospital-acquired infections, and in sportswear, where they combat odor-causing microbes. Recent studies highlight silver nitrate as an optimal antimicrobial agent, though challenges in durability and toxicity remain (Morais et al. 2016; Shukla et al. 2021; Vojnits et al. 2024) (Fig. 1.4).

1.11.5 Wastewater Bioremediation

Microbial consortia, including *Pseudomonas* and *Shewanella* species, degrade azo dyes and heavy metals in textile effluents. Using enzymes like azoreductases, these microbes break pollutants into non-toxic compounds, achieving up to 90% reduction in COD. Anaerobic bioreactors convert organic sludge into biogas, while biosurfactants enhance pollutant solubilization. Genetically engineered microbes (GEMs) further improve degradation efficiency, offering scalable solutions to meet environmental regulations (Vikrant et al. 2018; Kumar et al. 2021).

1.11.6 Biosurfactants and Sustainable Finishing

Biosurfactants like rhamnolipids from *Pseudomonas* species revolutionize textile processing by improving dye penetration and reducing energy use. These biodegradable agents replace synthetic surfactants, cutting water consumption in dye

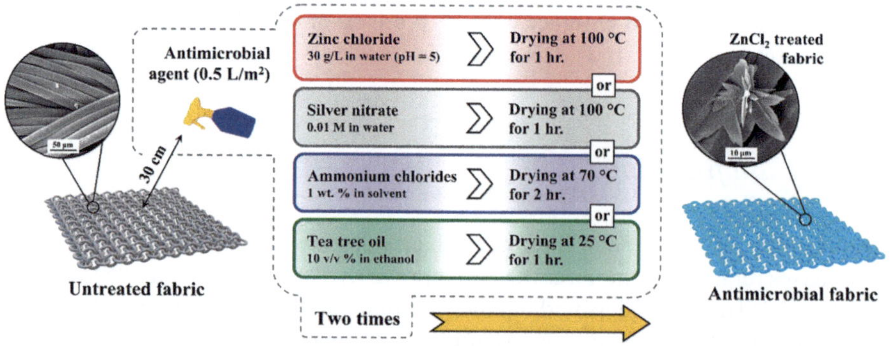

Fig. 1.4 Schematic representation of the whole methodology for antimicrobial fabric fabrication (Vojnits et al. 2024). (Reproduced under the terms and conditions of the Creative Commons Attribute (CC BY) License. Copyright © 2024 by the authors. Licensee MDPI, Basel, Switzerland)

baths by 40% and eliminating aquatic toxicity. For example, rhamnolipid-treated fabrics exhibit brighter colors and lower environmental impact, underscoring their role in sustainable production (Eras-Muñoz et al. 2022).

Microbial biotechnology is redefining the textile industry through innovations like enzyme-driven recycling of polyester waste and microbial synthesis of spider silk proteins. These advancements promise a future where textiles are high-performance, compostable, and produced within circular economy models. By harnessing microbial diversity and genetic engineering, the industry can phase out resource-intensive practices, mitigate climate impacts, and meet consumer demand for sustainability. This convergence of biology and technology positions microbial solutions as indispensable to a greener, more resilient textile sector.

1.12 Advantages of Microbial Biotechnology in Textiles

Microbial biotechnology has emerged as a cornerstone of sustainable innovation in the textile industry, offering transformative environmental, economic, and product quality benefits. By replacing resource-intensive conventional practices with eco-efficient microbial solutions, this technology addresses critical challenges while aligning with global sustainability goals.

1.12.1 Environmental Benefits

Microbial biotechnology significantly reduces the textile industry's environmental footprint by replacing resource-intensive practices with sustainable alternatives. Conventional processes, notorious for high water consumption and toxic chemical use, are mitigated through enzymatic treatments like bio-polishing with cellulases and chlorine-free bleaching with laccases, which operate under mild conditions to cut energy use by 30–50% and water consumption by up to 50%. Microbial bioremediation leverages strains such as *Pseudomonas* and white-rot fungi to degrade hazardous azo dyes and heavy metals in wastewater, reducing COD and protecting aquatic ecosystems. Additionally, bio-pigments from microbes like *Chromobacterium violaceum* and *Monascus purpureus* eliminate synthetic dye pollution, while biosurfactants derived from *Bacillus* spp. replace persistent synthetic surfactants. These innovations align with global sustainability goals, lowering greenhouse gas emissions and fostering circular production models (Vermelho et al. 2012).

1.12.2 Economic Advantages

Adopting microbial biotechnology enhances cost efficiency and market competitiveness for textile manufacturers. Enzymatic processes, such as low-temperature bio-polishing and dyeing, reduce energy and water expenses by 30–40%, while microbial biosurfactants improve dye penetration, minimizing chemical use. Agro-industrial waste is valorized into microbial enzymes and pigments, lowering raw material costs and supporting circular economy principles. Compliance with environmental regulations becomes more affordable, as bio-based effluent treatment systems cost 40–60% less than conventional methods. Furthermore, functional textiles such as antimicrobial fabrics coated with biogenic silver nanoparticles or UV-resistant microbial dyes command premium pricing, tapping into consumer demand for sustainable, high-performance products. These economic benefits, coupled with reduced wastewater treatment costs and regulatory fines, position microbial solutions as financially viable long-term strategies (Hatamie et al. 2015; Vikrant et al. 2018; Eras-Muñoz et al. 2022).

1.12.3 Product Quality Improvements

Microbial biotechnology elevates textile functionality, durability, and consumer appeal. Enzymatic bio-polishing removes microfibrils from cotton, producing smoother, pill-resistant fabrics with enhanced dye uptake and longevity. Microbial pigments, such as carotenoids from *Rhodotorula* spp., offer vibrant, fade-resistant colors with inherent antimicrobial properties, ideal for medical and sportswear textiles. Biosurfactants improve dye uniformity and brightness, while antimicrobial agents like chitosan coatings inhibit odor-causing bacteria, extending fabric freshness. These innovations result in textiles that are not only eco-friendly but also superior in performance, meeting modern demands for durability, comfort, and multifunctionality. By preserving fiber integrity and introducing novel properties, microbial biotechnology ensures textiles meet stringent quality standards while aligning with sustainability trends (Vermelho et al. 2012; Hatamie et al. 2015; Ahmed 2022).

1.13 Challenges and Limitations

While microbial biotechnology holds vast potential for the textile industry, its technological application remains in its infancy, leading to several constraints and drawbacks. Although microbial processes can enhance existing manufacturing methods, their effective implementation faces challenges due to technological barriers, economic considerations, and microbial pollution control standards. Key concerns within microbial biotechnology for textiles include the following:

1.13.1 Technical Challenges

Enzyme and Microbial Strain Development Enzymes such as cellulases, amylases, and proteases are widely used in textile processes like bio-polishing, scouring, and bleaching. However, the commercial production and development of these enzymes and microbial strains present significant hurdles. Cloning and expressing enzymes that remain stable and active under the harsh conditions of modern textile processing such as high temperatures, alkaline pH levels, and exposure to detergents pose major technical challenges. Additionally, industrial-scale enzyme purification can be costly and time-consuming. There is also a pressing need to engineer microorganisms tailored to the specific conditions of textile processing (Imran et al. 2015).

Substrate Specificity Microorganisms and their derived enzymes often exhibit narrow substrate specificity, limiting their application across diverse textile materials. For instance, an enzyme highly effective on cotton fibers may perform poorly on synthetics like polyester or nylon. Developing broad-spectrum microbial strains or specialized enzymes capable of efficiently processing a wide range of textile materials is critical to expanding their utility in the industry. Addressing substrate specificity is essential for ensuring compatibility with the varied fabrics used in textile production.

Process Control and Standardization The use of microbial cultures and enzymes in textiles often involves fermentation or biotechnological methods that depend heavily on factors such as pH, temperature, and nutrient concentration. Achieving reproducible performance of microbial systems across large-scale industrial operations remains challenging. The lack of standardized protocols and precise process control mechanisms often leads to unpredictable outcomes in microbial applications. Implementing automated monitoring and control systems for microbial processes is vital to overcoming this limitation.

Integration with Existing Systems Large-scale adoption of microbial biotechnology in textile manufacturing requires seamless integration with conventional production processes. Many traditional textile treatments, such as finishing and dyeing, rely on chemical reagents that may inhibit microbial activity. Retrofitting existing infrastructure to accommodate microbial systems demands substantial capital investment and strategic planning. Furthermore, alterations to established workflows could impact productivity and operational capacity, posing additional barriers to adoption.

1.13.2 Scalability Issues

A major obstacle to integrating microbial biotechnology in textiles is scaling up laboratory-proven processes to industrial production. While small-scale studies demonstrate promise, translating these solutions into commercial settings involves multifaceted challenges.

Production Scale-Up of Microbial Agents Microbial biotechnology often requires large-scale production of enzymes, bio-based dyes, or other microbial products. However, cultivating microorganisms in industrial-scale bioreactors involves optimizing growth conditions to maximize yield, efficiency, and product quality. Challenges such as nutrient distribution, oxygen transfer, and waste management in large bioreactors complicate scale-up efforts.

High Costs of Large-Scale Bioprocessing Despite the environmental and economic benefits of microbial biotechnology, the initial capital investment for bioreactors, fermenters, and downstream processing equipment is substantial. Ongoing operational costs, including raw material procurement and contamination prevention, further elevate expenses. These factors can render microbial solutions less competitive than conventional chemical processes. Reducing costs through advanced microbial strain development, fermentation optimization, and efficient product recovery is therefore critical (Hayat et al. 2017).

Transitioning from Lab-Scale to Industrial-Scale Systems Research typically begins with small-scale laboratory studies demonstrating the potential of microbial solutions. However, industrial-scale production introduces complexities: microbial strains must adapt to large fermentation systems, culture conditions require re-optimization, and production methods need adjustment. Bridging the gap between experimental validation and industrial implementation demands interdisciplinary collaboration and process innovation.

1.13.3 Regulatory Considerations

Regulatory frameworks governing the use of microorganisms and biotechnological products in textiles are rapidly evolving. Despite the sustainability promise of microbial biotechnology, compliance with regional and international regulations remains a critical consideration.

Safety Standards and Approvals The application of GMOs in textiles whether for enzyme production or antimicrobial treatments is subject to stringent safety evaluations. Regulatory bodies, such as the European Food Safety Authority (EFSA) and the U.S. Environmental Protection Agency (EPA), mandate rigorous risk assessments to ensure GMOs pose no threat to human health or ecosystems. Securing approvals can be time-consuming and costly, particularly for novel microbial applications.

Environmental Impact Assessment Introducing microbial technologies into textile production necessitates thorough environmental impact assessments (EIAs) to evaluate potential risks. Regulatory agencies in many countries require companies to assess microbial persistence, ecological interactions, and long-term environmental consequences before approving large-scale processes.

Consumer and Market Acceptance Consumer perception plays a pivotal role in the success of microbial-treated textiles. Concerns about GMO safety or the unfamiliarity of biotechnological products may hinder market acceptance. Stakeholders must prioritize transparent communication about the benefits, safety, and environmental advantages of microbial-based textiles to build public trust and drive adoption.

1.14 Future Prospects and Emerging Trends

The future of microbial biotechnology in the textile industry holds exciting possibilities as advances in genetic engineering, interdisciplinary research, and emerging technologies continue to progress. With growing demand for sustainable, eco-friendly, and innovative textile solutions, microbial biotechnology is poised to play a pivotal role in reshaping the industry's landscape.

1.14.1 Advances in Genetic Engineering

Synthetic Biology and Metabolic Engineering Recent achievements in genetic engineering, particularly in synthetic biology and metabolic engineering, have enabled the creation of microorganisms with enhanced capabilities for textile applications. By genetically modifying microbial strains, researchers can optimize their production of enzymes, dyes, and other bioproducts for superior performance. Engineered microorganisms can also be designed to thrive under extreme industrial conditions, such as high temperatures or alkaline environments. Future research may focus on refining strains to produce biodegradable dyes or sustainable materials, further aligning with global sustainability goals.

CRISPR-Cas9 and Genome Editing Advances in CRISPR-Cas9 technology have revolutionized genome editing, offering precise tools to tailor microbial strains for textile applications. CRISPR enables targeted modifications to microbial DNA, enhancing enzyme efficiency, stress resistance, or the ability to synthesize novel bio-based compounds. These innovations could significantly improve the precision and sustainability of textile manufacturing processes (Ahmar et al. 2023).

Engineered Microbial Consortia Emerging research explores the use of engineered microbial consortia which are interdependent communities of microorganisms where each species performs specialized functions. Such consortia could synergistically degrade textile effluents or produce diverse enzymes for fabric processing. By mimicking natural ecosystems, these systems could streamline production processes, reduce costs, and minimize environmental impact compared to single-strain approaches.

1.14.2 Integration with Other Technologies

Nanotechnology The integration of nanotechnology with microbial biotechnology offers innovative solutions for functional textiles. For example, bio-synthesized nanoparticles can impart water resistance, self-cleaning properties, or antibacterial features to fabrics without relying on harmful chemical treatments. Microbial processes enable the eco-friendly production of these nanoparticles, aligning with green manufacturing principles (Rahman and Tabassum 2024).

Artificial Intelligence and Machine Learning AI and machine learning are increasingly valuable for optimizing microbial processes in textiles. Predictive models can forecast the efficiency of enzymes or microorganisms under varying conditions (e.g., fermentation parameters, substrate types), accelerating process optimization and reducing trial-and-error experimentation. AI-driven automation also enhances real-time monitoring and control of microbial cultures, ensuring consistent outcomes in large-scale operations.

Circular Economy Models Microbial biotechnology aligns seamlessly with circular economy principles by promoting resource efficiency, waste reduction, and recycling. For instance, microorganisms can break down post-consumer synthetic fibers, recover chemicals from textile waste, or convert waste into bio-based materials. Such innovations could close resource loops, reduce reliance on virgin materials, and advance the industry's sustainability.

1.14.3 Potential New Applications

Biodegradable Fabrics and Packaging Microbial biotechnology enables the production of biodegradable textiles using bio-based polymers. Engineered microorganisms can synthesize biopolymers like polylactic acid (PLA) or polyhydroxyalkanoates (PHA), which degrade naturally and reduce microplastic pollution from synthetic fibers. These materials offer a sustainable alternative to conventional textiles, addressing environmental concerns associated with textile waste (Kopf et al. 2023).

Sustainable Textile Dyes The dyeing process, a major contributor to water consumption and chemical pollution, could be transformed by microbial-derived bio-dyes. Future advancements may yield vibrant, colorfast dyes produced sustainably through microbial fermentation. These bio-dyes, derived from renewable feedstocks, could minimize environmental impact while offering novel aesthetic possibilities for eco-conscious textile production.

1.15 Conclusion and Future Directions

Textile production has long been associated with significant environmental impacts due to its reliance on chemicals, water, and energy. Microbial biotechnology offers transformative potential by introducing sustainable alternatives for processes like dyeing, finishing, and waste management. Applications such as enzymatic treatments, microbial-derived dyes, antimicrobial fabrics, and bioremediation of wastewater highlight its capacity to reduce pollution and resource consumption. However, challenges persist in integrating these solutions into mainstream production. Developing microbial strains resilient to extreme industrial conditions (e.g., high temperatures, alkaline pH) remains technically demanding, while scaling lab-proven processes to industrial levels requires significant infrastructure investment. Regulatory hurdles, particularly for genetically modified organisms (GMOs), further delay commercialization.

Looking ahead, advancements in genetic engineering, such as CRISPR and synthetic biology, promise tailored microbes for bio-based fiber production, waste degradation, and novel functionalities. Integration with AI and nanotechnology could optimize processes and enhance textile properties like durability or antimicrobial performance. The circular economy framework aligns naturally with microbial biotechnology, enabling textile waste recycling and closed-loop systems. Concurrently, rising consumer demand for eco-friendly products and pressure on fast fashion to adopt sustainable practices will drive industry adoption. While challenges remain, interdisciplinary collaboration and technological innovation position microbial solutions as pivotal tools for achieving a sustainable, circular textile industry.

References

Agarwal P, Soni R, Kaur P, Madan A, Mishra R, Pandey J et al (2022) Cyanobacteria as a promising alternative for sustainable environment: synthesis of biofuel and biodegradable plastics. Front Microbiol 13:939347

Ahmar S, Hensel G, Gruszka D (2023) CRISPR/Cas9-mediated genome editing techniques and new breeding strategies in cereals–current status, improvements, and perspectives, vol 69. Biotechnol Adv, p 108248
Ahmed FJ (2022) Microbial biotechnology and bioremediation. Futuristic Biotechnol. https://doi.org/10.54393/fbt.v2i02.13
Branduardi P, Porro D (2012) Yeast: molecular and cell biology. Wiley, Hoboken, NJ, pp 347–370
De Sousa CS, Hassan SS, Pinto AC, Silva WM, De Almeida SS, Soares SDC et al (2018) Omics technologies and bio-engineering. Elsevier, Amsterdam, pp 3–20
Eltarahony M, El-Fakharany E, Abu-Serie M, Elkady M, Ibrahim A (2021) Statistical modeling of methylene blue degradation by yeast-bacteria consortium; optimization via agro-industrial waste, immobilization and application in real effluents. Microb Cell Factories 20:1–26
Eras-Muñoz E, Farré A, Sánchez A, Font X, Gea T (2022) Microbial biosurfactants: a review of recent environmental applications. Bioengineered 13:12365–12391
Gmoser R, Ferreira JA, Lennartsson PR, Taherzadeh MJ (2017) Filamentous ascomycetes fungi as a source of natural pigments. Fungal Biol Biotechnol 4:1–25
Hatamie A, Khan A, Golabi M, Turner AP, Beni V, Mak WC et al (2015) Zinc oxide nanostructure-modified textile and its application to biosensing, photocatalysis, and as antibacterial material. Langmuir 31:10913–10921
Hayat K, Menhas S, Bundschuh J, Chaudhary HJ (2017) Microbial biotechnology as an emerging industrial wastewater treatment process for arsenic mitigation: a critical review. J Clean Prod 151:427–438
Imran M, Crowley DE, Khalid A, Hussain S, Mumtaz MW, Arshad M (2015) Microbial biotechnology for decolorization of textile wastewaters. Rev Environ Sci Biotechnol 14:73–92
Ismail M, Akhtar K, Khan M, Kamal T, Khan MA, Asiri MA et al (2019) Pollution, toxicity and carcinogenicity of organic dyes and their catalytic bio-remediation. Curr Pharm Des 25:3645–3663
Jasińska A, Walaszczyk A, Paraszkiewicz K (2024) Omics-based approaches in research on textile dye microbial decolorization. Molecules 29:2771
Kapoor D, Sharma P, Sharma MMM, Kumari A, Kumar R (2020) Microbial diversity, interventions and scope. Springer, Cham, pp 259–299
Kehe J, Kulesa A, Ortiz A, Ackerman CM, Thakku SG, Sellers D et al (2019) Massively parallel screening of synthetic microbial communities. Proc Natl Acad Sci U S A 116:12804–12809
Kopf S, Åkesson D, Skrifvars M (2023) Textile fiber production of biopolymers–a review of spinning techniques for polyhydroxyalkanoates in biomedical applications. Polym Rev 63:200–245
Kumar D, Patel Z, Pandit P, Pandit R, Patel A, Joshi M et al (2021) Textile industry wastewaters from Jetpur, Gujarat, India, are dominated by Shewanellaceae, Bacteroidaceae, and Pseudomonadaceae harboring genes encoding catalytic enzymes for textile dye degradation. Front Environ Sci 9:720707
Lawson CE (2021) Retooling microbiome engineering for a sustainable future. Msystems 6. https://doi.org/10.1128/msystems.00925-00921
Mariyes TC, Ju E-J, Lee J-H (2024) Enhancing microbial resilience: the role of adaptive laboratory evolution in industrial biotechnology. J Life Sci 34:730–743
Mazotto AM, De Ramos SJ, LaA DB, Rocha NU, De Souza SA (2021) How can microbiology help to improve sustainability in the fashion industry? Environ Technol Innov 23:101760
Mccarty NS, Ledesma-Amaro R (2019) Synthetic biology tools to engineer microbial communities for biotechnology. Trends Biotechnol 37:181–197
Morais DS, Guedes RM, Lopes MA (2016) Antimicrobial approaches for textiles: from research to market. Materials 9:498
Navale GR, Dharne MS, Shinde SS (2021) Metabolic engineering and synthetic biology for isoprenoid production in Escherichia coli and Saccharomyces cerevisiae. Appl Microbiol Biotechnol 105:457–475
Nikel PI, Mattanovich D (2021) Microbial cell factories: a biotechnology journey across species. vol. 65. Portland Press Ltd., London. p 143–145

Parsaeimehr A, Ebirim RI, Ozbay G (2022) CRISPR-Cas technology a new era in genomic engineering. Biotechnol Rep 34:e00731

Rahman M, Tabassum Z (2024) Biotechnological approach to treat textile dyeing effluents: a critical review Analysing the practical applications. Text Leather Rev 7:124–152

Sharif A, Nasreen Z, Bashir R, Kalsoom S (2020) Microbial degradation of textile industry effluents: a review. Pure Appl Biol 9:2361–2382

Sherry A, Dell'agnese BM, Scott J (2023) Biohybrids: textile fibres provide scaffolds and highways for microbial translocation. Front Bioeng Biotechnol 11. https://doi.org/10.3389/fbioe.2023.1188965

Shukla A, Parmar P, Goswami D, Gehlot Y, Vala J, Parmar N et al (2021) Green chemistry for sustainable textiles. Elsevier, Amsterdam, pp 173–189

Singh R (2019) New and future developments in microbial biotechnology and bioengineering. Elsevier, Amsterdam, pp 107–114

Tang KHD (2023) State of the art in textile waste management: a review. Text 3:454–467

Timmis K, Hallsworth JE (2024) This is the age of microbial technology: crucial roles of learned societies and academies. Microb Biotechnol 17:e14450

Vermelho AB, Supuran CT, Guisan JM (2012) Microbial enzyme: applications in industry and in bioremediation. Enzyme Res 2012:1–2

Vikrant K, Giri BS, Raza N, Roy K, Kim K-H, Rai BN et al (2018) Recent advancements in bioremediation of dye: current status and challenges. Bioresour Technol 253:355–367

Vojnits K, Mohseni M, Parvinzadeh Gashti M, Nadaraja AV, Karimianghadim R, Crowther B et al (2024) Advancing antimicrobial textiles: a comprehensive study on combating ESKAPE pathogens and ensuring user safety. Materials 17:383

Wang S, Salmon S (2022) Progress toward circularity of polyester and cotton textiles. Sustain Chem 3:376–403

Woźniak E, Zimny T, Twardowski T (2020) Bioeconomy for sustainable development. Springer, Cham, pp 21–41

Yalcin-Enis I, Kucukali-Ozturk M, Sezgin H (2019) Nanoscience and biotechnology for environmental applications. Springer, Cham, pp 29–53

Chapter 2
Microbial Enzymes in Textile Bioprocessing

Fatima Azeem, Aleena Sumrin, Tazeen Malik, Musrat Shaheen, and Nazim Hussain

Abstract The textile industry faces significant challenges. The integration of enzymes offers a promising solution as various enzymes, including proteases, cellulases, and amylases, facilitate processes such as bio-polishing, de-sizing, and bleaching, thereby reducing reliance on harsh chemicals and labor-intensive methods. The use of enzymes not only improves fabric quality but also minimizes water and energy consumption while reducing waste generation. Specific enzymes serve distinct functions in textile processing. Cellulase improves the softness and absorbency of cotton fabrics, while amylase effectively removes starch-based sizing agents without fiber damage. Catalases break down residual hydrogen peroxide after bleaching, ensuring cleaner dyeing processes without extensive purification steps. Despite these benefits, challenges remain in scaling up enzyme applications for industrial use in the textile sector. Current limitations include integration into existing manufacturing infrastructure and cost-effective enzyme production. Ongoing research addresses these challenges through computational analysis of enzymatic behaviors and advances in microbial biotechnology. This chapter examines how microbial enzymes are redefining textile bioprocessing, aligning with sustainability goals and improving production methods. The development of more stable and specific enzymes promises to establish better practices in the textile industry while minimizing environmental damage.

Keywords Bioprocessing · Sustainable textiles · Enzymatic treatments · Bio-polishing · Cellulase · Amylase · Catalase · Green technology · Environmental sustainability · Enzyme optimization

F. Azeem · A. Sumrin · T. Malik · N. Hussain (✉)
Centre for Applied Molecular Biology (CAMB), University of the Punjab, Lahore, Pakistan
e-mail: aleena.camb@pu.edu.pk; nazim.camb@pu.edu.pk

M. Shaheen
Institute of Microbiology, Government College University Faisalabad, Faisalabad, Pakistan
e-mail: mussaratshaheen@gcuf.edu.pk

M. Khurshid et al. (eds.), *Microbial Biotechnology in Textile Innovation*, SDGs and Textiles, https://doi.org/10.1007/978-981-95-0347-6_2

2.1 Introduction

The use of enzymes in the textile industry has facilitated the development of innovative technologies for fiber processing, with significant implications for enhancing product quality. This approach exemplifies industrial or white biotechnology. Textile bioprocessing leverages bio-organisms and their enzymes to improve the productivity and sustainability of the entire industry. This methodology emerged as a response to the ecological challenges posed by conventional techniques, which often relied on hazardous materials. Enzymatic treatments in the textile industry aim to reduce environmental impact without compromising product quality (Duran and Duran 2000). The application of enzymes in textile processing dates back to ancient civilizations, where natural enzymes derived from dyeing fruits were used to soften fabrics. However, the use of microbial enzymes became prominent in the early twentieth century, marking a significant advancement in improving textile quality and performance. By the 1980s, enzyme-based methods had gained widespread adoption, extending beyond desizing to include processes such as bio-scouring, bio-finishing, and bio-polishing (Nigam 2013).

Cellulases, amylases, and laccases are among the enzymes widely utilised in various phases of textile manufacturing. Amylases play a significant role in desizing cotton textiles by breaking down starch, thereby allowing the fabric to be desized without compromising the integrity of the fibers. Cellulases are employed in denim finishing, where the hydrolysis of microfibers results in a smoother and more refined appearance. Additionally, laccases are utilized in dyeing processes, particularly for bleaching and decolorization (Duran and Duran 2000; Araujo et al. 2008). The adoption of enzymatic processes reduces energy consumption and wastewater production. For instance, compared to conventional techniques, biopolishing and bioscouring not only enhance fabric quality but also minimize effluent output. Advances in protein and genetic engineering have led to the development of more robust enzymes capable of withstanding harsh industrial conditions (Nigam 2013).

The use of microbial enzymes offers numerous benefits, including improved productivity, sustainability, and product quality in the textile industry. By replacing hazardous chemicals such as acids and alkalis, enzymes significantly reduce environmental pollution and waste generation. These enzymatic treatments are also energy-efficient, as they function optimally at lower energy requirements, thereby lowering carbon emissions. Moreover, enzyme-based processes improve textile durability and quality, for example, cellulases used in biopolishing enhance the texture and provide a smoother finish to fabrics (Raveendran et al. 2018).

Advancements in biotechnology, particularly in genetic engineering, have led to the development of robust enzymes capable of withstanding significant fluctuations in temperature and pH, thereby greatly expanding their industrial applications. It is predicted that the adoption of sustainable methods, driven by biotechnological advancements, will enable the textile industry to utilize more efficient enzymes to address emerging challenges. This chapter highlights the role of microbial enzymes in textile bioprocessing, focusing on their ability to enhance production while addressing environmental concerns, therefore, marking a pivotal transition towards environmentally friendly manufacturing. Furthermore, the chapter explores the types of enzymes, their diverse applications, and recent biotechnological innovations, alongside real-world case studies that illustrate their impact.

2.2 Mechanism of Microbial Enzymes

Enzymes are proteins that act as biocatalysts, facilitating metabolic and biological processes. Certain enzymes are also employed in industrial processes as organic catalysts. Microbial enzymes, derived from microorganisms, are particularly suited for commercial applications. Although enzymes from microbes were discovered in the twentieth century, extensive research has since been conducted on their isolation, production, and application in the bioindustry to advance knowledge and innovation (Pandey et al. 1999; Saleem et al. 2018). The use of enzymes in conventional catalytic processes are used either in immobilized or free form which mainly depends on their specificity. With ongoing advancements in biotechnology, it is now possible to design and engineer enzymes tailored to meet the specific needs of the textile industry. These tailored enzymes have been developed to carry out highly specific reactions. Innovations in fields such as metagenomics, protein engineering, and biochemical reaction engineering have enabled the invention of novel enzymes. Furthermore, molecular approaches are being utilised to enhance the quality and functionality of these enzymes, ensuring their effectiveness in various industrial applications (Chirumamilla et al. 2001).

2.2.1 Adaptable Microbial Enzymes for Industrial Use

Microbial enzymes are unique in their ability to function effectively and exhibit significant activity under optimal temperature and pH conditions. Three primary types of microbial enzymes, i.e. alkalophilic, thermophilic, and acidophilic have been identified based on their adaptability to specific environmental settings. Microbial contamination during large-scale industrial processes, which often persists over extended periods, can be mitigated using microorganisms that produce thermostable enzymes (Wang et al. 2012).

Thermostable enzymes exhibit improved penetration at elevated temperatures, facilitating enzyme digestion. Prolonged reaction times are often required for the complete hydrolysis and saccharification of large polysaccharides, but such time-intensive processes are frequently associated with contamination challenges. Due to their stability across a wide temperature range and activity under varying conditions, these enzymes are highly sought after (Zhang and Wu 2011).

One commercially valuable enzyme is thermophilic xylanase, widely used in brewing and mashing processes. Additionally, thermostable isoforms of laccase enzymes have applications in bioremediation, pulping, dyeing, and textile processing. These characteristics make them indispensable for various industrial applications.

2.2.2 Mechanisms of Enzyme Catalysis

Enzymes function by interacting with specific molecules known as substrates, accelerating biochemical processes. The mechanism by which enzymes act on substrates can be outlined as follows:

Substrate Binding The biocatalytic reaction begins when the substrate binds to a specific site on the enzyme known as the active site, which is complementary to the substrate. According to the lock and key model, substrates fit precisely into the active site, providing a clear explanation of how the enzyme functions. In textiles, substrates can be various types of fabrics and their molecular components. Additionally, the induced fit model suggests that when the enzyme and substrate bind, it induces a conformational change in the enzyme, leading to a better fit and enhancing the reaction (Frey and Hegeman 2007).

Catalysis After the substrate binds to the enzyme, an enzyme-substrate complex is formed, which accelerates the reaction and lowers the activation energy. Enzymes perform various forms of catalysis, such as covalent catalysis, where the enzyme forms a transient covalent bond to stabilize the transition state (Fig. 2.1). In acid-base catalysis, the enzyme donates or accepts protons to facilitate the reaction. In electrostatic catalysis, electrostatic interactions charge the transition state, aiding the catalytic process (Cook and Cleland 2007; Frey and Hegeman 2007).

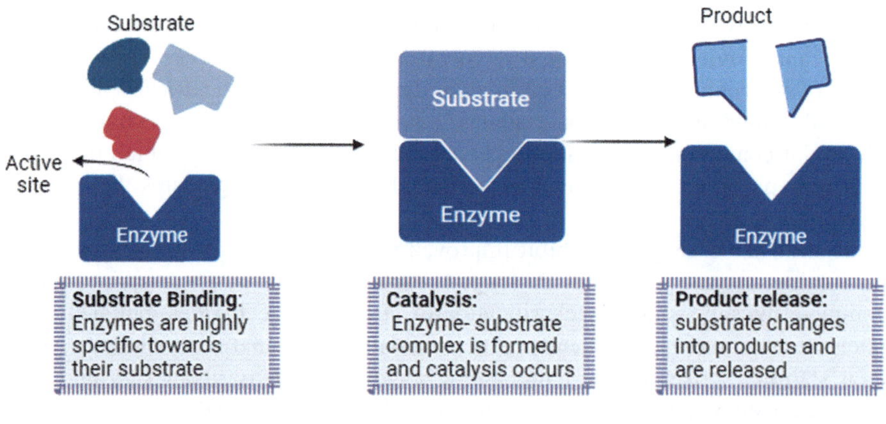

Fig. 2.1 Mechanism of enzyme action

Product Release The final step involves the release of the product from the active site. The enzyme then returns to its original state and can be reused in subsequent reactions.

2.3 Microbial Enzymes in Textile

Microbial enzymes play a crucial role in the textile industry due to their numerous benefits, including enhanced efficiency and improved sustainability. The key types of microbial enzymes used in textiles are discussed below and summarized in Table 2.1.

2.3.1 *Amylases: Sources and Applications*

Amylases are produced by various microorganisms, including both bacteria and fungi. The most common bacterial species are *Bacillus subtilis* and *B. licheniformis*, while the fungi *Aspergillus niger* and *Trichoderma reesei* are frequently used.

Table 2.1 The microbial sources of enzymes and their applications in textiles

Enzyme	Microbial sources	Use in textile
Protease	*Bacillus subtilis, Bacillus licheniformis, Bacillus pumilus, Aspergillus niger, Trichoderma reesei*	Protease not only increases the softness and smoothness of fabrics but also decreases shrinkage and are effective in breaking down protein stains in industrial applications
Amylases	*Bacillus subtilis, Bacillus licheniformis, Aspergillus niger, Trichoderma reese*	Amylases are eco-friendly and biodegradable. It operates under mild conditions (70–110 °C, pH 6.0–6.8), and in addition to this, it enhances the desizing process
Cellulases	*Trichoderma reesei, Humicola insolens, Aspergillus niger, Bacillus subtilis, Clostridium thermocellum*	-Cellulases reduce pilling and enhance fabric softness, so the dye uptake is improved. It is also environmentally friendly as it uses the residues of agricultural residues for enzyme production
Lipases	*Bacillus subtilis, Bacillus licheniformis, Bacillus pumilus, Pseudomonas aeruginosa, Aspergillus niger, Rhizopus oryzae*	Lipases enhance the uptake of dye and hydrophilicity of fibers like polyester. Moreover, it improves the vibrancy of color and reduces reliance on chemicals
Pectinases	*Aspergillus niger, Penicillium species, Trichoderma reesei, Bacillus subtilis, Erwinia chrysanthemi, Saccharomyces cerevisiae*	Pectinase removes impurities in fabric preparation to enhance dye absorption and reduce the need for harsh chemicals
Laccases	*Aspergillus species, Trametes versicolor, Phanerochaete chrysosporium, Pleurotus ostreatus, Bacillus subtilis, Klebsiella pneumoniae*	Laccase decolorizes synthetic dyes in textile effluents and improves overall fabric quality

Industrial-scale production of amylase typically involves the cultivation of these bacteria. Factors such as temperature, incubation time, growth medium, and pH influence enzyme synthesis. Research indicates that *Bacillus cereus* strains produce amylase with enhanced de-sizing ability, capable of completely de-sizing a sample in 72 h that is much slower than hydrochloric acid, which takes only 1 h (Amsamani and Umamaheswariand 2018; Rehman et al. 2023).

In textiles, amylases are usually used for the de-sizing process which is usually the washing of material made of wool that is free of sizing agents. This enzyme is well known to degrade the molecules of the starch and make textiles easier for further procedures such as dyeing and finishing. It also accelerates the breakdown of α-1,4-and/or α-1,6-glucosidic bonds and produces the units with floccose removed. These enzymes are useful for industrial applications because they perform well at high temperatures (70–110 °C) and a pH range of 6.0–6.8. (Tülek et al. 2021). Prior to the use of amylases, de-sizing was typically performed with harsh chemicals and high temperatures, which often weakened fabrics and compromised their quality. Amylase-based de-sizing allows for selective removal of sizing agents without damaging the fibers, thus enabling milder processing conditions (Amsamani and Umamaheswariand 2018).

The efficiency of amylases in textile de-sizing can be further enhanced through enzyme immobilization and stability optimization. Although immobilization can reduce enzymatic activity, some studies have demonstrated that immobilized α-amylase, particularly when attached to microspheres or zirconia-coated glass beads, exhibits improved washing performance in textile de-sizing (Tülek et al. 2021). Additionally, enzyme stability is critical, as the enzyme must remain functional under varying conditions such as pH, temperature, and chemical additives. Recent research is focused on developing amylases that can withstand these diverse conditions and maintain stability (Tülek et al. 2021).

The advantages of using amylases in the textile industry are many, with the main benefits outlined below:

1. *Environment friendly:* Amylases are biodegradable and operate under mild conditions, resulting in a lower environmental impact compared to conventional techniques. This makes them a more sustainable option for textile processing.
2. *Energy Efficiency:* Microbial amylases are effective at lower temperatures and pressures, reducing energy consumption and resource use during the de-sizing process.
3. *Specificity:* Amylases exhibit high specificity for starch, minimizing unwanted effects and ensuring better targeting of the substrate. This leads to enhanced quality of the final textile product.

2.3.2 Cellulases: Benefits and Applications

Cellulase is an enzyme that breaks down cellulose, a critical component of plant cell walls, and plays a significant role in catalysis. It is produced by various organisms, with fungi and bacteria being the most common sources. Each of these sources offers unique benefits, making them suitable for industrial applications.

Fungi, particularly filamentous fungi, are the most prominent producers of cellulase. *Trichoderma reesei* is especially recognized for its high cellulase production capacity, and it has been extensively studied for its commercial applications, particularly in industrial settings. Other fungal sources, such as *Humicola insolens* and *Aspergillus niger*, also produce cellulases, which are used in a variety of biotechnological applications, including the textile industry (Ikbal et al. 2024).

While fungal cellulases are more commonly utilized, certain bacteria also produce cellulase. Notable examples include *Bacillus subtilis* and *Clostridium thermocellum*, which produce cellulases that perform well at high temperatures. These bacterial cellulases are of particular interest for industrial applications, such as biofuel production (Uddin 2015; Ikbal et al. 2024).

The production of cellulase typically involves a fermentation process, during which microorganisms are cultivated in growth media suitable for their growth. The first crucial step in cellulase production is the selection and preparation of the substrate. Agricultural residues, such as bagasse, rice husks, and sugarcane, are commonly used due to their cost-effectiveness. Following substrate preparation, the selected microorganisms are inoculated into the substrate and incubated under controlled conditions. The temperature, pH, and aeration are carefully monitored to optimize enzyme production. Submerged fermentation is generally used for bacterial cellulase production, while solid-state fermentation is preferred for fungal cellulase production (Ikbal et al. 2024).

After the fermentation process, the next step involves processing the culture broth to extract the enzyme. This typically includes centrifugation to separate the biomass from the liquid. The supernatant, which contains cellulases, is then subjected to purification. Common purification techniques, such as chromatography and precipitation, are employed to isolate the active enzymes for use in the textile industry. Finally, the purified enzyme is converted into either liquid or powder form for practical use in textiles. This conversion process also stabilizes the enzyme, ensuring its enhanced activity (Korsa et al. 2023; Ikbal et al. 2024).

In the textile industry, cellulases are commonly used to reduce pilling, a process where fibers break and form small balls. Cellulases help break down these fibers, reducing the likelihood of pilling and thereby extending the lifespan of the fabric. Additionally, cellulases play a significant role in softening fibers. They are indispensable in textiles, primarily due to their roles in biopolishing and softening. Moreover, cellulases improve the appearance of fabrics, reduce pilling, and impart lasting softness to the material.

2.3.3 Pectinases: Benefits and Applications

Pectin is a polysaccharide found in the cell walls of plants and is degraded by pectinases. This group of enzymes plays a key role in various industrial applications, including the textile industry, where they are commonly used for fabric finishing and improving textile quality. Pectinase is produced by various microorganisms, including bacteria, fungi, and yeasts. Fungi such as *Aspergillus niger* and *Penicillium* species are notable producers of pectinases, particularly polygalacturonase and pectin lyase, which are widely used in industrial applications. *Trichoderma reesei* is another fungus known for producing pectinase, breaking down both cellulose and pectin. In addition to fungi, bacteria like *Bacillus subtilis* and *Erwinia chrysanthemi* also produce pectinases, which are primarily used in the fermentation process, with pectinolytic enzymes effectively degrading pectin. Although yeasts such as *Saccharomyces cerevisiae* also produce pectinase, they are less commonly used than fungal and bacterial sources (Alqahtani et al. 2022; Roman-Benn et al. 2023).

Pectinases are classified based on the mechanisms by which they act on pectin molecules. Polygalacturonases hydrolyze the α-1,4-glycosidic bonds in the molecule, breaking down polygalacturonic acid into oligosaccharides. Pectin lyases, on the other hand, cleave the pectin molecule via a β-elimination mechanism, producing methanol and pectate without generating toxic byproducts (Zheng et al. 2021). Pectin esterase removes methyl groups from pectin, converting it into pectic acid, which increases its reactivity and solubility (Roman-Benn et al. 2023).

In the textile industry, pectinases are used in fabric preparation, particularly for the pretreatment of natural fibers like cotton to remove impurities. This process enhances the fibers' viability, making them more receptive to dyeing. Pectinases improve dye uptake by breaking down the pectin, which may otherwise interfere with dye absorption. As a result, fabrics acquire a more vibrant and uniform color. Additionally, pectinases contribute to biopolishing by removing fuzz and protruding fibers, leading to a smoother finish (Alqahtani et al. 2022). This process is eco-friendly as it reduces the need for harsh chemicals, thereby improving the safety of the textiles. Furthermore, the wastewater generated during this procedure has a lower biological oxygen demand, making it easier to treat and less hazardous to the environment (Madhu and Chakraborty 2017; Roman-Benn et al. 2023).

2.3.4 Laccases: Benefits and Applications

Laccases are multicopper oxidase enzymes that are blue in color and catalyze the oxidation of various phenolic and non-phenolic compounds. These enzymes are found in a wide range of organisms and have gained significant importance in the textile industry due to their unique properties (Saleem et al. 2014, 2018; Arregui et al. 2019). *Scomycetes* and *Basidiomycetes* are the primary fungal sources of laccases, with notable species including *Aspergillus* species, *Trametes versicolor*, *Phanerochaete chrysosporium*, and *Pleurotus ostreatus* (Baldrian 2006). These fungi are known for their ability to degrade lignin and other organic materials, making them valuable sources of laccases for textile applications. Additionally, certain bacteria, such as *Bacillus subtilis* and *Klebsiella pneumoniae*, also produce laccases (Akram et al. 2022).

Laccases are characterized by their high substrate specificity and stability under varying environmental conditions, making them particularly suitable for use in the textile industry (Baldrian 2006; Arregui et al. 2019). The primary function of laccases is the oxidation of phenolic compounds using molecular oxygen as the electron acceptor, producing water as a byproduct. Their ideal substrate specificity enables them to oxidize various compounds, including phenols and aromatic amines, which are useful in textile processing (Baldrian 2006).

One of the most significant applications of laccases in the textile industry is their ability to decolorize synthetic dyes present in textile effluents. Laccases effectively degrade a wide range of chemical structures, making them an eco-friendly alternative to conventional bleaching agents, such as hydrogen peroxide. Through enzymatic bleaching, laccases not only decolorize the fabric but also modify its surface properties, enhancing dye uptake and improving overall fabric quality. This modification of surface properties results in better dye adhesion to fibers, leading to more vibrant and durable colors (Blánquez et al. 2019).

The use of laccases aligns with the increasing demand for environmentally sustainable practices in the textile industry. Manufacturers can reduce their environmental footprint while producing high-quality textiles, making laccases an attractive alternative to traditional chemical processes (Madhu and Chakraborty 2017; Akram et al. 2022).

2.3.5 *Proteases*

Proteases are enzymes that break down peptide bonds, leading to the degradation of proteins into amino acids. They can be sourced from microbes, plants, and animals; however, microbial sources are most commonly used in the textile industry. Among the microbial sources, *Bacillus* species are particularly important for producing alkaline proteases. Notable examples include *Bacillus subtilis*, *Bacillus licheniformis*, and *Bacillus pumilus*. These proteases are not only thermostable but also function well at high pH levels. Additionally, fungal sources, such as *Aspergillus niger* and *Trichoderma reesei*, are used in specific applications where a different enzymatic profile is required (Amory et al. 1999; Razzaq et al. 2019).

Proteases function by hydrolyzing proteins and can be classified according to their optimal pH levels. Alkaline proteases, which work optimally at pH levels between 8 and 12, are commonly used in laundry applications. In contrast, acidic proteases, which operate best at pH levels between 2 and 5, are less commonly used in the textile industry. The specificity of these enzymes allows them to target proteins in a selective manner without affecting other components of the fabric (Amory et al. 1999; Ikbal et al. 2024).

Proteases are also used to enhance the softness of fabrics by breaking down the cuticles on wool fibers. This process, known as shrink-resistance treatment, reduces shrinkage and improves fabric smoothness. In the case of wool, the breakdown of cuticle scales on the fibers results in a softer and smoother texture, with reduced shrinkage during washing (Ikbal et al. 2024). In laundry detergents, alkaline proteases play a significant role in breaking down protein-based stains. Their ability to function effectively at various temperatures makes them suitable for both cold and hot washes. Furthermore, these proteolytic enzymes can be used in biopolishing, which provides a smoother surface for fabrics. The use of proteases reduces the need for harsh chemicals and minimizes wastewater production (Madhu and Chakraborty 2017; Razzaq et al. 2019).

2.3.6 Lipases

These enzymes hydrolyze fats and oils into glycerol and fatty acids, playing a crucial role in the textile industry. They are used in fabric treatment, dyeing, and enhancing the quality of textiles. While numerous bacteria produce lipases, notable species include *Bacillus subtilis*, *Bacillus licheniformis*, and *Bacillus pumilus*, which are known for their alkaline lipases. Additionally, *Pseudomonas aeruginosa* produces thermotolerant lipases, which are significant in the textile industry. Important fungal sources of lipases include *Aspergillus niger* and *Rhizopus oryzae*, both of which exhibit notable lipolytic activity (Yao et al. 2021; Mesbah 2022).

Lipases function by hydrolyzing ester bonds, releasing fatty acids and glycerol. Like proteases, lipases can be classified based on their optimal pH levels. Alkaline lipases, which work best at pH levels between 8 and 10, are commonly used in detergent formulations. In contrast, acidic lipases operate optimally at pH levels between 4 and 5. Lipases are used to treat fibers, such as polyester, to improve their dyeability and hydrophilicity by breaking down ester linkages in polyethylene terephthalate (PET). This process introduces carboxylic and hydroxyl groups on the surface of the fibers, enhancing their moisture regain while causing minimal damage to the fabric (Mesbah 2022; Chatterjee et al. 2023).

Lipases also effectively remove natural oils in a degreasing process that is essential for better fabric finishing. This enzymatic treatment improves dyeing efficiency, as it allows better penetration of the dye, resulting in more vibrant colors. Moreover, the application of lipases in textile processing aligns with industry trends by reducing reliance on harsh chemical products (Mesbah 2022).

2.4 Applications of Microbial Enzymes

Microbial enzymes are environmentally friendly, and their enhanced sustainability in textile production offers many advantages. This chapter discusses some of the applications of microbial enzymes in textiles, along with their benefits and uses.

Table 2.2 Advantages and disadvantages of various denim de-sizing methods

Method	Advantages	Disadvantages
Physical methods	Simpler and cost-friendly	Time-consuming and effectiveness is not optimal for complete removal
Acid de-sizing	Remove sizing agents quickly	It may cause damage to cellulose fibers and require careful handling as it is corrosive in nature
Oxidative de-sizing	Very effective for many sizing agents	Exerts risk of cellulose damage and has environmental concerns due to the use of chemicals
Enzymatic de-sizing	Specificity to starch, eco-friendly, also preserves fabric strength and improves wettability	It is a slower process and may not be effective on all starch types (e.g., tapioca)

2.4.1 De-sizing

De-sizing is an important process in textile manufacturing aimed at removing sizing agents, primarily starch, from fabrics in preparation for finishing processes. Amylase is the enzyme typically used in this process. The fabric is first treated with a solution of amylase, typically at a concentration of 0.5–2%, along with wetting agents, under controlled temperatures of around 60–70 °C (Nigam 2013). The amylase then acts on the starch present, converting it into soluble forms. This reaction typically occurs over a period of hours to overnight, depending on the concentration of the enzyme and the specified conditions. After the enzyme has acted, the fabric is washed to remove the degraded starch products, leaving behind a clean substrate ready for dyeing.

This method is notable for its specificity, as it targets the starch without adversely affecting the cellulose fibers of cotton, thus preserving the integrity of the fabric (Rehman et al. 2023). In addition to enzymatic de-sizing, traditional de-sizing methods include physical and chemical approaches, such as acid de-sizing and oxidative de-sizing. Each of these methods has its advantages and disadvantages when compared to enzymatic de-sizing, as shown in Table 2.2.

2.4.2 Scouring

In textile production, scouring is an important pretreatment step aimed at removing contaminants such as proteins, lipids, and pectin from the fabric. Microbial enzymes like pectinases, lipases, and proteases are used in this process. Pectin is a polymer found in the cell walls of plants, and the pectinase enzyme acts on it, breaking it down into simpler molecules. This enzyme aids in the removal of pectin molecules from cotton fibers, which can interfere with fabric quality (Fig. 2.2). This procedure is environmentally friendly due to its better performance at neutral pH and low temperatures (Rehman et al. 2023).

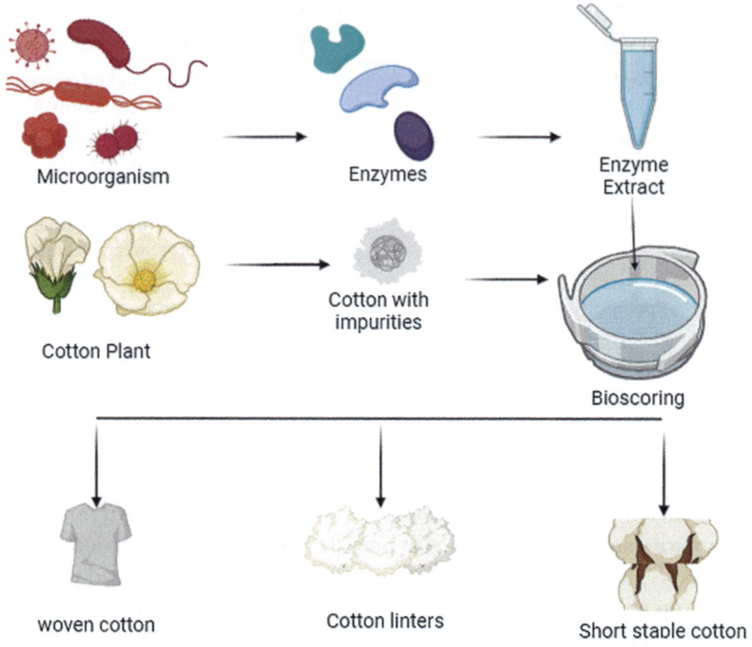

Fig. 2.2 Process of bioscouring (Enzymes such as cellulase, protease, and lipase, derived from microorganisms, are used to remove impurities like lipids, proteins, and pectins during the scouring process)

Additionally, lipases are enzymes that catalyze the breakdown of fats and oils, making them useful for cleaning lipid-based contaminants from textile products. This is particularly important for fabrics that have been exposed to oils during manufacturing, ensuring better finishing and dyeing results. Furthermore, proteases play a key role in scouring by breaking down protein-based contaminants. They effectively remove protein residues that could cause issues in subsequent processing, while also enhancing the overall quality and appearance of the fabric. These enzymes perform well under harsh conditions, leading to energy savings.

2.4.3 Bleaching

Laccases are particularly used for effective denim bleaching as they target the indigo dyes in jeans, facilitating color removal while minimizing damage. This process is also environmentally friendly, producing fewer harmful byproducts. It has the advantage of achieving various shades through controlled applications. Additionally, laccases also bleach cotton and other fibers, enhancing the whiteness of textiles while preserving the integrity of the fabrics (Pezzella et al. 2015).

During the bleaching process, laccases oxidize phenolic compounds, and this process is further enhanced using mediators that facilitate the oxidation reaction. The synergy between these processes improves enzyme diffusion, thereby enhancing the effectiveness of the bleaching.

2.4.4 Denim Processing

Enzymatic stone washing has revolutionized denim processing, offering a sustainable alternative to conventional methods. Traditionally, pumice stones are used in washing machines in the industry, where they are added to denim garments to create a faded look. The abrasion from the stones removes the indigo dye from the fabric's surface. In contrast, enzymatic denim processing consumes less water and produces no harmful effluents, as it eliminates the need for sodium hypochlorite. This process uses cellulose enzymes to fade the appearance of the fabric and remove surface fuzz as well (Rodríguez-Couto 2012).

2.4.5 Degumming of Silk

Degumming is a process that removes sericin from silk fibers, enhancing the sheen and color of the fabric. Sericin can make silk feel harsh and stiff, masking its natural luster. Traditionally, degumming was performed using alkaline solutions, such as sodium carbonate, combined with surfactants like sodium lauryl sulfate (Fig. 2.3). However, due to the various disadvantages of this method, it has been replaced by enzymatic processes. Enzymatic methods utilize enzymes such as papain or certain bacterial enzymes to hydrolyze sericin without damaging the fibers. Research has shown that natural enzymes extracted from papaya or pineapple skins can achieve better degumming results compared to traditional methods. After degumming, the silk fibers are unraveled from the cocoons, a process that can be done either manually or with automated machines. The resulting silk yarn can now be dyed, and due to its unique protein structure, it absorbs dye exceptionally well, leading to vibrant colors (Gaviria et al. 2023).

2.4.6 Wool Processing

Wool processing involves several stages to transform raw wool into a usable textile. The integration of enzymatic treatments has emerged as a sustainable method to reduce the environmental impact. Initially, impurities are removed from the fleece without the use of harsh chemicals. The cleaned wool is then carded, spun, and woven into fabrics. The final finishing operations include

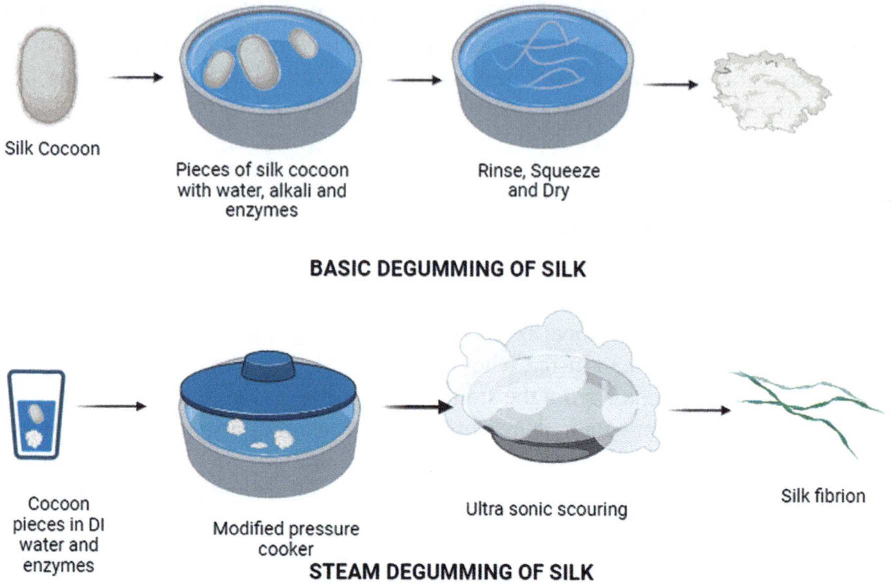

Fig. 2.3 Degumming of silk (Two methods are employed for silk degumming: basic degumming and steam degumming. Both methods utilize enzymes to remove sericin impurities from silk, enhancing its quality)

stabilizing dimensions and surface treatments. Enzymes play a crucial role, particularly in anti-felting treatments that enhance both durability and aesthetic appearance (Wang et al. 2018). Additionally, enzymes such as transglutaminase are used to cross-link the proteins, improving tensile strength and enhancing resistance to shrinkage. The surface properties of wool can also be modified, making it more receptive to dyes. This application not only enhances the quality of the fabric but also aligns with modern environmental standards in the textile industry. As research continues to evolve, these methods will become more prevalent, paving the way for greener practices in textile production (Popescu and Stanescu 2024).

2.5 Innovations in Microbial Enzyme Usage in Textiles and Their Applications

Genetic engineering plays a crucial role in enhancing enzyme efficiency within the textile industry. The following are key advancements and applications of genetic engineering techniques that improve enzyme production and functionality.

2.5.1 Advances in Genetic Engineering Techniques

Recombinant DNA technology allows for the insertion of specific genes into host organisms, enabling the high-yield production of targeted enzymes. For instance, genetic modification of *Bacillus amyloliquefaciens* has been shown to enhance the production of alkaline protease, which is essential for various textile applications. Additionally, site-directed genetic mutagenesis enables the modification of enzyme structures to improve their stability and activity (Jiang et al. 2022). This is particularly important in textile production, where enzymes must function efficiently under varying temperatures and pH conditions (Fig. 2.4). Another significant advancement is the integration of synthetic biology with genetic engineering, which allows the design of novel enzymes with tailored properties. A notable example of this is the collaboration between AMSilk and BRAIN Biotech, which focuses on developing high-performance, bio-based proteins for textiles. By leveraging these engineering techniques and bioinformatics, researchers aim to create enzymes that can outperform traditional methods. Furthermore, the development of biodegradable structural proteins with enhanced durability holds promise for replacing older materials (Jiang et al. 2022; Ndochinwa et al. 2024).

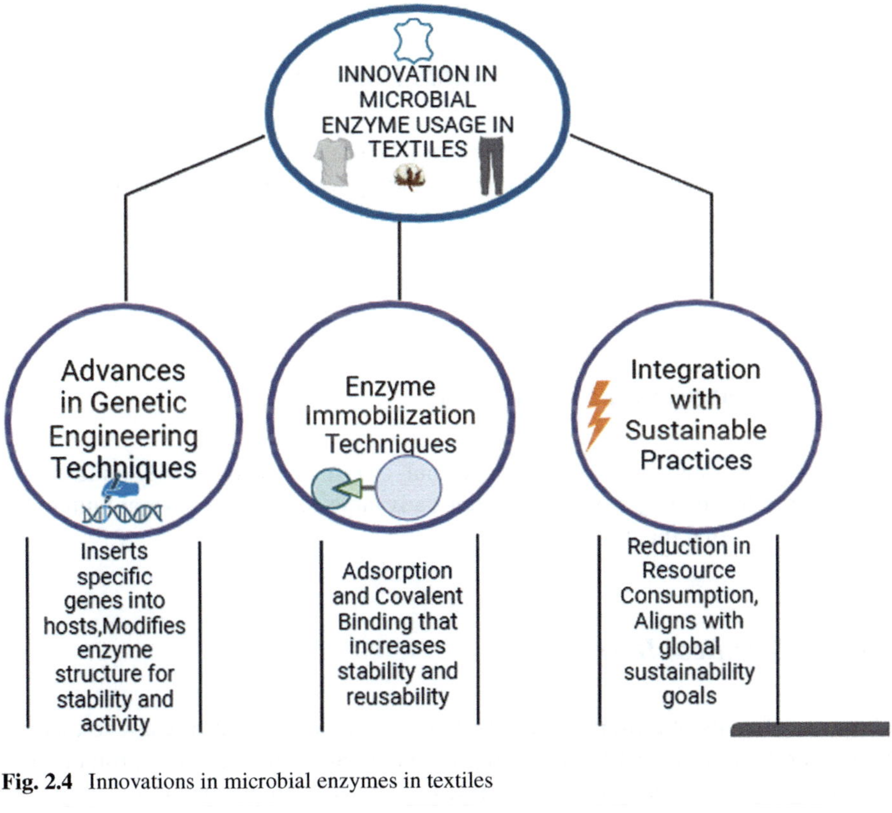

Fig. 2.4 Innovations in microbial enzymes in textiles

2.5.2 *Enzyme Immobilization Techniques*

Enzyme immobilization confines enzymes to textiles, enhancing their functionality in processes such as dye degradation, fabric treatment, and biopolishing. Two commonly used techniques for enzyme immobilization are adsorption and covalent binding. These methods utilize substrates, such as polyester, cotton, and other textile products, which act as enzyme carriers. One key advantage of using immobilized enzymes is their ability to be reused without a loss of activity (Maghraby et al. 2023). This process also allows for the separation of enzymes from the reaction mixture, leading to cleaner product isolation without the need for extensive purification steps (Fig. 2.4). Furthermore, immobilized enzymes exhibit better stability, which is crucial in textile applications where they must function under harsh conditions and exposure to chemicals. Immobilization can also extend the shelf life of enzymes by protecting them from degradation and inactivation, making long-term storage more feasible. In many cases, immobilization enhances catalytic activity by improving substrate accessibility. Techniques such as genetic engineering can be used to further customize the properties of immobilized enzymes. For example, engineered cellulase can be designed to have improved activity at extreme pH levels and varying temperatures (Homaei et al. 2013; Maghraby et al. 2023).

2.5.3 *Integration with Sustainable Practices*

The use of enzymes represents a significant advancement towards sustainable practices by reducing resource consumption, including water, energy, and raw materials. As enzyme technology and its applications continue to evolve through ongoing research, these biocatalysts play a crucial role in facilitating eco-friendly processes that align with global sustainability goals (Courth et al. 2021).

2.6 Challenges and Limitations

Although the implementation of microbial enzymes in the textile industry to enhance product quality has shown various advantages, there are several limitations and challenges that the industry faces.

2.6.1 *Technical Barriers*

Enzymes often exhibit limited stability, with denaturation occurring due to thermal stress during various textile processing methods. Strategies, such as protein engineering techniques, have been developed to increase stability, as discussed earlier.

However, a limited understanding of the structure-function relationship hinders the identification of optimal targets for site-directed mutations (Xie et al. 2014).

Additionally, the choice of support material for enzyme immobilization is critical, as it must provide mechanical strength while allowing substrate access. Some immobilization methods can lead to limitations where substrates cannot reach the enzymes due to diffusion barriers. Certain crosslinking agents used in the process can also cause enzyme inactivation (Homaei et al. 2013). Finally, the high degree of specificity of enzymes towards their substrates can limit their function in diverse industrial processes. For instance, enzymes used in textile applications must be able to interact effectively with a variety of chemicals (Fig. 2.5).

2.6.2 Economic and Regulatory Factors

The economic viability of microbial enzymes is significantly influenced by production costs. A techno-economic analysis revealed that the production cost of recombinant enzymes can reach approximately $316 per kg, which is substantially higher than that of traditional enzymes. This cost disparity is primarily due to the expenses associated with consumables, raw materials, and facilities, making it crucial to optimize the production process for cost-effective enzyme production (Sakhuja et al. 2021).

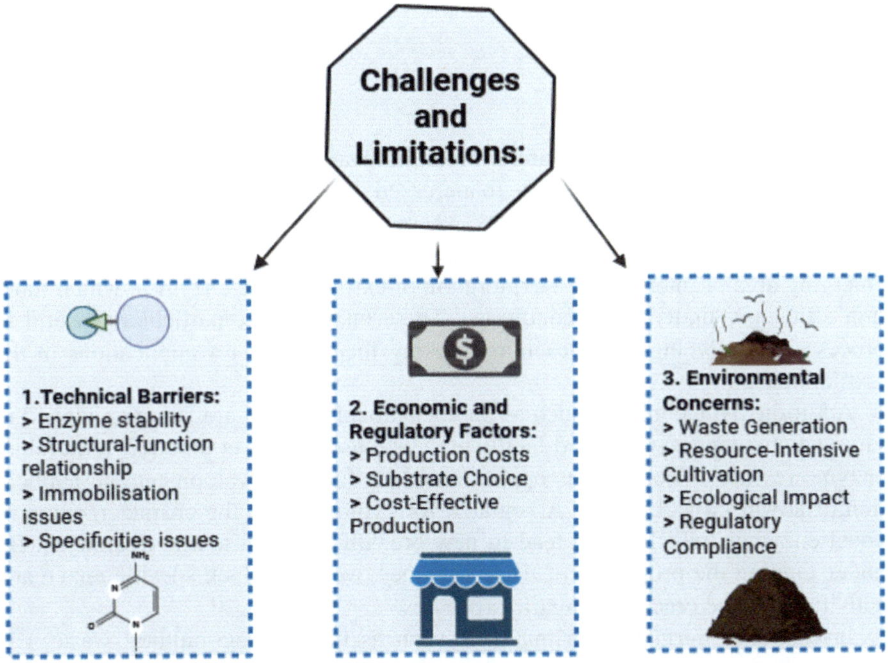

Fig. 2.5 Challenges and limitations of microbial enzymes in biotextiles

Another critical factor is the choice of substrate. While using high-purity substrates increases production costs, employing biomass as a substrate can reduce expenses while also addressing waste management concerns. Relying on inexpensive enzyme synthesis strategies may provide a viable approach to economically enhance the feasibility of the enzyme market (Sakhuja et al. 2021).

2.6.3 Environmental Concerns

Although there are various advantages to using enzymes, the processes of their production and purification can generate significant waste. A study revealed that the overall waste generated due to enzyme production can be as high as 37,835 g of waste per gram of enzyme produced. This raises concerns about environmental protection, as the cultivation of microbes also requires resource-intensive conditions. Temperature control and waste management must be carefully managed to optimize enzyme yields while minimizing environmental impact (Becker et al. 2021).

Furthermore, some enzymes that are considered safe may have undesirable consequences. For example, proteases, which are often considered advantageous, can negatively impact aquatic ecosystems due to their protein-degrading activity if not handled properly. As environmental regulations become stricter, textile industries must invest in compliance measures to ensure safe practices (Sakhuja et al. 2021).

2.7 Future Directions

The textile industry is witnessing various innovative trends driven by microbial enzymes, contributing significantly to increased sustainability. One prominent trend is the adoption of bio-based processes, where enzymes like cellulases enhance fabric quality while minimizing environmental impact (Silva et al. 2018). Another emerging area of interest is the exploration of extremophilic enzymes, which function efficiently under harsh conditions. These enzymes are particularly useful in processes like dyeing and bleaching, making them viable for applications in the textile industry.

Additionally, techniques such as photochemical grafting are being employed to improve enzyme retention and facilitate their reuse (Korsa et al. 2023). Microbial enzyme research also presents opportunities for further development and application of advanced techniques. A key area for exploration is the characterization of novel enzymes, which could lead to new breakthroughs in textile treatments. By understanding the properties of these enzymes, we can unlock specific enzymatic activities that are crucial for textile processes.

Investigating fermentation methods, such as utilizing agricultural waste, can enhance both the feasibility and sustainability of enzyme production. Comparing solid-state versus submerged fermentation techniques can also help maximize enzyme yields (Korsa et al. 2023). Furthermore, the integration of digital technologies, including machine learning and artificial intelligence, holds the potential to revolutionize enzyme applications in the textile industry. The future of textile bioprocessing is being shaped by the transformative potential of enzymes derived from microorganisms.

2.8 Conclusions

The revolutionization of the textile industry, particularly in biotextiles, is largely driven by microbial enzymes. These enzymes play a pivotal role throughout various stages, from fiber processing to finishing, significantly enhancing product quality while minimizing environmental impact. This chapter explores the importance, usage, limitations, and future prospects of microbial enzymes in textiles.

Enzymes such as cellulase, amylase, and protease, derived from microorganisms, have been extensively studied for their applications in textiles. For instance, cellulase is used to achieve stone-washed effects in denim, replacing the harmful pumice stones traditionally used. This enzymatic treatment not only enhances the fabric's appeal but also reduces water and energy consumption. Similarly, amylases efficiently remove starch-based agents, further improving resource efficiency.

The environmentally friendly nature of microbial enzymes contributes to non-toxic and biodegradable methods, which help reduce harmful waste generation. Enzymatic treatments also enhance the durability of fabrics, making them more appealing to consumers. As the pressure on textile manufacturers to adopt sustainable practices intensifies, the use of these enzymes aligns with global sustainability goals.

However, while the benefits of microbial enzymes are clear, certain limitations and challenges must be addressed. Enzyme stability is a major concern, as many enzymes are sensitive to extreme temperatures and pH, limiting their reusability and effectiveness. Immobilization techniques can increase enzyme stability and facilitate reuse, but they often raise production costs. Furthermore, the high specificity of enzymes can limit their versatility in diverse textile applications. Lastly, challenges related to safety and environmental compliance hinder the commercialization of new enzyme products.

In conclusion, microbial enzymes represent a sustainable component of the textile industry by converting synthetic fibers into monomers and transforming waste into valuable textiles. As consumer demand for eco-friendly products increases, enzyme-based processes offer significant potential. The non-toxicity of enzymes aligns with ecological goals, further boosting their market potential. Through continued research, innovation, and computational analysis to predict enzyme behavior, these enzymes hold great promise for revolutionizing the biotextile sector and fostering a more sustainable future.

References

Akram F, Ashraf S, Haq IU, Shah FI, Aqeel A (2022) Eminent industrial and biotechnological applications of laccases from bacterial source: a current overview. Appl Biochem Biotechnol:1–21

Alqahtani YS, More SS, Shaikh IA, Kj A, More VS, Niyonzima FN et al (2022) Production and purification of pectinase from Bacillus subtilis 15A-B92 and its biotechnological applications. Molecules 27:4195

Amory A, Clippe A, Konieczny-Janda G (1999) Use of alkaline proteases in industrial textile laundering processes. United States patent

Amsamani S, Umamaheswariand R (2018) Sustainability in fashion and apparels. WPI Publishing, pp 118–125

Araujo R, Casal M, Cavaco-Paulo A (2008) Application of enzymes for textile fibres processing. Biocatal Biotransformation 26:332–349

Arregui L, Ayala M, Gómez-Gil X, Gutiérrez-Soto G, Hernández-Luna CE, De Los H, Santos M et al (2019) Laccases: structure, function, and potential application in water bioremediation. Microb Cell Factories 18:1–33

Baldrian P (2006) Fungal laccases–occurrence and properties. FEMS Microbiol Rev 30:215–242

Becker M, Lütz S, Rosenthal K (2021) Environmental assessment of enzyme production and purification. Molecules 26:573

Blánquez A, Rodríguez J, Brissos V, Mendes S, Martins LO, Ball AS et al (2019) Decolorization and detoxification of textile dyes using a versatile Streptomyces laccase-natural mediator system. Saudi J Biol Sci 26:913–920

Chatterjee A, Puri S, Sharma PK, Deepa PR, Chowdhury S (2023) Nature-inspired enzyme engineering and sustainable catalysis: biochemical clues from the world of plants and extremophiles. Front Bioeng Biotechnol 11:1229300

Chirumamilla RR, Muralidhar R, Marchant R, Nigam P (2001) Improving the quality of industrially important enzymes by directed evolution. Mol Cell Biochem 224:159–168

Cook PF, Cleland WW (2007) Enzyme Kinetics and Mechanism (Garland Science, New York, USA). https://www.taylorfrancis.com/books/mono/10.4324/9780203833575/enzyme-kinetics-mechanism-paul-cookcleland

Courth K, Binsch M, Ali W, Ingenbosch K, Zorn H, Hoffmann-Jacobsen K et al (2021) Immobilization of peroxidase on textile carrier materials and their application in the bleaching of colored whey. J Dairy Sci 104:1548–1559

Duran N, Duran M (2000) Enzyme applications in the textile industry. Rev Prog Color Relat Top 30:41–44

Frey PA, Hegeman AD (2007) Enzymatic reaction mechanisms. Oxford University Press, Oxford

Gaviria A, Jaramillo-Quiceno N, Motta A, Restrepo-Osorio A (2023) Silk wastes and autoclaved degumming as an alternative for a sustainable silk process. Sci Rep 13:15296

Homaei AA, Sariri R, Vianello F, Stevanato R (2013) Enzyme immobilization: an update. J Chem Biol 6:185–205

Ikbal MS, Tisha FA, Asheque AI, Hasnat E, Uddin MA (2024) Eco-friendly biopolishing of cotton fabric through wasted sugarcane bagasse-derived enzymes. Heliyon 10:e26346

Jiang C, Ye C, Liu Y, Huang K, Jiang X, Zou D et al (2022) Genetic engineering for enhanced production of a novel alkaline protease BSP-1 in Bacillus amyloliquefaciens. Front Bioeng Biotechnol 10:977215

Korsa G, Konwarh R, Masi C, Ayele A, Haile S (2023) Microbial cellulase production and its potential application for textile industries. Ann Microbiol 73:13

Madhu A, Chakraborty JN (2017) Developments in application of enzymes for textile processing. J Clean Prod 145:114–133

Maghraby YR, El-Shabasy RM, Ibrahim AH, Azzazy HME-S (2023) Enzyme immobilization technologies and industrial applications. ACS Omega 8:5184–5196

Mesbah NM (2022) Industrial biotechnology based on enzymes from extreme environments. Front Bioeng Biotechnol 10:870083

Ndochinwa GO, Wang Q-Y, Okoro NO, Amadi OC, Nwagu TN, Nnamchi CI et al (2024) New advances in protein engineering for industrial applications: key takeaways. Open Life Sci 19:20220856

Nigam PS (2013) Microbial enzymes with special characteristics for biotechnological applications. Biomolecules 3:597–611

Pandey A, Selvakumar P, Soccol CR, Nigam P (1999) Solid state fermentation for the production of industrial enzymes. Curr Sci 77:149–162

Pezzella C, Guarino L, Piscitelli A (2015) How to enjoy laccases. Cell Mol Life Sci 72:923–940

Popescu C, Stanescu MD (2024) Eco-friendly processing of wool and sustainable valorization of this natural bioresource. Sustainability 16:4661

Raveendran S, Parameswaran B, Ummalyma SB, Abraham A, Mathew AK, Madhavan A et al (2018) Applications of microbial enzymes in food industry. Food Technol Biotech 56:16

Razzaq A, Shamsi S, Ali A, Ali Q, Sajjad M, Malik A et al (2019) Microbial proteases applications. Front Bioeng Biotechnol 7:110

Rehman A, Saeed A, Asad W, Khan I, Hayat A, Rehman MU et al (2023) Eco-friendly textile desizing with indigenously produced amylase from Bacillus cereus AS2. Sci Rep 13:11991

Rodríguez-Couto S (2012) Laccases for denim bleaching: an eco-friendly alternative. Sigma 1:10–12

Roman-Benn A, Contador CA, Li M-W, Lam H-M, Ah-Hen K, Ulloa PE et al (2023) Pectin: An overview of sources, extraction and applications in food products, biomedical, pharmaceutical and environmental issues. Food Chem Adv 2:100192

Sakhuja D, Ghai H, Rathour RK, Kumar P, Bhatt AK, Bhatia RK (2021) Cost-effective production of biocatalysts using inexpensive plant biomass: a review. 3 Biotech 11:280

Saleem R, Khurshid M, Ahmed S (2014) Xylanase, laccase and manganese peroxidase production from white rot fungi. Iran J Energy Environ 5:59–66

Saleem R, Khurshid M, Ahmed S (2018) Laccases, manganese peroxidases and xylanases used for the bio-bleaching of paper pulp: An environmental friendly approach. Protein Pept Lett 25:180–186

Silva C, Martins M, Jing S, Fu J, Cavaco-Paulo A (2018) Practical insights on enzyme stabilization. Crit Rev Biotechnol 38:335–350

Tülek A, Karataş E, Çakar MM, Aydın D, Yılmazcan Ö, Binay B (2021) Optimisation of the production and bleaching process for a new laccase from Madurella mycetomatis, expressed in Pichia pastoris: from secretion to yielding prominent. Mol Biotechnol 63:24–39

Uddin MG (2015) Effects of biopolishing on the quality of cotton fabrics using acid and neutral cellulases. Text Cloth Sustain 1:1–10

Wang X, Li D, Watanabe T, Shigemori Y, Mikawa T, Okajima T et al (2012) A glucose/O2 biofuel cell using recombinant thermophilic enzymes. Int J Electrochem Sci 7:1071–1078

Wang L, Yao J, Niu J, Liu J, Li B, Feng M (2018) Eco-friendly and highly efficient enzyme-based wool shrinkproofing finishing by multiple padding techniques. Polymers 10:1213

Xie Y, An J, Yang G, Wu G, Zhang Y, Cui L et al (2014) Enhanced enzyme kinetic stability by increasing rigidity within the active site. J Biol Chem 289:7994–8006

Yao W, Liu K, Liu H, Jiang Y, Wang R, Wang W et al (2021) A valuable product of microbial cell factories: microbial lipase. Front Microbiol 12:743377

Zhang S-B, Wu Z-L (2011) Identification of amino acid residues responsible for increased thermostability of feruloyl esterase a from Aspergillus niger using the PoPMuSiC algorithm. Bioresour Technol 102:2093–2096

Zheng L, Xu Y, Li Q, Zhu B (2021) Pectinolytic lyases: a comprehensive review of sources, category, property, structure, and catalytic mechanism of pectate lyases and pectin lyases. Bioresour Bioprocess 8:1–13

Chapter 3
Microbial Pigments as Bio-based Dyes in Textile

Fatima Qasim, Wajid Ali, Kinza Niaz, Muhammad Sajjad Ahmad, Heitor B. S. Bento, and Muhammad Irfan

Abstract The textile industry is increasingly seeking sustainable alternatives to synthetic dyes due to environmental and health concerns. Microbial pigments, derived from microorganisms such as bacteria, fungi, and algae, offer a promising bio-based solution. This review provides an in-depth analysis of the potential applications of microbial pigments in textiles, focusing on their production, properties, and advantages. These pigments are highly regarded for their sustainability, as they are produced from renewable resources and have a significantly lower ecological footprint compared to conventional synthetic dyes. Additionally, they are typically biodegradable, contributing to reduced environmental impact. Microbial pigments offer a diverse range of colors, potentially surpassing the palette of synthetic dyes and introducing novel hues to the textile industry. However, their practical application faces challenges related to stability, colorfastness, and scalability. These pigments often exhibit lower color retention and are susceptible to environmental factors, necessitating advancements in production methods and dyeing techniques to enhance their durability and performance. Recent research has made significant progress in microbial fermentation technologies and genetic modifications, improving pigment stability and yield.

Keywords Textile · Microbial pigments · Dyeing · Color · Ecofriendly

F. Qasim · W. Ali · K. Niaz · M. S. Ahmad · M. Irfan (✉)
Department of Biotechnology, University of Sargodha, Sargodha, Pakistan
e-mail: Irfan.ashraf@uos.edu.pk

H. B. S. Bento
Department of Bioprocess Engineering and Biotechnology, School of Pharmaceutical Sciences, São Paulo State University (UNESP), Araraquara, Brazil
e-mail: heitor.bento@unesp.br

M. Khurshid et al. (eds.), *Microbial Biotechnology in Textile Innovation*, SDGs and Textiles, https://doi.org/10.1007/978-981-95-0347-6_3

3.1 Introduction

Dyeing has been an integral part of human culture since the Bronze Age in Europe. The color of a fabric is its most captivating feature. Historically, natural dyes were used to color fabrics; however, they provided only a limited and often muted palette. Moreover, these dyes exhibited low color fastness, as they tended to fade upon exposure to washing and sunlight (Kant 2012). Some of the most well-known ancient dyes include madder, which produces a red hue derived from the roots of *Rubia tinctorium*, yellow obtained from the stigmas of the saffron plant (*Crocus sativus L.*), and indigo blue extracted from the leaves of *Indigofera tinctoria*.

The discovery of synthetic dyes in 1856 revolutionized the dyeing industry, leading to a significant decline in the use of natural dyes for textiles (Siva 2007). These artificial dyes provided a wider spectrum of vibrant and long-lasting colors, drastically improving color fastness (Alegbe and Uthman 2024). As synthetic dyes became more cost-effective, natural dyes were largely phased out by the early twentieth century. However, as the global demand for textiles continues to rise mainly driven by population growth and economic expansion, the industry faces significant environmental and resource-related challenges (Saravanan and Chandramohan 2011).

Currently, the textile industry consumes approximately 1.3 million tons of artificial colorants and their precursors annually, with nearly 200,000 tons lost as waste during dyeing and finishing processes. A typical textile processing unit generates about 125 L of wastewater per kilogram of fabric, often characterized by high chemical oxygen demand (COD), total dissolved solids (TDS), biological oxygen demand (BOD), and extreme pH fluctuations (Varadarajan and Venkatachalam 2016). Additionally, processes such as dyeing, finishing, and printing release toxic emissions, while yarn spinning and fabric weaving/knitting depend on fossil fuels, contributing to CO_2 emissions and particulate pollution. The industry's key environmental concerns include greenhouse gas emissions, toxic chemicals, excessive water consumption, and waste production, prompting a growing interest in eco-friendly dyes (Sandin and Peters 2018).

Although synthetic dyes have been widely used for centuries, they often contain harmful chemicals that can cause allergic reactions and other health complications. As a result, microbial pigments are emerging as sustainable alternatives for textile dyeing. These naturally derived pigments are biodegradable, non-toxic, and free from carcinogenic effects, making them increasingly popular in the textile industry. For instance, the red pigment prodigiosin (PG), derived from *Vibrio* spp. and *Serratia marcescens*, is used to dye materials such as silk, nylon, cotton, and acrylics, offering excellent wash and environmental resistance. Another widely used microbial dye is violacein, obtained from *Chromobacterium violaceum*, which effectively colors fabrics like viscose, cotton, polyester, and silk.

Additionally, violacein from *Pseudoalteromonas* spp. is commonly utilized in both textiles and toy manufacturing due to its eco-friendly properties. The pigment from *Janthinobacterium lividum* imparts a bluish-purple hue to silk, cotton, and wool, while appearing dark blue on nylon and vinylon. Similarly, the pigment from *Sclerotinia* spp. is frequently used to dye pure cotton (Di Salvo et al. 2023).

The widespread adoption of synthetic dyes was largely due to their affordability, ease of production, superior coloring properties, and minimal quantities required to achieve vibrant shades. However, many synthetic dyes were introduced without rigorous safety testing, raising significant health and environmental concerns (Downham and Collins 2000). For example, benzidine-based dyes have been linked to colorectal cancer, while carbon black, commonly used in printing inks, is classified as a potential carcinogen (Babitha 2009). These health risks have fueled a growing global demand for natural pigments (Manikprabhu and Lingappa 2013).

Plants and microbes serve as the primary sources of natural pigments. However, plant-derived pigments face limitations such as seasonal availability, stability issues, and solubility constraints. In contrast, microbial pigments present a promising alternative, as they can be rapidly produced in low-cost culture media, remain unaffected by climatic conditions, and exhibit high compatibility with various textiles. Given their economic feasibility, widespread availability, and non-toxic nature, microbial pigments are poised to replace hazardous synthetic dyes in the textile industry (Albarakaty 2022).

3.2 Microbial Pigments

The demand for renewable and non-polluting biological products is increasing. Microorganisms serve as a valuable source of bio-based products, including microbial pigments, which have applications across various industries. These pigments provide a sustainable alternative to synthetic dyes. Biopigments can be obtained from terrestrial and aquatic ecosystems, encompassing animals, plants, and microorganisms (Fig. 3.1). Although most bio-pigments are currently derived from land plants, the aquatic environment which harbors nearly 80% of all biotic forms represents a rich and underutilized source of pigments for textile dyeing (Ramesh et al. 2019). Microorganisms produce stable pigments such as carotenoids and flavonoids, which exhibit a stronger affinity for textiles compared to other natural colorants. Key microbial pigments include carotenoids, tetrapyrroles, flavonoids, and xanthophylls, such as astaxanthin (Malik et al. 2012). In addition to their ease of extraction, these pigments offer various bioactive properties, including antibacterial, anticancer, and antioxidant effects.

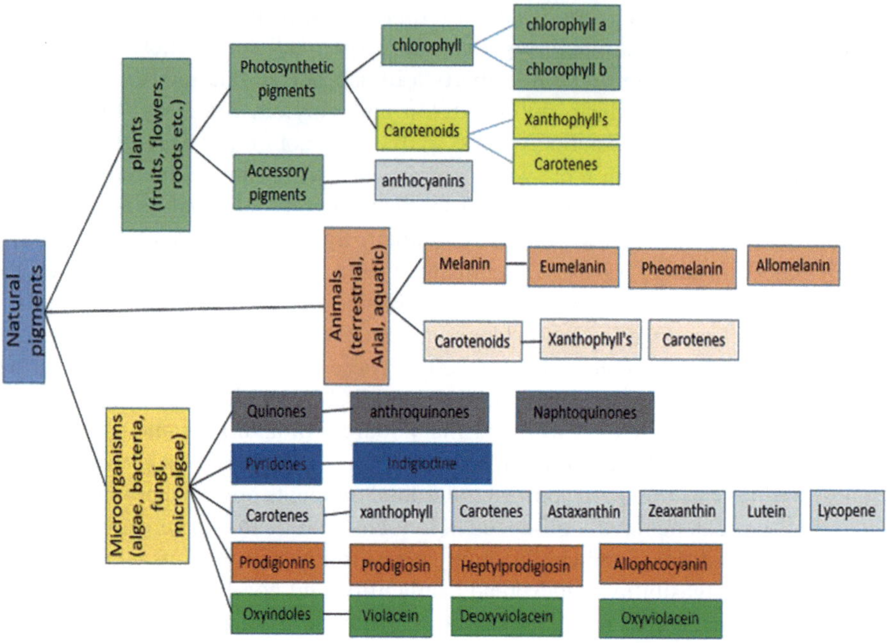

Fig. 3.1 Bio-based pigments and their applications in the pharmaceutical and textile industries

3.3 Sources of Microbial Pigments

3.3.1 Fungal Pigments

Fungi are incredibly diverse and widespread organisms, inhabiting a wide range of environments, including extreme conditions. This diversity enables fungi to produce unique substances known as secondary metabolites (Ramesh et al. 2019). Many fungal genera, such as *Thermomyces, Alternaria, Chlorociboria, Cunninghamella, Fusarium, Monascus, Acrostalagmus, Aspergillus, Phymatotrichum, Talaromyces, Bisporomyces, Trichoderma, Cordyceps, Curvularia, Penicillium,* and *Scytalidium*, are known to produce biopigments suitable for dyeing textiles (Lagashetti et al. 2019). For instance, *Scytalidium cuboideum* produces red pigments known as naphthoquinones, *Scytalidium ganodermophthorum* yields yellow pigments, and *Chlorociboria aeruginosa* synthesizes green pigments. These pigments can effectively dye various textile components, including bleached cotton, acrylics, wool, polyester, and polyamides.

Filamentous fungi produce a wide range of pigments, including anthraquinones, quinones, violacein, flavins, phenazines, melanins, indigo, and carotenoids (Weber et al. 2014). Among fungal pigments, carotenoids and polyketides are particularly significant (Dufossé et al. 2005, 2014). Over 200 fungal species have been identified as carotenoid producers, with these pigments being synthesized via the isoprenoid pathway. Carotenoids are known for their vivid colors and biological activities, including their role as precursors to vitamin A and their antimicrobial, antioxidant, and membrane-stabilizing properties (Pfander 1992). Various fungal classes synthesize carotenes, a major group of biopigments, including *Basidiomycetes* (e.g., *Rhodosporidium, Sclerotium, Sclerotinia, Sporidiobolus, Ustilago*), *Ascomycetes* (e.g., *Aspergillus, Aschersonia, Cercospora, Penicillium*), and *Zygomycetes* (e.g., *Blakeslea, Mucor, Phycomyces*) (Ferahtia et al. 2021). Carotenoids exist in two primary forms: carotenes (hydrocarbons) and xanthophylls (oxygenated derivatives of carotenes) (Avalos and Carmen Limón 2015). They are widespread in nature and typically range in color from yellow to orange-red (Venil et al. 2020a, b).

Several fungi produce specific carotenoids used in commercial applications. For instance, *Ashbya gossypii* synthesizes lactoflavin (riboflavin), while *Blakeslea trispora* produces β-carotene and lycopene, which are also used in the commercial production of the colorant Arpink Red. Anthraquinones, another group of fungal pigments, typically exhibit a yellow color but can also appear in shades of orange, violet, red, and brown, depending on their chemical substituents (Gessler et al. 2013). These pigments are widely valued for their vibrant hues and applications in the cosmetics, food, textile, and printing industries. The complexity of anthraquinones, combined with various chemical and environmental factors such as light, temperature, pH, oxygen transfer, and nutrient availability, significantly influences their production (Xiao-Mei et al. 2011). Notably, *Fusarium oxysporum* is a well-known producer of anthraquinones used in wool dyeing (Venil et al. 2020a, b).

Fungal polyketides, which exhibit a broad spectrum of colors, include anthraquinones, azaphilones, naphthoquinones, and hydroxyanthraquinones (Nagia and El-Mohamedy 2007). These pigments not only provide vivid coloration but also possess significant bioactive properties, including anticancer, antimicrobial, immunosuppressive, and antibiotic activities (Mapari et al. 2010). Azaphilones, a specific class of polyketide-derived pigments, are produced by fungi such as *Penicillium, Aspergillus, Monascus,* and *Chaetomium. Monascus* fungi, for example, synthesize six major polyketide pigments: monascorubramine, monascorubin, rubropuntamine, rubropunctatin (red), monascin (yellow), and ankaflavin (orange). However, these pigments are sensitive to light, heat, and pH changes. Additionally, *Monascus* pigments can interact with environmental compounds such as amino acids and proteins to form hydrophilic pigments (Mapari et al. 2009; Venil et al. 2020a, b).

Melanin pigments, produced by *Lasiodiplodia theobromae* isolated from wood, have been successfully used to dye poplar wood veneer, enhancing both its aesthetic appeal and environmental resistance. In black fungi such as *Cryptococcus neoformans, Cladosporium cladosporioides,* and *Aspergillus versicolor,* melanin pigments function as radioprotectors and energy transducers, providing resistance to ionizing radiation (Robinson et al. 2012). Due to their stability and consistent quality, fungal pigments have been proposed as sustainable alternatives to synthetic dyes in the textile industry (Meruvu and Dos Santos 2021). Moreover, these pigments can absorb ultraviolet light, offering potential UV protection when applied to textiles (Nambela et al. 2020).

3.3.2 Bacterial Pigments

Bacteria offer several advantages over fungi for pigment synthesis, primarily due to their shorter life cycle and ease of genetic modification (Venil et al. 2013, 2014). These pigment-producing bacteria are found across a wide variety of environments, from soil and rhizospheres to marine habitats, freshwater, and desert sands. They are capable of surviving extreme conditions, such as high temperatures, saline environments, and even as endophytes in plants (Narsing Rao et al. 2017). Like fungi, bacteria produce a diverse array of pigments, including carotenoids, prodigiosin, melanin, zeaxanthin, violacein, pyocyanin, and actinorhodin (Ahmad et al. 2012a, b).

These pigments can interact with textile fibers through ionic interactions, which are impacted by the pH of the dye bath used during the coloring process (Kramar et al. 2014). In addition to providing color, bacterial pigments exhibit important biological activities, such as antimicrobial properties against various pathogens, anticancer effects, antioxidant activity, and UV protection (Fig. 3.2). These features make bacterial pigments highly promising for use as functional dyes, offering applications beyond their visual impact.

Although these pigments are often referred to as "bacterial pigments" or "biopigments," they function more like dyes from a textile science perspective. They form chemical bonds with the functional groups on textile fibers, despite not being water-soluble (Kramar and Kostic 2022). Notable bacteria involved in pigment production include *Serratia*, *Pseudomonas*, and *Streptomyces*, which generate a broad spectrum of colors such as pink, blue, red-orange, yellow, and green (Malik et al. 2012; Aman Mohammadi et al. 2022). It is important to note that the shade of an isolated pigment does not always match the final hue seen on dyed fabric or fibers. The final

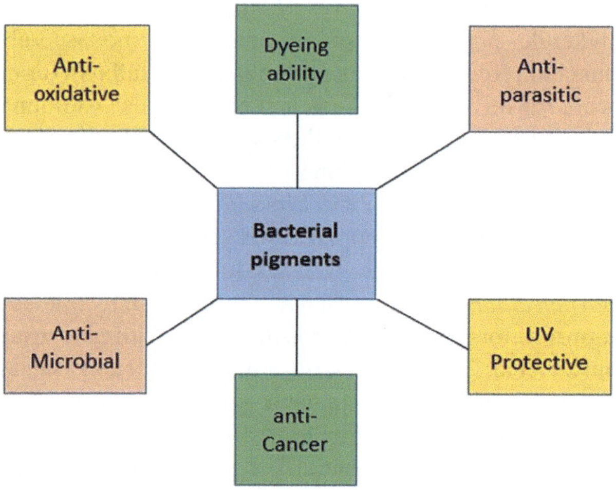

Fig. 3.2 Key applications of bacterial pigments

color depends on the type of substrate used; the same pigment can yield different shades on different fabrics. Additionally, dyeing conditions such as pH, temperature, and the use of mordants can also affect the resulting color (Alihosseini et al. 2010).

Recent advancements in genetic engineering have made it possible to modify bacteria to produce specific pigments. For example, *Streptomyces coelicolor*, which naturally produces a blue pigment, can be genetically altered to produce radiant yellow (kalafungin), orange, or yellow-red (anthraquinones) pigments (Narsing Rao et al. 2017). rodigiosin, a red pigment first isolated from *Vibrio psychoerythrus* and *Serratia marcescens*, is well known for its antibacterial, antineoplastic (anticancer), and anti-malarial properties (Khanafari et al. 2006). *seudomonas aeruginosa* produces pyocyanin, a blue pigment with antibacterial and antifungal properties that also serves as a biocontrol agent (Vinotha et al. 2019). Violacein, a violet-colored pigment discovered in *Chromobacterium violaceum* from the Amazon River, has a broad range of biological applications in the textile, food, and healthcare industries (Yada et al. 2008; Kamla et al. 2012). Carotenoids, which range from yellow to orange-red in color, are produced by bacteria like *Serratia* and *Streptomyces*, with other high-carotenoid producers including *Agrobacterium*, *Myxococcus*, *Sulfolobus*, and *Mycobacterium* (Browning et al. 2003).

3.3.3 Algal Pigments

Algae are a diverse group of pigmented eukaryotic organisms found in both freshwater and marine environments. Depending on the species, algae can alter the color of the water, producing shades of blue, green, or brown. Algae are classified into six major phyla based on their dominant pigments, which in turn determine their coloration (Menaa et al. 2021; Almoulki and Akkaya 2022). The Chlorophyta (Green Algae) are green due to their high content of chlorophyll pigments, specifically chlorophyll a, chlorophyll b, carotenes, and xanthophylls. Phaeophyta (Brown Algae) includes brown algae, which contain both xanthophylls and chlorophyll a, contributing to their characteristic brown color. Rhodophyta (Red Algae) are red because of the presence of phycoerythrin, chlorophyll a, phycocyanin, xanthophylls, and carotenes. Euglenophyta (Euglena) species contain carotenes, chlorophyll a, chlorophyll b, and xanthophylls. Chrysophyta (Diatoms and Golden-Brown Algae) possess chlorophyll a and carotenes, which contribute to their golden-brown hue. Finally, Pyrrophyta (Dinoflagellates) have carotenes, chlorophyll a, and xanthophylls (Menaa et al. 2021).

Each group of algae is characterized by its unique pigment combination, which affects its color and ecological role. Algae-based dyes offer several advantages for the textile industry due to their non-toxic nature and the absence of allergens. Algal products contribute to enhanced energy and water efficiency and promote safer working conditions by eliminating the need for hazardous chemicals (Bose and Rajasekaran 2016). Despite these benefits, the use of algal pigments in textiles is still limited due to insufficient research on their application in fabric refining and dye separation methods.

Studies have demonstrated that *Sargassum muticum* produces vibrant color on cotton fabric when dyed at 75 °C for 30 min in an alkaline solution with 4.0 g of NaCl per 100 mL. Cotton fabrics dyed with *Colpomenia sinuosa* and *Lithomyrtus obtusa* show optimal color strength under specific conditions: *Colpomenia sinuosa* is most effective at a pH of 11, with a dyeing time of 60 min at 60 °C and 8.0 g of NaCl per 100 mL, while *Lithomyrtus obtusa* performs best at 30 °C for 45 min, at a pH of 7.0, using 6.0 g of NaCl per 100 mL (Azeem et al. 2019). A study investigated pigments extracted from red, brown, and green algae to assess their ability to inhibit common pathogens and their adherence to textiles while retaining color (Mona et al. 2019). In Britain, scientists were pioneers in using algae to color textiles and artworks. Diatoms, single-celled organisms with shiny, crystalline shells, possess the unique ability to change color depending on the arrangement of pores in their shells. This color change occurs without altering the fabric's chemical composition. As the paint hardens, algae align horizontally on the surface, creating consistent light reflection, which results in a strong visual impact even with translucent paints and silica shells. These characteristics suggest that algae-based dyes could be a viable, cost-effective, and safe alternative for textile dyeing. Algae demonstrate the effectiveness of using sustainable resources (Ayyanar et al. 2023).

Algal pigments are commonly analyzed using advanced techniques such as spectrophotometry and high-performance liquid chromatography, which are essential for water quality assessment and phytoplankton research. Important pigments like chlorophylls, xanthophylls, and carotenoids have commercial value and are used in the pharmaceutical, textile, cosmetic, and food industries. Additionally, there is increasing interest in using phycoerythrin, a red pigment from the macroalga Gracilaria sp., for dyeing cotton fabrics (Njiru et al. 2018). Chlorophylls provide a green color, carotenoids yield yellow or orange hues, and phycobilins, which are water-soluble, include phycoerythrin (red) and phycocyanin (blue) (Tunáková and Militký 2016).

3.4 Technologies for Producing Pigments from Bacteria

3.4.1 Strain Development

In the past, random mutagenesis followed by repeated rounds of selection were the primary methods used to enhance strains; these techniques remain highly effective today (Sonderegger and Sauer 2003; Portnoy et al. 2011). The development of gene deletion techniques over the past decade has enabled the effective inactivation of genomic DNA, significantly advancing bacterial metabolic engineering years (Murphy et al. 2000; Bloor and Cranenburgh 2006). As a result, the focus of strain development has shifted toward a more integrated and systematic approach. Wild strains often produce insufficient pigment content for commercial applications, and as a result, the evolution of industrial strains is driven by economic factors. Isolating bacteria that can generate colors in shorter fermentation times has become crucial. A common factor across all commercial fermentation processes is the enhancement of microbes responsible for the mass production of industrial products. Common mutagens, such as 1-methyl-3-nitro-1-nitrosoguanidine (NTG), ethyl methane sulfonate (EMS), and ultraviolet light (UV), can easily enhance strains, resulting in multiple times more pigmentation than would otherwise occur (Tripathi et al. 2001; Chen et al. 2003).

3.4.2 *Fermentation*

Today, the biotechnological synthesis of colors using microorganisms is an important field of study. In relation to pigment production, microbial fermentation and gene editing have been extensively researched. The two main strategies for biotechnologically producing natural colors are: (1) identifying new color sources and enhancing their color production capacity, and (2) improving existing strains and optimizing process parameters to achieve higher and more stable yields from well-established, reliable colorant sources (Fig. 3.3). Another significant challenge is the investment required for manufacturing facilities and research and development.

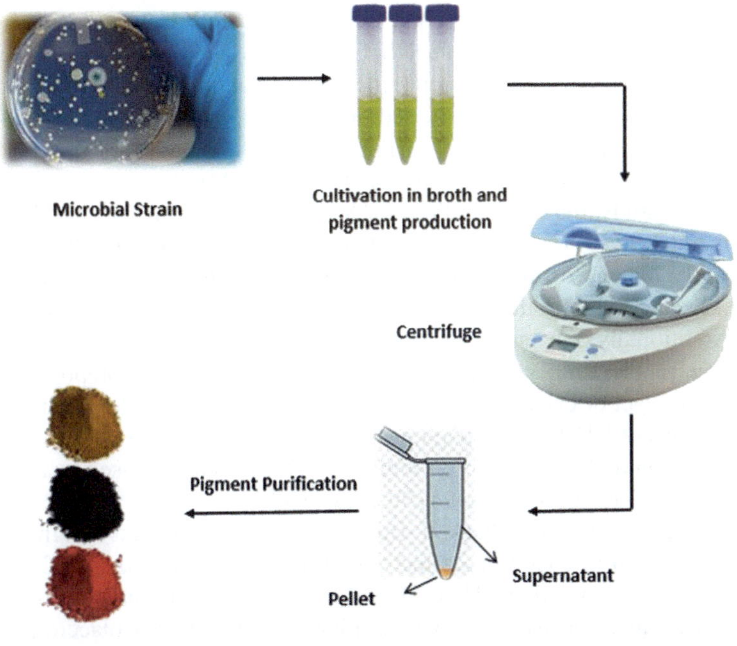

Fig. 3.3 Biotechnological production of natural pigments using microbial strains

When fermentation physiology and metabolic engineering are applied correctly, it is possible to produce colorants in large quantities from bacteria (Nielsen and Olsson 2002). Advances in gene technology have led to efforts to create cell factories for pigment production through the heterologous expression of biosynthetic pathways from both known and novel pigment-producing microorganisms. A key strategy to achieve high yields of valuable fermentative compounds is the optimization of fermentation processes (Parekh et al. 2000).

Medium optimization is one of the most critical steps in achieving the highest pigment yields. This process considers various factors, including medium composition, operating parameters, pH, temperature, agitation, and airflow. While traditional optimization methods often focus on adjusting one factor at a time, this approach has limitations, such as requiring numerous trial runs and being time-consuming. Response Surface Methodology (RSM) has become increasingly important for process and material improvements in pigment production. This method allows the simultaneous analysis of multiple variables, discovered through carefully planned experiments (Gharibzahedi et al. 2012). The primary advantage of RSM is that it reduces the number of trials required to evaluate several interacting variables, making it quicker and less labor-intensive compared to traditional techniques (Ghasemlou et al. 2012; Gharibzahedi et al. 2013).

Several studies have employed the Plackett–Burman design to identify key factors involved in specific processes (Yu et al. 1997). SM has also been effectively used to optimize fermentation conditions and medium components while requiring fewer tests (Mamatha et al. 2008; Singh and Bishnoi 2012). Additionally, RSM is used to examine quadratic effects and interactions between process variables. For instance, using statistical experimental techniques, a study optimized canthaxanthin synthesis from *D. natronolimnaea* HS-1 grown on cheese whey, a cost-effective substrate. Sequential optimization (including fractional experimental design, response surface analysis, and center composite design) led to an increased production of canthaxanthin (2872 ± 77 g/L) in response to changes in pH, cheese whey concentration, and yeast extract (Khodaiyan et al. 2008).

In another study, the ideal medium for *S. marcescens* to produce prodigiosin was determined Cultivation with glycine and sugar as energy and carbohydrate sources significantly increased prodigiosin biosynthesis. When glycine and sucrose were added to the medium, prodigiosin production increased 2.13 and 2.25 times, respectively. Additionally, the inorganic supplement KH_2PO_4 enhanced prodigiosin formation by accelerating cell growth, while other inorganic components had no significant impact or even reduced production. RSM analysis identified the optimal concentrations of KH_2PO_4 and peptone. Experimental confirmation showed only an 8.76% variation in the predicted prodigiosin yield, which increased by almost 8.52-fold (from 283.3 to 2623.39 mg/L) (Su et al. 2011).

Various studies used the Plackett–Burman and Box–Behnken designs to investigate how cultivation conditions affect the ability of *Duganella sp.* B2 to produce violacein. They found that pH, flask volume, and the amounts of beef extract, L-tryptophan, and potassium nitrate significantly influenced violacein production. Under optimal conditions, *Duganella sp.* B2 produced 1.82 g/L of violacein, a

4.6-fold increase, which was 3.6 times greater than the best yield reported for *C. violaceum* (Wang et al. 2009).

The cost of bioprocesses significantly impacts the substrates used for bioproduct synthesis. To produce bioproducts economically, it is essential to select low-cost and efficient substrates. Pigments have been successfully produced from various agricultural by-products, including corn cob (Velmurugan et al. 2011), sugarcane bagasse (Silveira et al. 2013), grape waste (Silveira et al. 2008), seed from jackfruit (Babitha et al. 2006), maize steep liquor (Hamano and Kilikian 2006), wheat materials (Domínguez-Espinosa and Webb 2003), and cassava (Yongsmith et al. 1993). By using low-cost feedstocks to reduce the cost of microbial fermentation, the production of canthaxanthin using carotenogenic strains can be more economical and productive from an industrial perspective (Bhosale 2004).

Agro-industrial residues have recently been explored as carbohydrate and nutrient sources for carotenoid pigment production. These include cheese whey (Frengova et al. 1995), molasses from sugarcane (Bhosale and Gadre 2001), sucrose syrup (Buzzini and Martini 2000), peat hydrolysate (Martin et al. 1993), cellobiose (Cruz and Parajo 1998), and beetroot molasses (Aksu and Eren 2005). Beet molasses, a by-product of beet sugar production, contains approximately 50% (w/w) sucrose and is rich in growth factors such as inositol, glutamic acid, pantothenic acid, betaine, biotin, and trace metals. This makes beet molasses a valuable, low-cost nutrient source for microbial fermentation (Xu and Xu 2014; Atayeter 2015). This material, being a major waste product of the beetroot sugar industry, has the potential to serve as a cost-effective substrate for the microbial fermentation process that produces carotenoids (Asadi 2006).

One of the most notable characteristics of the *Rheinheimera baltica* group is its ability to produce an intense blue pigment, although its chemical properties and production dynamics remain poorly understood, and its ecological significance is still unclear. Recent studies indicate that environmental factors, such as interactions with other microorganisms, play a significant role in pigment formation (Bode 2006). For instance, *Pseudomonas aeruginosa* was found to be encouraged to produce pyocyanin, a blue pigment with antibiotic properties, when cultured alongside an *Enterobacter* species (Pup14B) (Angell et al. 2006). Several small signaling molecules have been identified, and numerous microbial systems have been molecularly characterized. These findings suggest that environmental factors, including the composition of the growth medium and species interactions, affect pigment production. A derivative of violacein is also produced by the *Janthinobacterium lividum* S9601 strain (deposited as FERM P15894), isolated from discarded cocoons or stained silk strands. This dye is noted for its exceptional color tone, fabric texture, and dyeing fastness (JP 10113169). Simultaneous production of protease and violacein occurs when protein fibers, such as animal hair or fabric, are dipped in the culture, significantly enhancing the feel of the fibers while also causing them to degrade (JP 6341069) (Bhaskaran 2014). The liquid cultures of *Chromobacterium* and *Janthinobacterium* species contain degraded materials, including animal hair. To optimize the production of bacterial pigments, large-scale culture techniques must be developed using readily available, low-cost materials for commercial processes. Standardization of the culture medium and fermentation conditions is crucial to developing a method that maximizes pigment yields. By using statistical experimental design techniques, process variability can be minimized, development time shortened, and the output response enhanced to ensure more consistent and higher pigment production.

3.4.3 Recovering and Separating Pigments from Bacteria

The development of new research fields focusing on the physiological and pharmacological aspects of bacterial pigments has led to improvements in both the quantity and quality standards for these pigments. However, significant challenges remain in the separation and purification processes, which hinder the large-scale production of prodigiosin. Traditional methods for separating and purifying prodigiosin have relied on organic solvent extraction (Montaner and Pérez-Tomás 2001). Since most of the prodigiosin produced by *Serratia marcescens* is bound to the bacterial envelopes, with only a small amount being released into the culture broth, the process is laborious and time-consuming (Wang et al. 2004). Additionally, a large volume of solvent is consumed, and only small quantities of highly pure prodigiosin are obtained.

Several natural macromolecules, such as organic acids, peptides, proteins, nucleic acids, and other substances, have been isolated and purified using non-ionic adsorbed resins (Hamilton et al. 2000; Hicketier and Buchholz 2002). These resins offer high loading capacities, which enable large-scale compound separation. The desired product can be directly adsorbed from the culture broth onto the chosen resin. This new approach eliminates the need for cell separation and prodigiosin extraction stages. By using fewer solvents for extraction and recyclable adsorbents, this method can significantly reduce operational costs.

An efficient adsorption technique for the immediate separation and purification of prodigiosin from culture medium was developed resulting in excellent quantitative recovery using non-ionic adsorbents. Compared to the 52% recovery achieved by silica-gel chromatography and traditional extraction methods, the overall recovery rate of this process (85%) was notably higher. Moreover, the adsorption resin (X-5) used in this procedure was highly loadable and easily regenerable. These advantages make it possible to scale up the separation and purification of bacterial pigments effectively (Wang et al. 2004).

3.5 Factors Affecting Bacterial Pigment Production

Various factors influence the production of pigments by microbes (Fig. 3.4), which are briefly discussed below.

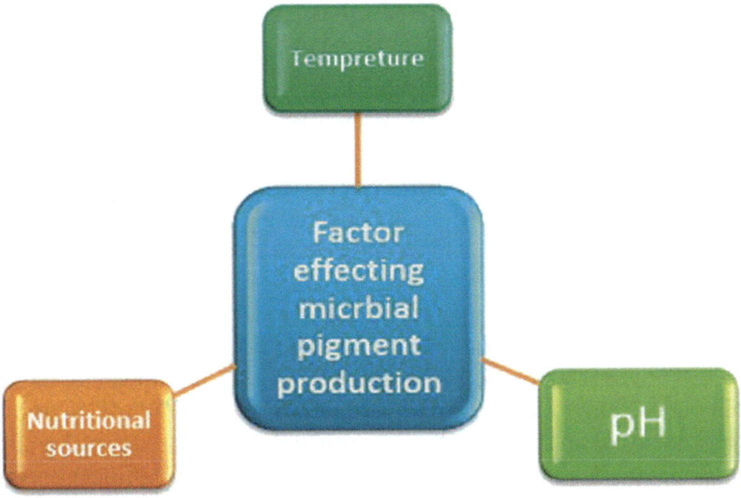

Fig. 3.4 Factors influencing microbial pigment production

3.5.1 Temperature

The temperature during incubation significantly impacts bacterial pigment synthesis. A temperature range of 23–30 °C is optimal for bacterial growth and pigment production. If the temperature increases beyond this range, pigment synthesis decreases, indicating an inverse correlation between temperature and pigment production (Kumar et al. 2015).

3.5.2 pH

The color of bacterial pigments is influenced by pH; bacteria will only produce pigments in a pH-neutral to slightly alkaline environment (Kumar et al. 2015).

3.5.3 Nutritional Sources

The primary variable in pigment production is the source. Some organisms require a higher carbon source, while others need nitrogen or additional minerals to support their growth. However, certain minerals and nitrogen sources have a significant impact on bacterial pigment production. Therefore, the amount of the source should be carefully regulated based on the bacterial strain, as excessive quantities can inhibit their growth.

3.6 Innovative Methods for Generating Bacterial Pigments

To remain globally competitive, biotechnology companies are constantly searching for new bacterial pigments and ways to increase the output of valuable products. These objectives are typically achieved through conventional techniques such as media formulation, bioreactor and process design, microbial strain selection, and culture enhancement. However, these approaches have notable shortcomings, including the time required to achieve desired results, high costs, and often a low success rate.

Two innovative methods for producing industrially desired bacterial products (e.g., enzymes and pigments) in quantities exceeding 5 years of research have been introduced. These strategies are based on the way bacteria respond to their environment (elicitation) and to nearby bacteria (quorum sensing). While quorum sensing has been extensively studied, much of the research has focused on laboratory-scale applications in the biomedical fields. Nevertheless, there is significant potential for industrial use of this bacterial interaction mechanism, as evidenced by the range of physiological functions influenced by quorum sensing (Raina et al. 2011).

Quorum sensing systems are known to regulate the synthesis of bacterial pigments, such as prodigiosin and violacein (McClean et al. 1997; Thomson et al. 2000). By accumulating signaling molecules, a bacterial cell using quorum sensing can determine the density of its neighbors. For example, when cultivated on G agar plates, *Pantoea agglomerans* has been reported to produce a deep blue pigment at high cell concentrations. The biotechnology and medical sectors could benefit from quorum sensing-induced bacterial pigment production, as this method could facilitate the large-scale production of bioproducts that are commercially attractive through bacterial communication. These compounds are promising candidates for industrial-scale production since they require low cultural inputs when biotic elicitors and quorum-sensing molecules are applied, significantly boosting the yield of the desired products (Raina et al. 2011).

3.7 Applications of Microbial Pigments in Textile Industry

The textile industry uses approximately 1.3 million tons of dyes, pigments, and dye precursors, with an annual market value of around 2.3 billion US dollars. Predominantly synthesized, these dyes have significant drawbacks: (i) their production requires hazardous chemicals, raising significant worker safety concerns; (ii) they generate hazardous wastes; and (iii) they are detrimental to the environment (Venil et al. 2013). Bacterial and fungal pigments offer a promising alternative source of natural colorants, characterized by rapid growth, straightforward manufacturing, and independence from weather conditions.

Each year, 200,000 tons of dyes are discharged as effluents during dyeing and finishing processes. These dyes often escape traditional wastewater treatment methods and remain in the environment due to their exceptional stability against light, water, temperature, detergents, soap, bleach, chemicals, and perspiration (Ogugbue and Sawidis 2011). In response to the increasing focus on eco-friendly dyes, pigments produced by bacteria, fungi, yeast, and algae provide sustainable alternatives for textile dyeing (Chadni et al. 2017). These pigments are distinguished by their wide range of colors and chemical structures (Metwally et al. 2021). Beyond their aesthetic qualities, they offer notable bioactive properties, including anticancer, antimicrobial, antioxidant, insect repellent, and ultraviolet protective effects (dos Santos Silva et al. 2020; Haji and Naebe 2020; Venil et al. 2020a, b; Kramar and Kostic 2022). Some of these properties are shown in Fig. 3.5.

Microbial pigments have demonstrated versatility in fabric dyeing applications. Prodigiosin, sourced from *Vibrio* species, has shown effectiveness in dyeing dry wool, silk, nylon, and acrylic fabrics (Alihosseini et al. 2008). In comparison, pigments extracted from *Serratia marcescens*, when used in conjunction with tamarind as a mordant, exhibit the capability to color a wide array of fabrics, including polyesters, acrylics, polyester microfibers, silk, and cotton (Venil et al. 2021). The cultured *Serratia marcescens* producing prodigiosin is shown in Fig. 3.6.

Moreover, the crude gel of prodigiosin pigment, harvested directly from the cultured medium, offers significant economic and environmental benefits, positioning it as a viable replacement for synthetic dyes in the textile industry. This is due to prodigiosin's superior biodegradability and lower toxicity (Metwally et al. 2021). In contrast, synthetic dyes, which are frequently sourced from non-renewable or

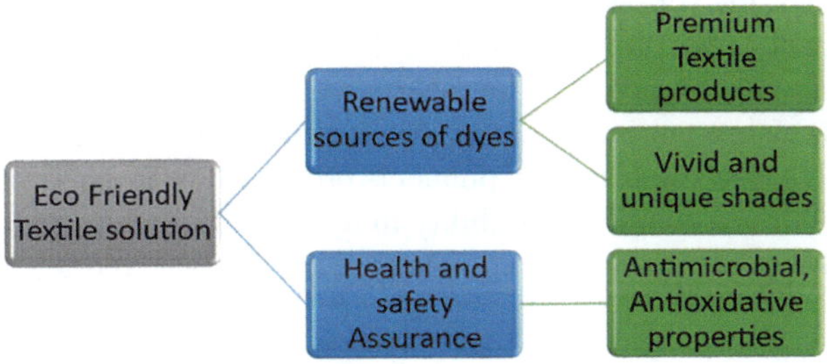

Fig. 3.5 Overview of the benefits of microbial pigments

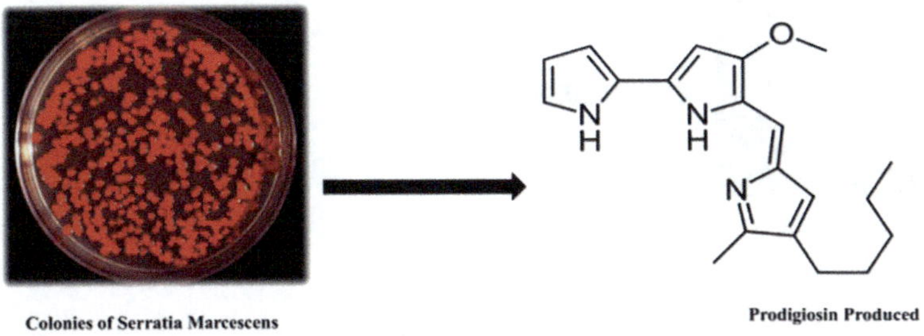

Fig. 3.6 *Serratia marcescens* as a microbial source of the red pigment prodigiosin

petrochemical origins, present considerable ecological and health risks due to their harmful and toxic effects. The environmental friendliness and safety profile of prodigiosin underscore its potential to replace these conventional dyes, contributing to more sustainable and eco-conscious textile practices.

Furthermore, prodigiosin pigment production via fermentation offers significant environmental and economic advantages, particularly when utilizing natural substrates such as agro-industrial byproducts, which are both cost-effective and environmentally friendly (Nguyen et al. 2022). Investigations have also revealed that prodigiosin offers excellent fastness properties and vibrant coloration in textiles, suggesting its potential as a natural replacement for synthetic dyes in the textile industry (Metwally et al. 2021; Kramar and Kostic 2022; Mouro et al. 2023).

Similarly, anthraquinones from *Fusarium oxysporum* (Fig. 3.7) have been applied to dye dry wool fabrics (Nagia and El-Mohamedy 2007). The studies have investigated the use of pigments from *Penicillium minioluteum* for dyeing wet blue goat nappa skin. Furthermore, a red pigment extracted from *Talaromyces verruculosus* has demonstrated effective coloration for cotton fabrics, providing a suitable color tone without cytotoxicity (Chadni et al. 2017).

Microbial pigments impart varied color tones depending on the type of textile. Colorfastness testing indicates that dyed fabrics retain their color across a range of external factors, including perspiration, multiple washes, and friction or rubbing (Venil et al. 2013). Similar textile dyeing was observed with the pigment derived from *Janthinobacterium lividum*, which yields a blue-purplish hue on silk, wool, and cotton, while resulting in a dark blue shade on nylon and vinylon (Hiroshi 2000). Similarly, the yellow pigment sourced from *Thermomyces* was evaluated for its dyeing efficacy on cotton, silk, and wool fabrics. It was found that silk showed a significantly greater attraction to the *Thermomyces* pigment compared to the other textiles, resulting in a more vibrant and pronounced coloration on silk (Poorniammal et al. 2013). The deep blue and red pigments produced by *Streptomyces* strains NP2 and NP4 exhibit considerable variability in dyeing efficacy depending on the type of material. These pigments resulted in intense and vibrant coloration on polyamide

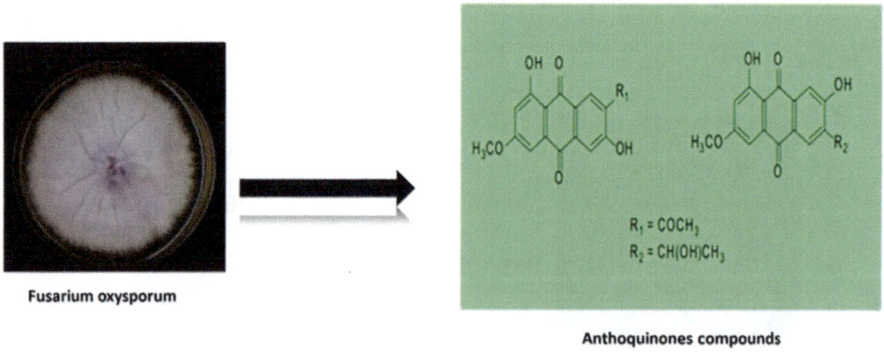

Fig. 3.7 Anthraquinone compounds extracted from *Fusarium oxysporum* grown on PDA, used as natural dyes for wool

and acrylic fibers, whereas they produced faint staining on cotton and other cellulosic fibers (Kramar et al. 2014).

The dyeing process employed a straightforward method, either immersing the fabric in pigment extract or heating it with bacterial cells. By modifying the immersion time and adjusting the temperature of the dye bath, color variations were achieved (Venil et al. 2013). Moreover, textiles dyed with microbial pigments have demonstrated antimicrobial properties. Specifically, fabrics colored with prodiginines derived from *Vibrio* species exhibited antibacterial activity against *Staphylococcus aureus* and *E. coli* (Alihosseini et al. 2008). Moreover, textiles dyed with microbial pigments demonstrate antimicrobial properties. Specifically, fabrics colored with prodiginines (structures shown in Fig. 3.8), derived from *Vibrio* species, exhibited antibacterial activity against *Staphylococcus aureus* and *E. coli* (Borić et al. 2011).

In recent years, the production of colorants (natural dyes) for textile use has garnered considerable attention. Currently, most colorants are derived from non-renewable sources such as fossil oils. Despite the cost-effectiveness and advanced technical sophistication of synthetic colorant production, which provides a comprehensive spectrum of colors, the shift toward natural dyes is gaining momentum. However, synthetic colorants face several significant challenges, including reliance on non-renewable oil sources, sustainability concerns, environmental toxicity, and potential health risks associated with certain dyes.

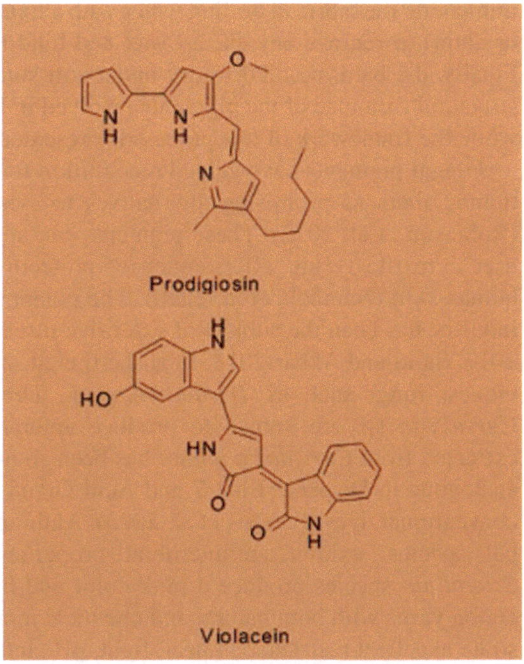

Fig. 3.8 Chemical structures of prodiginine derivatives with antimicrobial properties

As a viable alternative, the biosynthesis of pigments via fermentation offers significant potential. These biosynthesized pigments can serve as fundamental chromophores for subsequent chemical alterations, potentially resulting in colorants with a wide range of hues. Microorganisms synthesize a wide array of stable pigments, including carotenoids, flavonoids, quinones, and rubramines. The fermentation process not only achieves higher pigment yields but also generates fewer residues compared to other traditional plant and animal sources (Hobson and Wales 1998). Additionally, certain natural colorants, particularly anthraquinone-type compounds, have demonstrated significant antibacterial activity in addition to producing vibrant hues, making them functional dyes for antimicrobial textiles (Frandsen et al. 2006).

Another application of bacterial pigments is their use in the creation of batik, where prodigiosin and violacein are employed for dyeing. Batik, a culturally significant attire typically worn by women in Southeast Asia, is a traditional gown distinguished by its intricate motifs, which often include representations of leaves, flowers, and geometric patterns. The chosen design was first drawn by an artist using pencil, following the "Batik-Tulis" technique. Subsequently, a mixture of beeswax and paraffin wax was melted and applied to the crafted motifs using a technique known as "canting." The beeswax helped secure the fabric, while the paraffin wax contributed to the characteristic cracking observed in batik. In the areas where the wax had soaked into the fabric, the dye could not penetrate. After the waxing procedure, the fabrics were coloured with bacterial pigment using the brushing method. The hue was modified by incorporating either ethyl acetate (for red and purple pigments) or acetone (for yellow pigment). The next step involved immersing the fabric in boiling water with a fixative (copper sulphate, iron sulphate, or alum) to remove any excess wax and bind the bacterial pigment to the fabric. Finally, the batik was left to dry under soft sunlight (Ahmad et al. 2012a, b). The structural formulae of the pigments involved in Batik making are shown in Fig. 3.9, while the framework of this process is presented in Fig. 3.10.

Fungal pigments have gained recognition for their stability and reliability, positioning them as promising alternatives to synthetic dyes in the textile industry (Robinson et al. 2012). These pigments can absorb ultraviolet light, which, when used in textile dyeing, offers enhanced protection against harmful UV radiation for human skin (Nambela et al. 2020). The potential of fungal pigments in the textile industry has been the subject of extensive research (Kumar et al. 2015; Caro et al. 2017; Sajid and Akbar 2018; Lagashetti et al. 2019). Using biotechnological techniques, fungi such as *Trichoderma* sp., *Drechslera* sp., *Aspergillus* sp., and *Curvularia* sp. are known to produce anthraquinones. Specifically, cynodontin extracted from *Curvularia lunata* has been shown to yield two anthraquinone dyes analogous to Disperse Blue 7 and Acid Green 28, with properties comparable to conventional dyes (Stevens et al. 2023). Anthraquinones, produced by various fungal species, exhibit antimicrobial properties. Anthraquinones isolated from *Sclerotinia* species produce a pink color and have been successfully used to dye cotton yarns with both natural and chemical mordants. The resulting fibers demonstrate excellent resistance to heat, light, pH, and temperature fluctuations (Perumal et al. 2009).

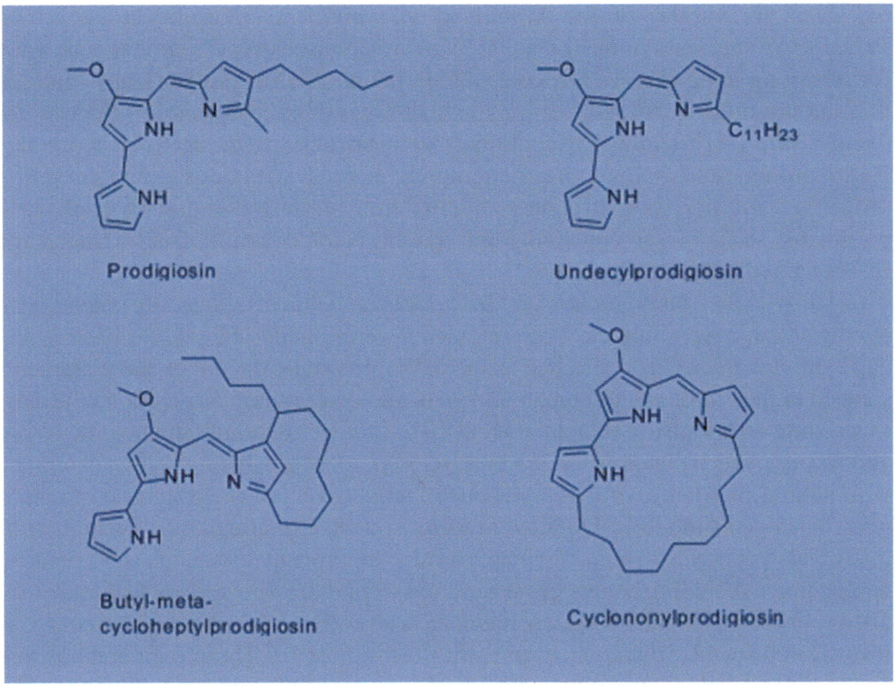

Fig. 3.9 Microbial pigments used in Batik fabric dyeing

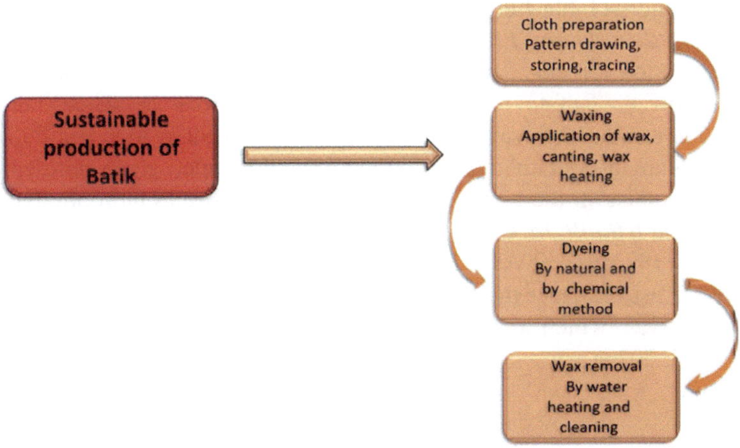

Fig. 3.10 Framework for Batik production process

The study reported the presence of red and yellow pigments in *T. australis* and *P. murcisnum*, which show a strong affinity for wool fibers and are water-soluble. These dyes maintain their integrity under temperature and pH changes, making them suitable for industrial applications. Chemical analysis has revealed that the red dyes are similar to the Monascus-type pigments (Visagie et al. 2014). *Penicillium* species synthesize ankaflavin, a Monascus-type pigment renowned for its strong ionic affinity for wool (Afshari et al. 2015). Another study reported that anthraquinone extracted from *Fusarium oxysporum* is highly effective for dyeing wool, providing exceptional colorfastness and high dye uptake. Anthraquinones are known for producing vivid colors with outstanding durability properties. Furthermore, structural modification of these compounds could be a promising strategy for enhancing the synthesis of dye molecules (Nagia and El-Mohamedy 2007).

A study isolated a pigment from *Trichoderma* sp. and evaluated its potential for dyeing silk and wool fabrics. These pigments were found to be safe for human skin and exhibited microbial resistance features. Wool samples dyed with these pigments showed higher color values compared to silk and demonstrated exceptional fastness to washing and abrasion (Gupta et al. 2019). Another study investigated the use of yellow pigments isolated from *Thermomyces* sp. in textile dyeing. When combined with natural mordants, yellow pigments effectively reduce the influence of pathogens. The dyed silk fabric exhibited moderate overall colorfastness. Due to its antimicrobial properties, this pigment shows significant potential for medical applications, including bandages, masks, and wound dressings. (Poorniammal et al. 2013). In a study, reddish brown pigment derived from *Alternaria alternata* was investigated for its efficacy in pigmenting cotton fabrics. These pigments showed commendable color durability against both sweat and abrasion, marking an initial investigation into the use of *Alternaria alternata* pigments for cotton dyeing (Devi and Karuppan 2015; Venil et al. 2020a, b). Meanwhile, Velmurugan et al. assessed a range of water-soluble fungal pigments from *Monascus purpureus*, *Isaria farinosa*, *Emericella nidulans*, *Fusarium verticillioides*, and *Penicillium purpurogenum* for dyeing cotton threads. Their findings demonstrated that pre-mordanting with alum and ferrous sulfate resulted in notable variations in shades and colors (Velmurugan et al. 2010; Venil et al. 2020a, b). In addition to scientific advancements with pigments obtained from fungi and bacteria, microalgae also demonstrate significant promise in the textile industry. A study utilized chlorophylls extracted from *Caulerpa lentillifera*. Considering the extensive availability of microbial pigments, their affinity for diverse textiles, and their non-toxic properties, these pigments are likely to enhance their market appeal. They present a promising alternative to synthetic dyes, which are harmful to both the environment and human health (Barreto et al. 2023).

3.8 Advantages of Using Microbial Pigments

Microorganisms represent a viable alternative source of natural colorants, with fungi, in particular, offering a broader spectrum of pigments that can be produced economically and with high efficacy (Poorniammal et al. 2021) (Table 3.1). Bio-based colorants derived from plants, animals, and microorganisms are

Table. 3.1 Pigments produced by different microorganisms, modified from (Malik et al. 2012; Venil et al. 2020a, b)

Microorganisms	Name of species	Pigment/molecule	Color
Bacteria	Corneobacterium insidiosin	Indigodine	Blue
	Rugamonas rubra, Streptoverticillium rubrireticuli, Alteromonas rubra	Prodigiosin	Red
	Pseudomonas aeruginosa	Pyocyanin	Blue-green
	Janthinobacterium lividum	Violacein	Purple
	Bradyrhizobium spp	Canthaxanthin	Orange
	Streptomyces echinorube	Rubrolone	Red
	Flavobacterium spp.	Zeaxanthin	Yellow
	Xanthomonas oryzae	Xanthomonadin	Yellow
	Paracoccus carotinifaciens	Astaxanthin	Pink-red
	Staphylococcus aureus	Staphyloxanthin zeaxanthin	Golden yellow
	Bradyrhizobium sp	Canthaxanthin	Dark-red
Fungi	*Blakeslea trispora*	β-carotene	Cream
	Penicillium oxalicum	Anthraquinone	Arpink red
	Alternaria alternata	Anthraquinone	Reddish-Brown
	Aspergillus sp. AN01	Asperyellone	Yellow
	Talaromyces australis	Carotenoids	Red
	Phoma herbarum	Magenta pigment	Magenta
	Monascus purpureus	Rubropunctamine	Red
	Blakeslea trispora	Lycopene	Red
	Cordyceps unilateralis	Naphtoquinone	Deep-blood-red
	Ashbya gossip	Riboflavin	Yellow
	Penicillium murcianum	Carotenoids	Yellow
	Talaromyces australis	Carotenoids	Red
	Chlorociboria aeruginosa	Quinones	Green
	Scytalidium cuboideum	Quinones	Red
Yeast	*Phaffia rhodozyma*	Astaxanthin	Pink-red
	Rhodotorula sp. *Rhodotorula glutinis*	Torularhordin	Orange-red
Algae	*Dunaliella salina*	β-carotene	Red
	Chlorococcum	Lutein	
	Hematococcus	Canthaxanthin	
Actinomycetes	*Streptoverticillium rubrireticuli*	Prodigiosin	Red
	Streptomyces echinoruber	Rubrolone	Red

environmentally friendly and typically exhibit low or no toxicity (Dufosse et al. 2014; Gupta et al. 2019). However, the mass production of pigments from plants and animals is limited by several drawbacks. In contrast, microbial pigments offer substantial potential across various fields due to their numerous advantages (Panesar et al. 2015; Tirumale and Wani 2018).

Although the production cost of β-carotene is significantly higher, its natural and safe characteristics make it a viable competitor to synthetic dyes (Venil et al. 2013; Nigam and Luke 2016). Microbial cells capable of generating pigments are known as pigment producers. These synthesize a wide array of colors, most of which are water-soluble (Joshi et al. 2003; Chattopadhyay et al. 2008; Lin and Xu 2020). Natural pigments are primarily utilized as color additives and serve as antibiotics and antioxidants. Due to the widespread use of synthetic dyes and conflicting safety reports associated with them, there is a pressing need to find safe, natural alternatives. Microbial dyes offer several advantages over plant- or animal-derived colorants, including higher production efficiency, economic viability, enhanced durability, and simplified post-processing (Velmurugan et al. 2010; Tuli et al. 2015).

Fungi represents a beneficial and accessible alternative source of organic dyes, offering notable advantages over plant-derived pigments (Indra Arulselvi et al. 2014; Caro et al. 2017; Gmoser et al. 2017; Narsing Rao et al. 2017; Ramesh et al. 2019). They ensure consistent pigment production throughout the year, exhibit rapid growth in cost-effective media, and generate pigments in a diverse range of shades. These pigments are both stable and soluble, simplifying the manufacturing process (Joshi et al. 2003; Manikprabhu and Lingappa 2013).

Various fungal families, including *Monascaceae, Trichocomaceae, Nectriaceae, Hypocreaceae, Pleosporaceae, Cordycipitaceae, Xylariaceae, Chaetomiaceae, Sordariaceae, Chlorociboriaceae, Hyaloscyphaceae, Hymenochaetaceae, Polyporaceae, Ophiostomataceae, Tremellaceae, Herpotrichiellaceae*, and *Tuberaceae*, have been identified as significant pigment producers (Dufossé 2006; Tuli et al. 2015; Akilandeswari and Pradeep 2016; Caro et al. 2017; Vega Gutierrez and Robinson 2017; Ramesh et al. 2019). These fungi synthesize a wide array of pigments as secondary metabolites, including carotenoids, melanin, flavins, phenazines, quinones, monascins, violacein, and indigo (Mortensen 2006; Kirti et al. 2014).

A strain of *Talaromyces verruculosus* was isolated for its ability to produce a red pigment. This pigment was extracted and applied to cotton fabric using a mordant to achieve effective dyeing. The study assessing this process revealed that the resulting color was satisfactory and retained stability even after laundering. Additionally, a cytotoxicity evaluation confirmed that the pigment-dyed fabrics were safe for human use (Chadni et al. 2017). Fungal pigments present new opportunities for textile production, particularly for medical applications. They offer a promising avenue for exploring natural, eco-friendly pigments that align with public demand for sustainable products. Moreover, biotechnological advancements enable the large-scale production of these pigments at minimal cost, with high yields and efficient extraction processes free from mycotoxins. The development of fungal pigments represents a significant step toward promoting sustainable and environmentally friendly dyes (Poorniammal et al. 2021).

However, several challenges must be addressed to popularize microbial pigments for industrial dyeing, including significant capital investment, improvements in extraction processes, higher production efficiency, and enhanced color permanence. Despite these obstacles, the market for microbial pigments is expanding, offering innovative and sustainable alternatives for the textile industry. This growth opens new avenues for biotechnological advancements and eco-friendly solutions.

3.9 Future Directions

Synthetic pigments have significantly harmed both human health and the ecosystem. In comparison, pigments derived from microorganisms such as bacteria, algae, microalgae, fungi, and yeast have proven to be more valuable and sought after due to their eco-friendly properties. Given the substantial environmental and health risks associated with synthetic dyes, along with the implementation of stringent environmental regulations such as the EU Green Deal, the search for viable alternatives has become imperative.

Biogenic dyes derived from microorganisms have emerged as a promising solution in the textile dyeing industry. These naturally sourced dyes offer a diverse range of chemical structures and essential color options, making them a compelling alternative to synthetic counterparts (Fried et al. 2022). Microorganisms produce a wide variety of pigments, yielding a spectrum of vibrant colors. Notably, fungal pigments, produced through cost-effective processes, demonstrate high colorfastness, positioning them as a viable alternative (Dufosse et al. 2014). The authorized use of fungal pigments varies globally, influenced by local regulations and traditional practices. For instance, a patent has been granted for the use of *Monascus*-like pigments derived from *Penicillium* species on various substrates, including textiles, cotton, wool, leather, paint, and paper (Mapari et al. 2011).

The primary challenge in enhancing both pigment and biomass yield lies in their negative correlation. To achieve controlled pigment production, the relationship between biomass accumulation and pigment synthesis must be thoroughly investigated. Genetic engineering techniques have been employed to regulate pigment production, with recombinant DNA technology being used to modify the function of enzymes involved in carotenoid biosynthesis (Gmoser et al. 2017). However, the commercial manufacturing of fungal pigments, such as *Monascus* pigments, faces difficulties due to their limited water solubility and instability when exposed to varying pH levels, heat, and light. These challenges have been addressed by developing methods to replace oxygen with nitrogen from amino groups during pigment production (Gmoser et al. 2017). Additionally, the co-production of toxic compounds alongside pigments presents a significant hurdle, limiting their application and hindering regulatory approval. The commercial potential of valuable fungal metabolites is further restricted due to the concurrent production of mycotoxins. Concerns regarding the toxicity of fungal pigments have also emerged, necessitating further research to ensure their safety for industrial use.

Regardless of whether they originate from terrestrial or marine sources, introducing versatile and commercially viable microbial pigment products requires substantial investment and extensive experimental research. Several critical factors must be investigated for various biotechnological applications, including color stability under different conditions such as heat (thermolability at varying temperatures and autoclaving), light (photolysis), pH, agitation, aeration, and dissolved oxygen levels. Additionally, optimizing solubility characteristics (e.g., lipolytic, hydrosoluble, and oxidized forms), refining production processes, and conducting rigorous toxicological evaluations are essential (Ramesh et al. 2019).

Apart from these factors, microbial pigment productivity in fermentation is influenced by the type and design of bioreactors such as traditional bioreactors, stirred-tank bioreactors, airlift bioreactors, and trickle-bed reactors as well as the fermentation mode (batch, fed-batch, or continuous) and the physicochemical and biological conditions during the fermentation process (Venil et al. 2014). Although many microbial pigments have the potential for commercialization, high production costs due to expensive synthetic media often pose a major challenge. To address this issue, cheaper agro-industrial residues are being explored as alternative substrates (Ramesh et al. 2019).

The increasing use of synthetic dyes and chemicals in the dyeing industry has significant effects on both human health and the environment. To mitigate these issues, it is crucial to develop efficient and eco-friendly water systems for dyeing processes. Potential solutions include the use of natural dyes and mordants, biodegradable chemicals and enzymes, optimization of the dyeing process, and the adoption of water-free systems. These methods can be implemented individually or in combination with traditional techniques to create more sustainable dyeing practices. The complexity of dye degradation and the wide variety of dyes used for cellulosic fibers highlight the need for further research and development. Most current studies focus on natural dyes and their application to protein fibers. Additionally, a more thorough investigation into the environmental impact and behavior of dyes and their by-products during the dyeing process is necessary. This knowledge will contribute to technological advancements and more sustainable applications in the dyeing industry (Khattab et al. 2020).

3.10 Conclusion

Microbial pigments present a promising advancement in the pursuit of sustainable textile dyeing strategies. Their potential advantages include reduced environmental impact, biodegradability, and a diverse range of colors which make them viable alternatives to synthetic dyes. However, challenges such as stability, colorfastness, and large-scale production must be addressed to enhance their practical application. Current research focused on optimizing microbial fermentation processes and improving pigment stability is essential for overcoming these limitations. With continued innovation and refinement, microbial pigments have the potential to significantly contribute to more sustainable practices in the textile industry, aligning with broader environmental goals and growing consumer demand for eco-friendly products.

References

Afshari M, Shahidi F, Mortazavi SA, Tabatabai F, Es' Haghi Z (2015) Investigating the influence of pH, temperature and agitation speed on yellow pigment production by Penicillium aculeatum ATCC 10409. Nat Prod Res 29:1300–1306

Ahmad WA, Wan Ahmad WY, Zakaria ZA, Yusof NZ, Ahmad WA, Ahmad WYW et al (2012a) Application of bacterial pigments as colorant. Springer, Cham

Ahmad WA, Wan Ahmad WY, Zakaria ZA, Yusof NZ, Ahmad WA, Ahmad WYW et al (2012b) Application of bacterial pigments as colorant: the Malaysian perspective. Springer, Cham

Akilandeswari P, Pradeep B (2016) Exploration of industrially important pigments from soil fungi. Appl Microbiol Biotechnol 100:1631–1643

Aksu Z, Eren AT (2005) Carotenoids production by the yeast Rhodotorula mucilaginosa: use of agricultural wastes as a carbon source. Process Biochem 40:2985–2991

Albarakaty FM (2022) Pigments synthesis by microorganisms an overview. Teikyo Med J 45:3965–3979

Alegbe EO, Uthman TO (2024) A review of history, properties, classification, applications and challenges of natural and synthetic dyes. Heliyon 10:e33646

Alihosseini F, Ju KS, Lango J, Hammock BD, Sun G (2008) Antibacterial colorants: characterization of prodiginines and their applications on textile materials. Biotechnol Prog 24:742–747

Alihosseini F, Lango J, Ju KS, Hammock BD, Sun G (2010) Mutation of bacterium Vibrio gazogenes for selective preparation of colorants. Biotechnol Prog 26:352–360

Almoulki T, Akkaya E (2022) Algae-based dyes for the textile industry. In: 6th EurAsia Waste Management Symposium

Aman Mohammadi M, Ahangari H, Mousazadeh S, Hosseini SM, Dufossé L (2022) Microbial pigments as an alternative to synthetic dyes and food additives: a brief review of recent studies. Bioprocess Biosyst Eng 45:1–12

Angell S, Bench BJ, Williams H, Watanabe CM (2006) Pyocyanin isolated from a marine microbial population: synergistic production between two distinct bacterial species and mode of action. Chem Biol 13:1349–1359

Asadi M (2006) Beet-sugar handbook. John Wiley & Sons, Hoboken, NJ

Atayeter S (2015) The availability of sugar beet molasses as a substitutive medium for cultivation of Nannochloropsis sp. Hibberd, 1981 (Eustigmatophyta). Iran J Fish Sci 14:1089–1097

Avalos J, Carmen Limón M (2015) Biological roles of fungal carotenoids. Curr Genet 61:309–324

Ayyanar I, Ravindran A, Rajendran R, Arulprakasam A, Vellaisamy B (2023) Algae materials. Elsevier, Amsterdam, pp 231–246

Azeem M, Iqbal N, Mir RA, Adeel S, Batool F, Khan AA et al (2019) Harnessing natural colorants from algal species for fabric dyeing: a sustainable eco-friendly approach for textile processing. J Appl Phycol 31:3941–3948

Babitha S (2009) Biotechnology for agro-industrial residues utilisation: utilisation of agro-residues. Springer, Dordrecht, pp 147–162

Babitha S, Soccol CR, Pandey A (2006) Jackfruit seed—a novel substrate for the production of Monascus pigments through solid-state fermentation. Food Tech Biotech 44:465–471

Barreto JVDO, Casanova LM, Junior AN, Reis-Mansur MCPP, Vermelho AB (2023) Microbial pigments: major groups and industrial applications. Microorganisms 11:2920

Bhaskaran K (2014) Process for the production of violacein and its derivative deoxyviolacein containing bioactive pigment from Chromobacterium sp. (MTCC5522)

Bhosale P (2004) Environmental and cultural stimulants in the production of carotenoids from microorganisms. Appl Microbiol Biotechnol 63:351–361

Bhosale P, Gadre R (2001) β-carotene production in sugarcane molasses by a Rhodotorula glutinis mutant. J Ind Microbiol Biotechnol 26:327–332

Bloor AE, Cranenburgh RM (2006) An efficient method of selectable marker gene excision by Xer recombination for gene replacement in bacterial chromosomes. Appl Environ Microbiol 72:2520–2525

Bode HB (2006) No need to be pure: mix the cultures! Chem Biol 13:1245–1246

Borić M, Danevčič T, Stopar D (2011) Prodigiosin from Vibrio sp. DSM 14379; a new UV-protective pigment. Microb Ecol 62:528–536

Bose P, Rajasekaran A (2016) Decolourization of textile dyes using bacterial consortium. Int J Sci Res Environ Sci 4:0017–0022

Browning DF, Whitworth DE, Hodgson DA (2003) Light-induced carotenogenesis in Myxococcus xanthus: functional characterization of the ECF sigma factor CarQ and antisigma factor CarR. Mol Microbiol 48:237–251

Buzzini P, Martini A (2000) Production of carotenoids by strains of Rhodotorula glutinis cultured in raw materials of agro-industrial origin. Bioresour Technol 71:41–44

Caro Y, Venkatachalam M, Lebeau J, Fouillaud M, Dufossé L (2017) Fungal metabolites. Springer, Berlin, pp 499–568

Chadni Z, Rahaman MH, Jerin I, Hoque K, Reza MA (2017) Extraction and optimisation of red pigment production as secondary metabolites from Talaromyces verruculosus and its potential use in textile industries. Mycology 8:48–57

Chattopadhyay P, Chatterjee S, Sen SK (2008) Biotechnological potential of natural food grade biocolorants. Afr J Biotechnol 7:2972–2985

Chen Y, Li D, Lu W, Xing J, Hui B, Han Y (2003) Screening and characterization of astaxanthin-hyperproducing mutants of Haematococcus pluvialis. Biotechnol Lett 25:527–529

Cruz J, Parajo J (1998) Improved astaxanthin production by Xanthophyllomyces dendrorhous growing on enzymatic wood hydrolysates containing glucose and cellobiose. Food Chem 63:479–484

Devi S, Karuppan P (2015) Reddish brown pigments from Alternaria alternata for textile dyeing and printing. Indian J Fibre Text Res 40:315–319

Di Salvo E, Lo Vecchio G, De Pasquale R, De Maria L, Tardugno R, Vadalà R et al (2023) Natural pigments production and their application in food, health and other industries. Nutrients 15:1923

Domínguez-Espinosa RM, Webb C (2003) Submerged fermentation in wheat substrates for production of Monascus pigments. World J Microbiol Biotechnol 19:329–336

Dos Santos Silva PM, Fiaschitello TR, De Queiroz RS, Freeman HS, Da Costa SA, Leo P et al (2020) Natural dye from Croton urucurana Baill. bark: extraction, physicochemical characterization, textile dyeing and color fastness properties. Dyes Pigments 173:107953

Downham A, Collins P (2000) Colouring our foods in the last and next millennium. Int J Food Sci Technol 35:5–22

Dufossé L (2006) Microbial production of food grade pigments. Food Tech Biotech 44:313–321

Dufossé L, Galaup P, Yaron A, Arad SM, Blanc P, Murthy KNC et al (2005) Microorganisms and microalgae as sources of pigments for food use: a scientific oddity or an industrial reality? Trends Food Sci Technol 16:389–406

Dufossé L, Fouillaud M, Caro Y, Mapari SA, Sutthiwong N (2014) Filamentous fungi are large-scale producers of pigments and colorants for the food industry. Curr Opin Biotechnol 26:56–61

Ferahtia A, Mohammed Tahar HF, Mimeche BE (2021) Surface water quality assessment in semi-arid region (El Hodna watershed, Algeria) based on water quality index (WQI). Stud Univ Babeş-Bolyai Chem 66:127–142

Frandsen RJ, Nielsen NJ, Maolanon N, Sørensen JC, Olsson S, Nielsen J et al (2006) The biosynthetic pathway for aurofusarin in Fusarium graminearum reveals a close link between the naphthoquinones and naphthopyrones. Mol Microbiol 61:1069–1080
Frengova G, Simova E, Beshkova D (1995) Effect of temperature changes on the production of yeast pigments co-cultivated with lacto-acid bacteria in whey ultrafiltrate. Biotechnol Lett 17:1001–1006
Fried R, Oprea I, Fleck K, Rudroff F (2022) Biogenic colourants in the textile industry–a promising and sustainable alternative to synthetic dyes. Green Chem 24:13–35
Gessler N, Egorova A, Belozerskaya T (2013) Fungal anthraquinones. Appl Biochem Microbiol 49:85–99
Gharibzahedi SMT, Mousavi SM, Hamedi M, Khodaiyan F, Razavi SH (2012) Development of an optimal formulation for oxidative stability of walnut-beverage emulsions based on gum arabic and xanthan gum using response surface methodology. Carbohydr Polym 87:1611–1619
Gharibzahedi SMT, Mousavi SM, Hamedi M, Khodaiyan F (2013) Application of response surface modeling to optimize critical structural components of walnut–beverage emulsion with respect to analysis of the physicochemical aspects. Food Bioprocess Technol 6:456–469
Ghasemlou M, Khodaiyan F, Gharibzahedi SMT (2012) Enhanced production of Iranian kefir grain biomass by optimization and empirical modeling of fermentation conditions using response surface methodology. Food Bioprocess Technol 5:3230–3235
Gmoser R, Ferreira JA, Lennartsson PR, Taherzadeh MJ (2017) Filamentous ascomycetes fungi as a source of natural pigments. Fungal Biol biotechnol 4:1–25
Gupta N, Poddar K, Sarkar D, Kumari N, Padhan B, Sarkar A (2019) Fruit waste management by pigment production and utilization of residual as bioadsorbent. J Environ Manag 244:138–143
Haji A, Naebe M (2020) Cleaner dyeing of textiles using plasma treatment and natural dyes: a review. J Clean Prod 265:121866
Hamano PS, Kilikian BV (2006) Production of red pigments by Monascus ruber in culture media containing corn steep liquor. Braz J Chem Eng 23:443–449
Hamilton GE, Luechau F, Burton SC, Lyddiatt A (2000) Development of a mixed mode adsorption process for the direct product sequestration of an extracellular protease from microbial batch cultures. J Biotechnol 79:103–115
Hicketier M, Buchholz K (2002) Fluidized bed adsorption of Cephalosporin C. J Biotechnol 93:253–268
Hiroshi K (2000) Isolation of bacteria producing bluish-purple pigment and use for dyeing. Jpn Agric Res Q 34:131–140
Hobson DK, Wales DS (1998) Green dyes. J Soc Dyers Colour 114:42–44
Indra Arulselvi P, Umamaheswari S, Ranandkumar Sharma G, Karthik C, Jayakrishna C (2014) Screening of yellow pigment producing bacterial isolates from various eco-climatic areas and analysis of the carotenoid produced by the isolate. J Food Process Technol 5:2
Joshi V, Attri D, Bala A, Bhushan S (2003) Microbial pigments. Indian J Biotechnol 2:362–369
Kamla M, Jayanti T, Sneh G (2012) A review on microbial pigment. Int J Microbial Res Technol 1:361–365
Kant R (2012) Textile dyeing industry an environmental hazard. Nat Sci 4:22–26
Khanafari A, Assadi MM, Fakhr FA (2006) Review of prodigiosin, pigmentation in Serratia marcescens. Online J Biol Sci 6:1–13
Khattab TA, Abdelrahman MS, Rehan M (2020) Textile dyeing industry: environmental impacts and remediation. Environ Sci Pollut Res 27:3803–3818
Khodaiyan F, Razavi SH, Mousavi SM (2008) Optimization of canthaxanthin production by Dietzia natronolimnaea HS-1 from cheese whey using statistical experimental methods. Biochem Eng J 40:415–422
Kirti K, Amita S, Priti S, Mukesh Kumar A, Jyoti S (2014) Colorful world of microbes: carotenoids and their applications. Adv Biol 2014:837891
Kramar A, Kostic MM (2022) Bacterial secondary metabolites as biopigments for textile dyeing. Text 2:252–264
Kramar A, Ilic-Tomic T, Petkovic M, Radulović N, Kostic M, Jocic D et al (2014) Crude bacterial extracts of two new Streptomyces sp. isolates as bio-colorants for textile dyeing. World J Microbiol Biotechnol 30:2231–2240

Kumar A, Vishwakarma HS, Singh J, Dwivedi S, Kumar M (2015) Microbial pigments: production and their applications in various industries. Int J Pharm Chem Biol Sci 5:203–212
Lagashetti AC, Dufossé L, Singh SK, Singh PN (2019) Fungal pigments and their prospects in different industries. Microorganisms 7:604
Lin L, Xu J (2020) Fungal pigments and their roles associated with human health. J Fungi 6:280
Malik K, Tokkas J, Goyal S (2012) Microbial pigments: a review. Int J Microbial Res Technol 1:361–365
Mamatha S, Ravi R, Venkateswaran G (2008) Medium optimization of gamma linolenic acid production in Mucor rouxii CFR-G15 using RSM. Food Bioprocess Technol 1:405–409
Manikprabhu D, Lingappa K (2013) γ Actinorhodin a natural and attorney source for synthetic dye to detect acid production of fungi. Saudi J Biol Sci 20:163–168
Mapari SA, Meyer AS, Thrane U, Frisvad JC (2009) Identification of potentially safe promising fungal cell factories for the production of polyketide natural food colorants using chemotaxonomic rationale. Microb Cell Factories 8:1–15
Mapari SA, Thrane U, Meyer AS (2010) Fungal polyketide azaphilone pigments as future natural food colorants? Trends Biotechnol 28:300–307
Mapari SA, Meyer AS, Frisvad JC, Thrane U (2011) Google Patents
Martin AM, Lu C, Patel TR (1993) Growth parameters for the yeast Rhodotorula rubra grown in peat extracts. J Ferment Bioeng 76:321–325
Mcclean KH, Winson MK, Fish L, Taylor A, Chhabra SR, Camara M et al (1997) Quorum sensing and Chromobacterium violaceum: exploitation of violacein production and inhibition for the detection of N-acylhomoserine lactones. Microbiology 143:3703–3711
Menaa F, Wijesinghe U, Thiripuranathar G, Althobaiti NA, Albalawi AE, Khan BA et al (2021) Marine algae-derived bioactive compounds: a new wave of nanodrugs? Mar Drugs 19:484
Meruvu H, Dos Santos JC (2021) Colors of life: a review on fungal pigments. Crit Rev Biotechnol 41:1153–1177
Metwally RA, El Sikaily A, El-Sersy NA, Ghozlan HA, Sabry SA (2021) Antimicrobial activity of textile fabrics dyed with prodigiosin pigment extracted from marine Serratia rubidaea RAM_Alex bacteria. Egypt J Aquat Res 47:301–305
Mona S, Yazhini M, Shaukat F, Sekarenthiran C, Maya S (2019) Extraction of algal pigments and their suitability as natural dyes. J Algal Biomass Util 10:1–8
Montaner B, Pérez-Tomás R (2001) Prodigiosin-induced apoptosis in human colon cancer cells. Life Sci 68:2025–2036
Mortensen A (2006) Carotenoids and other pigments as natural colorants. Pure Appl Chem 78:1477–1491
Mouro C, Gomes AP, Costa RV, Moghtader F, Gouveia IC (2023) The sustainable bioactive dyeing of textiles: a novel strategy using bacterial pigments, natural antibacterial ingredients, and deep eutectic solvents. Gels 9:800
Murphy KC, Campellone KG, Poteete AR (2000) PCR-mediated gene replacement in Escherichia coli. Gene 246:321–330
Nagia F, El-Mohamedy R (2007) Dyeing of wool with natural anthraquinone dyes from fusarium oxysporum. Dyes Pigments 75:550–555
Nambela L, Haule LV, Mgani Q (2020) A review on source, chemistry, green synthesis and application of textile colorants. J Clean Prod 246:119036
Narsing Rao MP, Xiao M, Li W-J (2017) Fungal and bacterial pigments: secondary metabolites with wide applications. Front Microbiol 8:1113
Nguyen TH, Wang S-L, Nguyen VB (2022) Recent advances in eco-friendly and scaling-up bioproduction of prodigiosin and its potential applications in agriculture. Agronomy 12:3099
Nielsen J, Olsson L (2002) An expanded role for microbial physiology in metabolic engineering and functional genomics: moving towards systems biology. FEMS Yeast Res 2:175–181
Nigam PS, Luke JS (2016) Food additives: production of microbial pigments and their antioxidant properties. Curr Opin Food Sci 7:93–100
Njiru CW, Makonde HM, Kahindo JM, Matoka EO, Matoka CM (2018) Bio-prospecting of macro-algae for potential industrial dyes. Afr J Biotechnol 17:804–810

Ogugbue CJ, Sawidis T (2011) Bioremediation and detoxification of synthetic wastewater containing triarylmethane dyes by Aeromonas hydrophila isolated from industrial effluent. Biotechnol Res Int 2011:967925

Panesar R, Kaur S, Panesar PS (2015) Production of microbial pigments utilizing agro-industrial waste: a review. Curr Opin Food Sci 1:70–76

Parekh S, Vinci V, Strobel R (2000) Improvement of microbial strains and fermentation processes. Appl Microbiol Biotechnol 54:287–301

Perumal K, Stalin V, Chandrasekarenthiran S, Sumathi E, Saravanakumar A (2009) Extraction and characterization of pigment from Sclerotinia sp. and its use in dyeing cotton. Text Res J 79:1178–1187

Pfander H (1992) Carotenoids: an overview. Methods Enzymol 213:3–13

Poorniammal R, Parthiban M, Gunasekaran S, Murugesan R, Thilagavathi G (2013) Natural dye production from Thermomyces sp fungi for textile application. Indian J Fibre Text Res 38:276–279

Poorniammal R, Prabhu S, Dufossé L, Kannan J (2021) Safety evaluation of fungal pigments for food applications. J Fungi 7:692

Portnoy VA, Bezdan D, Zengler K (2011) Adaptive laboratory evolution—harnessing the power of biology for metabolic engineering. Curr Opin Biotechnol 22:590–594

Raina S, Murphy T, Vizio D, Reffatti P, Keshavarz T (2011) Novel strategies for over-production of microbial products. Chem Eng Trans 24:847–852

Ramesh C, Vinithkumar NV, Kirubagaran R, Venil CK, Dufossé L (2019) Multifaceted applications of microbial pigments: current knowledge, challenges and future directions for public health implications. Microorganisms 7:186

Robinson SC, Tudor D, Cooper PA (2012) Utilizing pigment-producing fungi to add commercial value to American beech (Fagus grandifolia). Appl Microbiol Biotechnol 93:1041–1048

Sajid S, Akbar N (2018) Applications of fungal pigments in biotechnology. Pure Appl Biol 7:922–930

Sandin G, Peters GM (2018) Environmental impact of textile reuse and recycling–a review. J Clean Prod 184:353–365

Saravanan P, Chandramohan G (2011) Dyeing of silk with ecofriendly natural dye obtained from barks of Ficus Religiosa L. Univers J Environ Res Technol 1:268–273

Silveira ST, Daroit DJ, Brandelli A (2008) Pigment production by Monascus purpureus in grape waste using factorial design. LWT - Food Sci Technol 41:170–174

Silveira ST, Daroit DJ, Sant'anna V, Brandelli A (2013) Stability modeling of red pigments produced by Monascus purpureus in submerged cultivations with sugarcane bagasse. Food Bioprocess Technol 6:1007–1014

Singh A, Bishnoi NR (2012) Optimization of ethanol production from microwave alkali pretreated rice straw using statistical experimental designs by Saccharomyces cerevisiae. Ind Crop Prod 37:334–341

Siva R (2007) Status of natural dyes and dye-yielding plants in India. Curr Sci 92:916–925

Sonderegger M, Sauer U (2003) Evolutionary engineering of Saccharomyces cerevisiae for anaerobic growth on xylose. Appl Environ Microbiol 69:1990–1998

Stevens C, Bechtold TM, Manian AP, Pham T (2023) Handbook of Natural Colorants. John Wiley & Sons. https://doi.org/10.1002/9781119811749

Su W-T, Tsou T-Y, Liu H-L (2011) Response surface optimization of microbial prodigiosin production from Serratia marcescens. J Taiwan Inst Chem Eng 42:217–222

Thomson N, Crow M, Mcgowan S, Cox A, Salmond G (2000) Biosynthesis of carbapenem antibiotic and prodigiosin pigment in Serratia is under quorum sensing control. Mol Microbiol 36:539–556

Tirumale S, Wani NA (2018) Fungi and their role in sustainable development: current perspectives. Springer, Cham, pp 413–426

Tripathi U, Venkateshwaran G, Sarada R, Ravishankar G (2001) Studies on Haematococcus pluvialis for improved production of astaxanthin by mutagenesis. World J Microbiol Biotechnol 17:143–148

Tuli HS, Chaudhary P, Beniwal V, Sharma AK (2015) Microbial pigments as natural color sources: current trends and future perspectives. J Food Sci Technol 52:4669–4678
Tunáková V, Militký J (2016) Advanced textiles with enhanced electrical conductivity. Vlak Text 23:202–210
Varadarajan G, Venkatachalam P (2016) Sustainable textile dyeing processes. Environ Chem Lett 14:113–122
Vega Gutierrez PT, Robinson SC (2017) Determining the presence of Spalted wood in Spanish Marquetry woodworks of the 1500s through the 1800s. Coatings 7:188
Velmurugan P, Kim M-J, Park J-S, Karthikeyan K, Lakshmanaperumalsamy P, Lee K-J et al (2010) Dyeing of cotton yarn with five water soluble fungal pigments obtained from five fungi. Fibers Polym 11:598–605
Velmurugan P, Hur H, Balachandar V, Kamala-Kannan S, Lee K-J, Lee S-M et al (2011) Monascus pigment production by solid-state fermentation with corn cob substrate. J Biosci Bioeng 112:590–594
Venil CK, Zakaria ZA, Ahmad WA (2013) Bacterial pigments and their applications. Process Biochem 48:1065–1079
Venil CK, Aruldass CA, Dufossé L, Zakaria ZA, Ahmad WA (2014) Current perspective on bacterial pigments: emerging sustainable compounds with coloring and biological properties for the industry–an incisive evaluation. RSC Adv 4:39523–39529
Venil CK, Devi PR, Ahmad WA (2020a) Valorisation of agro-industrial residues–volume I: biological approaches. Springer, Cham, pp 149–162
Venil CK, Velmurugan P, Dufossé L, Devi PR, Ravi AV (2020b) Fungal pigments: potential coloring compounds for wide ranging applications in textile dyeing. J Fungi 6:68
Venil CK, Dufossé L, Velmurugan P, Malathi M, Lakshmanaperumalsamy P (2021) Extraction and application of pigment from Serratia marcescens SB08, an insect enteric gut bacterium, for textile dyeing. Text 1:21–36
Vinotha M, Prabhavathi P, Vijayaraghavan R, Kumar D (2019) Bacterial pigments and their application in textile industries using mordants. J Adv Sci Res 10:139–145
Visagie C, Houbraken J, Frisvad JC, Hong S-B, Klaassen C, Perrone G et al (2014) Identification and nomenclature of the genus Penicillium. Stud Mycol 78:343–371
Wang X, Tao J, Wei D, Shen Y, Tong W (2004) Development of an adsorption procedure for the direct separation and purification of prodigiosin from culture broth. Biotechnol Appl Biochem 40:277–280
Wang H, Jiang P, Lu Y, Ruan Z, Jiang R, Xing X-H et al (2009) Optimization of culture conditions for violacein production by a new strain of Duganella sp. B2. Biochem Eng J 44:119–124
Weber G, Chen HL, Hinsch E, Freitas S, Robinson S (2014) Pigments extracted from the wood-staining fungi Chlorociboria aeruginosa, Scytalidium cuboideum, and S. Ganodermophthorum show potential for use as textile dyes. Color Techno 130:445–452
Xiao-Mei Y, Jun-Sheng L, Qi-Qian L, Guo-Xia H, Liu-Juan Y (2011) Evaluation of the potential toxicity of Anthraquinone derivatives in Chinese herbal medicines by the resonance light scattering Spectrum. Asian J Chem 23:3631–3634
Xu K, Xu P (2014) Betaine and beet molasses enhance L-lactic acid production by Bacillus coagulans. PLoS One 9:e100731
Yada S, Wang Y, Zou Y, Nagasaki K, Hosokawa K, Osaka I et al (2008) Isolation and characterization of two groups of novel marine bacteria producing violacein. Mar Biotechnol 10:128–132
Yongsmith B, Tabloka W, Yongmanitchai W, Bavavoda R (1993) Culture conditions for yellow pigment formation by Monascus sp. KB 10 grown on cassava medium. World J Microbiol Biotechnol 9:85–90
Yu X, Hallett S, Sheppard J, Watson A (1997) Application of the Plackett-Burman experimental design to evaluate nutritional requirements for the production of Colletotrichum coccodes spores. Appl Microbiol Biotechnol 47:301–305

Chapter 4
Innovative Applications of Microbial Biosurfactants for Textile Cleaning and Finishing

Ghulam Mustafa, Muhammad Islam, Abdur Rehman, Waqas Ahmad, Apridah Cameliawati Djohan, Ika Rahmatul Layly, Marcelo Franco, and Muhammad Irfan

Abstract The textile industry generates significant environmental pollution through its widespread use of synthetic and traditional chemical surfactants in cleaning and finishing processes. Bio-surfactants offer an eco-friendly and sustainable alternative for textile refining. This chapter examines the characteristic features, types, and mechanisms of action of bio-surfactants, focusing on their production through microorganisms and their applications in textile processing. These compounds, characterized by their hydrophilic heads and hydrophobic tails, have diverse applications across the textile industry. The chapter explores how genetic engineering and advanced genome editing tools can enhance bio-surfactant production by leveraging the natural genetic capacity of microorganisms. Applications discussed include soil and dirt removal, fabric treatment, oil removal, and final textile refining. Compared to traditional toxic chemicals that significantly contribute to environmental pollution, bio-surfactants offer both environmental sustainability and economic feasibility, making them an optimal choice for the textile industry.

Keywords Textile processing · Environmental sustainability · Microbial production · Eco-friendly textiles · Fabric treatment · Sustainable manufacturing

---

G. Mustafa · M. Islam · A. Rehman · W. Ahmad · M. Irfan (✉)
Department of Biotechnology, University of Sargodha, Sargodha, Pakistan
e-mail: Irfan.ashraf@uos.edu.pk

A. C. Djohan · I. R. Layly
Research Center for Applied Microbiology, National Research and Innovation Agency (BRIN), Cibinong Bogor, Indonesia
e-mail: apri004@brin.go.id; ikar003@brin.go.id

M. Franco
Department of Exact Science, State University of Santa Cruz, Ilheus, Brazil
e-mail: mfranco@uesc.br

M. Khurshid et al. (eds.), *Microbial Biotechnology in Textile Innovation*, SDGs and Textiles, https://doi.org/10.1007/978-981-95-0347-6_4

4.1 Introduction

Surfactants are amphipathic molecules containing both hydrophobic and hydrophilic components that allow segregation at physical interfaces. The non-polar part is typically made up of hydrocarbon chains, while the polar component can be anionic, amphoteric, nonionic, or cationic. Surfactants reduce interfacial and surface tensions due to the combination of their hydrophobic and hydrophilic moieties, which form microemulsions and enable the solubilization of hydrocarbons in water (Farias et al. 2021).

Textile products are essential in daily life, and their safety during both production and use is of utmost importance. Various factors, including the materials used, manufacturing processes, and finishing chemicals, can impact our health (Salerno-Kochan and Kowalski 2020). In recent years, there has been a noticeable trend toward the development of new-generation textiles with additional functions, which often serve as key selling points for consumers (Mittal and Bahners 2017). Given their unique properties, surfactants play a significant role in the textile industry. They enhance dye adherence to fibers, ensuring better color fastness, and facilitate deep dye penetration into fabrics. Moreover, surfactants prevent dye clumping, which could lead to streaks or patches during the dyeing process. During the scouring stage, surfactants act as emulsifiers, aiding in the removal of contaminants, such as oils, waxes, and sizing agents, from textile fibers. This helps make the fibers cleaner and more receptive to subsequent processing stages (Sivaramakrishnan 2013).

The textile finishing process requires significant amounts of water, and pretreating textile inputs is crucial for producing high-quality fabrics. Effective friction management during manufacturing involves eliminating fiber admixtures and lipophilic compounds, which are often used as lubricants (Kesting et al. 1996). Biosurfactants are increasingly used in textile finishing to mitigate environmental pollution, owing to their emulsification, solubilization, dispersion, wetting, and detergency properties (Mittal and Bahners 2017). These biosurfactants also play a role in textile dyeing and are now commonly found in commercial detergents for cleaning garments. Dye solubility is a significant challenge in the textile industry, as dyes with low water solubility can lead to uneven dispersion and surface buildup on fibers. Adding surfactants to dye mixtures increases the water solubility of the dyes, resulting in more uniform color distribution and improved fiber penetration (Quagliotto et al. 2006).

Due to their hydrophobic (non-polar) and hydrophilic (polar) components (Fig. 4.1), biosurfactants can reduce the surface tension between two phases. The hydrocarbon chain of β-hydroxy fatty acids forms the hydrophobic tail, while the hydrophilic head contains polar functional groups such as carboxylic acid, amino acids, sugars, alcohols, or phosphates (Banat et al. 2010). he amphiphilic nature of these compounds leads to aggregation at liquid/liquid (oil/water), liquid/gas, and liquid/solid interfaces, reducing surface tension and forming emulsions. Surfactants significantly influence daily life, being integral to many commonly used products (Gürkök and Özdal 2021).

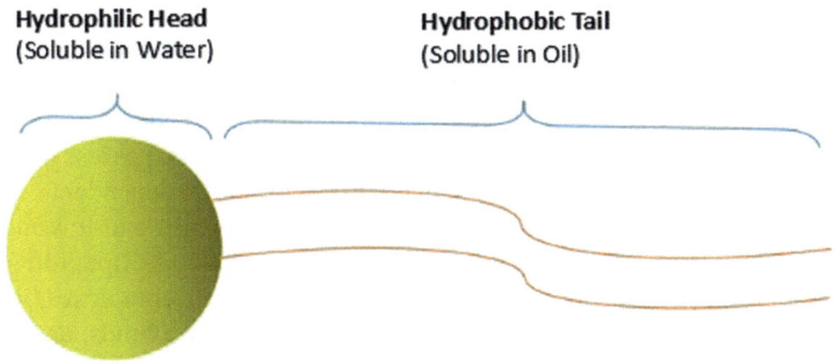

Fig. 4.1 Structure of surfactant/biosurfactant

Synthetic surfactants, derived from the petrochemical industry, offer high production and low manufacturing costs. However, this production method is often viewed as unsustainable, conflicting with national goals for a green economy, consumer demands for sustainably produced products, and the United Nations' sustainable development goals. Additionally, synthetic surfactants can pose environmental challenges due to their toxicity and poor biocompatibility (Johnson et al. 2021). In contrast, microbial surfactants are metabolic byproducts of fungal or bacterial fermentation and are considered more eco-friendly. They exhibit lower toxicity compared to synthetic surfactants (Banat et al. 2010; Farias et al. 2021). Microorganisms produce these substances when cultivated on both soluble and insoluble substrates, such as carbohydrates, oils, residues, and hydrocarbons (Silva et al. 2014; Naughton et al. 2019).

Switching from synthetic to biosurfactants could reduce CO_2 emissions by 1.5 million tons annually, contributing to an 8% reduction in lifetime CO_2 emissions (Rocha et al. 2019; Farias et al. 2021). With 10 million tons of surfactants produced globally each year, biosurfactants currently account for around 10% of the market share. These natural surfactants find applications in diverse industries, including agriculture, medicine, pharmaceuticals, civil sectors (e.g., sewage and waste treatment), and food and petroleum industries (for emulsifiers, moisturizers, creams, and medicines) (Ribeiro et al. 2020a, b, c). Additionally, biosurfactants, produced from renewable substrates, offer financial benefits and enhanced biodegradability (Sarubbo et al. 2022).

4.2 Types of Biosurfactants

Surfactants are classified into amphoteric, cationic, anionic, and nonionic types, based on the charge present on their hydrophilic heads, which influences their solubility. Common anionic surfactants, such as linear alkylbenzene sulfonate (LAS) and sodium dodecyl sulfate (SDS), are widely used in laundry detergents, personal and household cleaners, cosmetics, pharmaceuticals, and in the removal of soil hydrocarbons (Wang et al. 2020; Xu et al. 2021). The bacteriostatic properties of catalytic surfactants, commonly found in detergents, softeners, and conditioners, make them valuable in the production of both cleaning products and antibiotics (Fredell 2019). Amphoteric surfactants, such as lauryl betaine, possess both anionic and cationic characteristics, which are pH- and environment-dependent.

Biosurfactants are categorized into high molecular-weight and low molecular-weight groups. High molecular-weight biosurfactants include lipoproteins, lipopolysaccharides, proteins, polysaccharides, and biopolymer complexes, while low molecular-weight biosurfactants consist of lipopeptides, glycolipids, and phospholipids (Rosenberg and Ron 1999; Drakontis and Amin 2020). Based on their molecular structure, biosurfactants from microorganisms can be classified as lipoproteins, lipopeptides, glycolipids, phospholipids, or polymeric surfactants (Table 4.1). Other biosurfactants include fatty acids, particulate matter, and neutral lipids.

Glycolipid biosurfactants are generally divided into four categories: trehalolipids, sophorolipids, rhamnolipids, and mannosylerythritol lipids. Pseudomonas aeruginosa, for instance, produces rhamnolipids, which are valued for their ability to emulsify hydrocarbons as well as their antibacterial and antifungal properties (Rahman et al. 2002; Benincasa et al. 2004; Abdel-Mawgoud et al. 2010). Trehalolipids, produced by Rhodococcus and Arthrobacter species, are utilized for their hydrocarbon-solubilizing and emulsifying qualities, and are also applied in microbial-enhanced oil recovery, bioremediation, and various food and cosmetic applications. These surfactants are also known for their immunomodulatory and anticancer properties (Paściak et al. 2010; Bages-Estopa et al. 2018).

Sophorolipids, commonly produced by yeast species such as *Torulopsis sp.* and *Candida sp.*, have numerous applications in medicine, cosmetics, and oil/water emulsion stabilization. They are valued for their anti-inflammatory, antibacterial, and immune system-regulating capabilities (Asmer et al. 1988; Gaur et al. 2019a, b). Notable lipopeptide biosurfactants include viscosin from *P. fluorescens*, gramicidin from *B. brevis*, serravettin from *Serratia marcescens*, lichenisin from *B. licheniformis*, surfactin and iturin from *Bacillus subtilis*, and polymyxin from *B. polymyxa*. Among these, surfactin is particularly known for its thrombolytic, antiviral, antibacterial, anti-inflammatory, and antimycoplasma properties (Mireles et al. 2001; Carrillo et al. 2003).

Biosurfactants, which include phospholipids and fatty acids, are produced by microorganisms when they oxidize alkanes (Rehm and Reiff 2005). Phospholipids, such as lecithin and lysolecithin, are often used in gene carrier systems due to their role as membrane constituents, while fatty acids find applications in the food industry (Mcclements and Gumus 2016). Prominent polymeric biosurfactants, such as emulsan, liposan, and mannoprotein, possess potent bioemulsifying properties. Emulsan, derived from *Acinetobacter calcoaceticus*, is one of the most potent stabilizers available today (Rosenberg et al. 1979; Cameron et al. 1988).

Table 4.1 Types of biosurfactants

Class	Subclass	Microbes used	References
Glycolipids	Rhamnolipids	*Pseudomonas ssp.*	de Araújo Padilha et al. (2019)
		Pseudomonas cepacia	Silva et al. (2014)
		Pseudomonas aeruginosa	De Sousa et al. (2011), de Araújo Padilha et al. (2019), Salazar-Bryam et al. (2021)
		Serratia rubidaea	Nalini and Parthasarathi (2018)
		Lysinibacillus sphaericus	Gaur et al. (2019a, b)
	Trehalolipids	*Nocardia farcinica*	Christova et al. (2015)
		C. bombicola	Pinto et al. (2018)
		Rhodococcus sp.	Kuyukina et al. (2015)
	Sophorolipids	*Candida sphaerica*	Luna et al. (2011)
		Cutaneotrichosporon mucoides	Marcelino et al. (2019)
		Starmerella bombicola	Kurtzman et al. (2010), Liu et al. (2019)
	Mannosylerythritol lipids	*Pseudozyma aphidis*	Niu et al. (2019)
	Not informed	*Marinobacter hydrocarbonoclasticus*	Zenati et al. (2018)
		Saccharomyces cerevisiae	Ribeiro et al. (2020a, b, c)
		Meyerozyma guilliermondii	Camargo et al. (2018)
		Candida utilis	Ribeiro et al. (2020a, b, c)
Lipopeptides	Surfactin	*Bacillus nealsonii*	Phulpoto et al. (2020)
		Bacillus subtilis	Jamal et al. (2011), Datta et al. (2018), Wu et al. (2019), Yang et al. (2020)
	Lichenysin	*Bacillus licheniformis*	Joshi et al. (2008)
	Not informed	*Virgibacillus salarius*	Elazzazy et al. (2015)
		Bacillus velezensis	Nair et al. (2020)
		Pseudomonas azotoformans	Das and Kumar (2018)
		Bacillus pumilus	Marchut-Mikolajczyk et al. (2018)
		Bacillus cereus	Durval et al. (2020)
		Bacillus pseudomycoides	Li et al. (2016)
		Halomonas sp.	Wu et al. (2019)
Phospholipids	Not informed	*K. pneumoniae*	Jamal et al. (2011)
		Thiobacillus thiooxidans	Vandana and Singh (2018)
Polymeric surfactants	Emulsan	*Acinetobacter lwoffi*	Wittgens and Rosenau (2018)
	Rufisan		Rufino et al. (2013)
	Liposan	*Candida lipolytica*	Vandana and Singh (2018)
	Alasan	*Acinetobacter radioresistens*	Navon-Venezia et al. (1995)

4.3 Properties of Surfactants

Synthetic surfactants and biosurfactants share similar characteristics, including reduced foaming capacity, solubility, emulsification, stability, surface tension, and detergency (Santos et al. 2016; Drakontis and Amin 2020; Shakeri et al. 2021). However, the performance of biosurfactants generally exceeds that of synthetic surfactants, with additional benefits that make them more preferable. A surfactant's efficacy is commonly characterized by its critical micelle concentration (CMC). The CMC of biosurfactants typically ranges from 1 to 2000 mg/L, depending on their chemical structure (Bergström 2015). An effective biosurfactant can reduce water surface tension from approximately 72 mN/m to below 35 mN/m, and interfacial tension between oil and water from around 40 mN/m to 1 mN/m (Twigg et al. 2021). Compared to synthetic surfactants, biosurfactants have lower CMC values and more favorable interfacial and surface tension properties.

Biosurfactants are produced by microorganisms in various environments and remain effective under extreme conditions. They can maintain their physicochemical properties at pH values ranging from 3 to 12 and can tolerate extreme temperature conditions. Furthermore, biosurfactants can endure saline concentrations of up to 10% (w/v), while synthetic surfactants are typically inactivated at NaCl concentrations ≥2% (Santos et al. 2016). Santos et al. (2017) observed that the surface tension reduction and emulsification capacity of biosurfactants produced by *C. lipolytica* remained consistent for 120 days in the presence of NaCl (1–5%), pH values ranging from 5 to 9, and temperatures between 40 and 50 °C. Additionally, research on *Streptomyces sp.* revealed that the biosurfactant produced by this microorganism remained efficient across a wide range of temperatures (4–120 °C), pH levels (2–12), salt concentrations (2–12%), and heating durations at 90 °C (10–120 min) (Santos et al. 2019).

The diverse molecular structures of biosurfactants lead to functional activities that are often specific to their particular architectures. These biosurfactants can self-organize into various micelle morphologies, such as spherical, rod-like, and worm-like structures, enhancing their specificity. This property is particularly beneficial for demulsifying industrial emulsions, detoxifying contaminants, and in applications within the pharmaceutical, food, and cosmetic industries (Santos et al. 2016). Simple modifications to the molecular structures of surfactants can significantly impact their functionality. For example, sophorolipid surfactants with the same fatty acid side chains may not all be lactonic or acidic. The lactonic form possesses strong antibacterial properties, whereas the acidic form lacks these characteristics.

In general, biosurfactants are more biocompatible and biodegradable than their synthetic counterparts. Studies have shown that biosurfactants can degrade spontaneously in the absence of additional microbial support, even in diverse environments with varying temperatures, pH levels, and durations (Rodríguez-López et al. 2020). They also enhance the solubility of toxins, promoting their degradation. A study found that oil breakdown rates in sand and seawater exceeded 90% in the presence of biosurfactants (Silva et al. 2018). Another study observed that a biosurfactant derived from *Candida sphaerica* facilitated the breakdown of hydrocarbons in saltwater by promoting the proliferation of surrounding bacteria (Luna et al. 2015). Moreover, it was demonstrated that biosurfactants improved the degradation of motor oil in contaminated soil (Chaprão et al. 2018).

Biosurfactants are also valuable due to their natural glycolipidic and lipoprotein structures, making them suitable for use in pharmaceuticals, food, and cosmetics (Adu et al. 2020). In contrast, synthetic surfactants, commonly used in wastewater treatment and pollution elimination, may end up in natural water sources. When used intentionally or accidentally, synthetic surfactants can negatively impact aquatic ecosystems (Yuan et al. 2014). The extent of the damage depends on the concentration of these surfactants, which may increase over time. High concentrations can be harmful to aquatic life and, through the food chain, may pose risks to humans (Ivanković and Hrenović 2010). In contrast, biosurfactants are much less harmful to aquatic organisms and degrade more easily in water and soil, as they are naturally occurring and produced by bacteria (Ivanković and Hrenović 2010; Yuan et al. 2014). This makes them a more environmentally sustainable alternative for industries looking to replace synthetic surfactants.

4.4 Production of Biosurfactants

4.4.1 Microorganisms and Growing Media

Many bacteria, including *Pseudomonas*, *Corynebacterium*, *Candida*, *Bacillus*, and *Rhodococcus*, are commonly used for biosurfactant production. Among these, *Pseudomonas aeruginosa* is particularly well-known for producing rhamnolipids, as evidenced by various studies conducted between 2015 and 2021 ($n = 36$) (Martins and Martins 2018). Due to the high demand for biosurfactants in the market, biotechnological approaches are employed to identify production pathways and develop strains or recombinant mutants that produce the highly sought-after products. For instance, the unique *S. bombicola* mutant (ΔatΔsble) produces bolaforms, a type of sophorolipid with two sophorose units on either side of a hydrophobic tail. This structure exhibits greater tolerance to higher pH levels, making it more suitable for applications such as drug delivery in biomedicine (Soetaert et al. 2017; Van Renterghem et al. 2018).

Glucose and glycerol are commonly used substrates for biosurfactant production; however, they can also be synthesized using various other substances. The widespread industrial use of glucose, a product derived from food sources, increases the cost of biosurfactant production (Ejike Ogbonna et al. 2021). As an alternative, several reviews have discussed the use of different substrates for biosurfactant production (Wongsirichot et al. 2021). When complex substrates are employed, contaminants may remain in the final product. For example, the fermentation of kerosene by *Serratia sp.* to produce non-cytotoxic biosurfactants can result in significant contamination in the downstream phase (Borah et al. 2019). Biosurfactant production media may require the addition of yeast extract or macronutrients such as chlorine, sulfate, or phosphate salts of various metals (e.g., Ca, Na, K, Mg, Fe). Recent research indicates that mineral fermentation media are commonly used (in 40% of the cases examined), with modest variations in content depending on the specific biosurfactant being produced. To reduce costs, low-cost by-products and waste materials are increasingly being explored as alternative substrates and media supplements for sophorolipid production (Ingham and Winterburn 2022).

4.4.2 Fermentation Process

Solid-state fermentation (SSF) and submerged fermentation (SmF) are the primary bioprocesses used for biosurfactant production. SSF is a cost-effective process due to its use of industrial and agro-industrial waste, the removal of substrate restrictions, and lower energy requirements. However, SSF presents challenges, including substrate heterogeneity, difficulties in operational monitoring, and complex downstream processing (Sánchez et al. 2015; Kumar et al. 2021).

Biosurfactant production volumes typically range from 100 to 1200 mL at the laboratory scale. Few studies report larger volumes, often at the bench scale (approximately 3 L), and even fewer detail scale-up operations to the pilot scale (20 and 50 L). Yields at these scales range from 0.83 to 27 g/L. The highest SmF yield observed was approximately 26 g/L from 1.2 L of *P. aeruginosa* UCP 0992, which used vegetable oil waste and maize steep liquor as substrates (Silva et al. 2018). The lowest yield at the same scale was 0.83 g/L from 500 mL of surfactin produced with *B. velezensis* MHNK1 (Jakinala et al. 2019). At 50 L production volumes, higher sophorolipid yields (21 and 27 g/L) were obtained using *Candida* species and high-fat-containing residues (Luna et al. 2015; da Rocha Junior et al. 2019).

When comparing different processes, it is important to evaluate not only the production scale but also the biosurfactant purity and efficiency. Highly productive strains, optimal fermentation conditions, and cost-effective substrates are crucial for reducing costs and increasing biosurfactant output for environmental applications (Silva et al. 2018; Cazals et al. 2020; Durval et al. 2020). As a result, alternatives such as inexpensive substrates (e.g., organic waste and by-products) and SSF are being explored to optimize the production process (Luna et al. 2015; Jiménez-Peñalver et al. 2019).

For example, in a scale-up method using *Candida sphaerica* UCP 0995, maize steep liquor and groundnut oil refinery byproducts were used in 20-L bioreactors to produce a biosurfactant that dispersed 90% of oil droplets in saltwater. The process yielded 21 g/L in 144 h, which was comparable to lab-scale results. Similarly, molasses and winterized oil cake were used as low-cost substrates in an SSF process for *S. bombicola* in 22-L and 100-L reactors, yielding 0.19 gSL/gDM and 0.14 gSL/gDM, respectively, demonstrating the microorganism's production flexibility (Sánchez et al. 2015; Rodríguez et al. 2021).

Fermentation conditions vary depending on the microorganisms and biosurfactants being produced. Fermentation temperatures typically range from 25 °C to 40 °C, with 30 °C being the most commonly used. At this temperature, genera such as *Pseudomonas* and *Bacillus* are frequently studied. Fermentation times can vary widely, ranging from 48 to 288 h. For instance, *Wickerhamomyces anomalus* CCMA 0358 is known for its rapid biosurfactant production in under 24 h, which helps reduce production costs. While faster rates are occasionally reported, agitation speeds generally range from 135 to 250 rpm. Although high-speed agitation has been shown to reduce yeast cell numbers, it has also been observed to enhance biosurfactant synthesis in certain conditions (Souza et al. 2018).

Biosurfactant production is more expensive than chemical surfactant manufacturing, with downstream processing accounting for up to 60% of the total cost (Sarubbo et al. 2015; Souza et al. 2018). In submerged fermentation, biomass is typically removed via centrifugation before downstream processing, although this step is not always detailed in reviews. Common purification methods include extraction, precipitation, and crystallization, with precipitation typically occurring under acidic conditions. Extraction often involves organic solvents, such as methanol-chloroform, methanol-n-hexane-water, or ethyl acetate-n-hexane mixtures (Jiménez-Peñalver et al. 2016; Shen et al. 2019). Solvent extraction is widely used to recover biosurfactants in SSF. During crystallization, buffers such as phosphate and phthalate are employed to purify the biosurfactant. One of the latest innovations is the use of zero-valent iron nanoparticles for purification, along with integrated gravity separation to improve sophorolipid recovery (Dolman et al. 2017; Sahebnazar et al. 2018).

Biosurfactants are known to reduce surface tension to as low as 40 mN/m (Ayed et al. 2015). Techniques like TLC, LC-MS, and FTIR are commonly used for the purification and structural characterization of biosurfactants (Jiménez-Peñalver et al. 2020). In conclusion, recent efforts in biosurfactant production have focused on utilizing new feedstocks, particularly industrial waste and low-cost materials, which, when combined with SSF, can make the process more cost-effective. However, the high costs of purification and the lack of detailed downstream processing knowledge have led to the production of crude biosurfactant extracts. Therefore, economic evaluations that consider purification methods and future applications are essential for the advancement of biosurfactant production. The steps involved in biosurfactant production are outlined in Fig. 4.2.

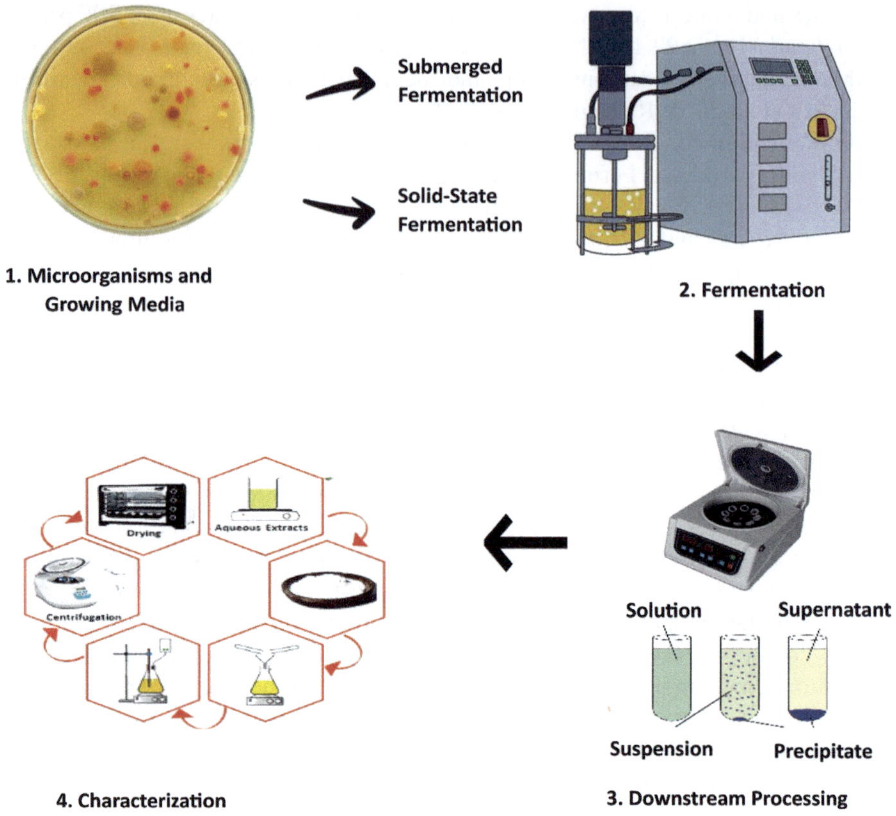

Fig. 4.2 Biosurfactant production process

4.5 Employing Genetic Engineering for Increased Bio-surfactant Production

Biosurfactants have a wide range of industrial applications across various fields. Genetic engineering and advanced genome editing technologies have significantly boosted the production of biosurfactants, making them increasingly important in industrial settings in recent years (Ambaye et al. 2021). Microorganisms inherently possess genetic instructions for producing biosurfactants, and cloning and other techniques leverage this genetic information to enhance biosurfactant production. Researchers are continually exploring new microbial strains with immense potential for producing these valuable compounds. Genetically engineered microbial strains can produce much larger quantities of biosurfactants compared to their wild-type counterparts. Recombinant DNA technology and other genome-editing techniques enable the production of biosurfactants at a much lower cost, with altered chemical properties that improve their tolerance to extreme salt, pH, and temperature conditions (Satpute et al. 2016).

Studies have shown that *Escherichia coli* strains carrying the *srfA* and *BioS* genes can produce significantly higher amounts of biosurfactants from substrates like olive oil compared to non-engineered wild-type strains. Similar findings from various researchers have demonstrated the increased biosurfactant production using genetically engineered microbial strains (Sekhon 2012; Kandasamy et al. 2019). In fact, biosurfactant production has been shown to double in genetically engineered strains (Ambaye et al. 2021). Low productivity of biosurfactants remains a major challenge for large-scale industrial applications. However, metabolically engineered strains have addressed this issue by significantly increasing biosurfactant production. Their titer has been enhanced up to 12.8 g/L, with a yield of 65.0 mmol/mol of sucrose.

4.6 Surfactants: Mechanism of Action

Biosurfactants have remarkable applications across various industries due to their unique structural and chemical properties. When in contact with water, they form micelles or aggregates (Fig. 4.3). These structures consist of a hydrophilic head, which remains exposed to the surrounding liquid, and a hydrophobic tail, which is oriented toward the center. Different types of aggregates can form, including lipid bilayers, cylindrical micelles, and spherical micelles, depending on the chemical structure of the biosurfactant.

The dual hydrophilic and hydrophobic nature of biosurfactants enhances their ability to remove oils and dirt, making them highly effective for both small- and large-scale industrial applications. Surfactant components adsorb oils, surrounding and removing them while preventing redeposition.

The toxicity and biodegradation behavior of biosurfactants vary depending on their type. Studies have indicated that they can be toxic to aquatic life to some extent, although the impact depends on the sensitivity of individual species. While some species can tolerate high concentrations, others are sensitive even to low levels. Both anionic and non-anionic biosurfactants exhibit negative effects, but non-anionic types tend to be more toxic.

Recent research has demonstrated that anionic surfactants, including alcohol ethoxylates and phenol ethoxylates, are biodegradable under both anaerobic and aerobic conditions. However, their aerobic degradation is limited, achieving a maximum removal rate of up to 20%. Complete removal typically requires extended periods. Factors such as alkyl chain length, water solubility, molecular weight, and the degree of ethoxylation significantly influence their biodegradability.

Biosurfactants offer a wide range of applications, including dispersing, cleaning, and emulsifying, and both anti-foaming and foaming activities. They are also used as lubricants in car engines, for corrosion prevention, and in the production of aerosols. Additionally, they are employed in porous rocks to enhance oil flow (Sivaramakrishnan 2013).

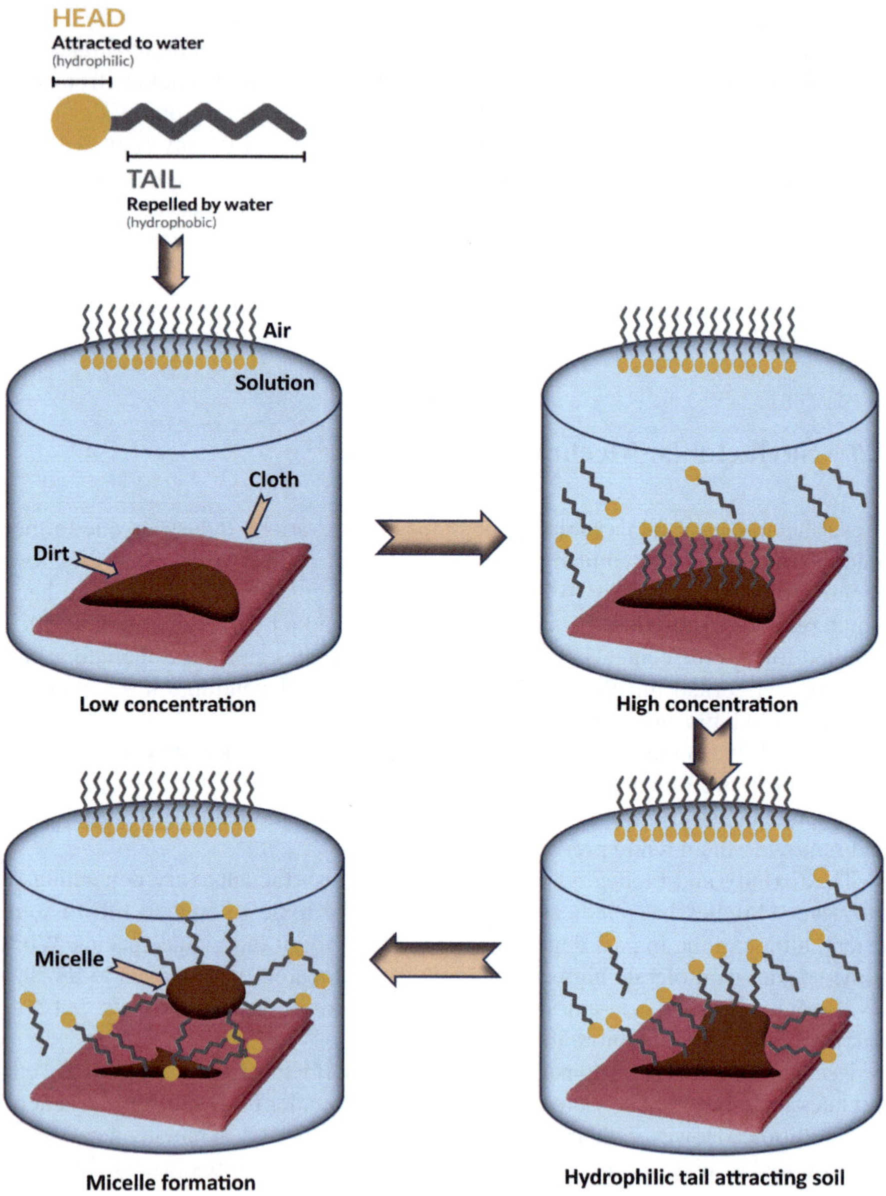

Fig. 4.3 Mechanism of action of surfactants

4.6.1 Classification of Surfactant

Surfactants are classified into four major categories based on their nature: cationic, zwitterionic, anionic, and non-ionic (Fig. 4.4).

4.6.1.1 Anionic Surfactants

Anionic surfactants contain negatively charged components at their hydrophilic heads, such as phosphate, sulfate, carboxylates, and sulfonates. A prominent example of an anionic surfactant is sodium stearate, which accounts for more than 50% of global usage. These surfactants are primarily used in emulsion polymerization. Other notable examples include linear alkyl benzene sulfonates (LABs), dioctyl sodium sulfosuccinate (DOSS), and alkyl-aryl ether phosphates. Anionic surfactants always contain at least one negatively charged group, while counterions such as Na^+, K^+, and quaternary ammonium ions carry a positive charge (Hibbs 2006; Sivaramakrishnan 2013).

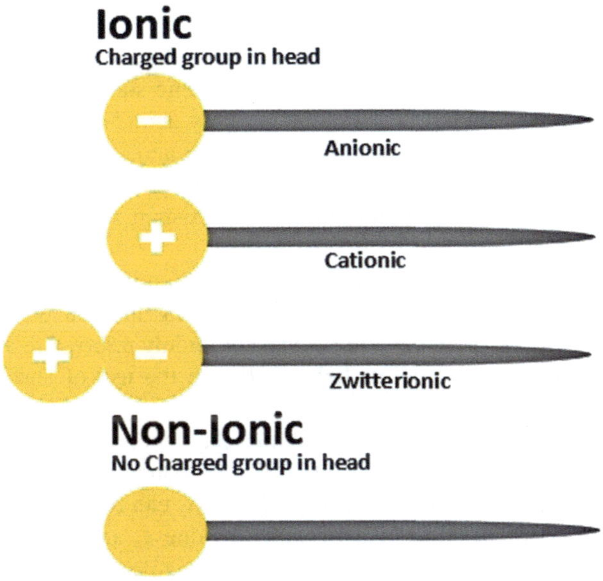

Fig. 4.4 Ionic and non-ionic surfactants/classification of surfactants

4.6.1.2 Cationic Surfactants

Cationic surfactants carry a positive charge and exhibit antifungal and antimicrobial properties. Due to their cationic nature, they can disrupt viral and bacterial cell membranes. Examples of cationic surfactants include alkyl pyridinium salts ($R_3PyN^+Cl^-$) and quaternary ammonium salts ($R_4N^+Cl^-$) (Sivaramakrishnan 2013; Fait et al. 2019).

4.6.1.3 Zwitterionic Surfactants

Zwitterionic surfactants carry both negative and positive charges, with anionic and cationic components attached to the same molecule. These charges typically neutralize each other under normal environmental conditions. The cationic or anionic behaviour of these surfactants can also vary depending on the pH of the environment. Examples of zwitterionic surfactants include amino acids (Seredyuk et al. 2001; Sivaramakrishnan 2013).

4.6.1.4 Non-anionic

Non-ionic surfactants are a category of surfactants that do not undergo ionization when dissolved in water. They remain in a non-ionic state, which enhances their stability. These surfactants can tolerate both highly acidic and alkaline conditions. They are soluble in both organic solvents and water and exhibit excellent compatibility with other substances. Non-ionic surfactants possess both hydrophilic and hydrophobic groups, making them stable and resistant to ionization in aqueous environments.

Their solubility in water is a result of hydrogen bonding. However, as temperature increases, hydrogen bonding decreases, leading to reduced solubility. This results in the formation of cloudy emulsions, commonly referred to as cloud points. The determination of cloud points helps optimize the use of surfactants at high temperatures.

Examples of non-ionic surfactants include alkyl (poly)glycosides (R-O-(Glyc) x), alkyl polyalcohols ($R\text{-}O\text{-}CH_2\text{-}CHOH\text{-}CH_2\text{-}OH$), and alkyl polyamines ($R\text{-}NH\text{-}(CH_2\text{–}CH_2\text{–}NH)x\text{-}H$). Their widespread applications can be attributed to their eco-toxicological properties, resistance to water hardness, and effectiveness at low temperatures (Moore 1997; Broze 1999). High molecular weight non-ionic surfactants are particularly important for large-scale industrial applications, such as in the textile industry. These properties collectively drive their extensive use in various industries.

4.7 Applications of Biosurfactants in Textile Cleaning

Fabric is an essential component of daily life, with the human body in direct contact with fabric materials during use (Salerno-Kochan and Kowalski 2020). Textile cleaning is a resource-intensive process, as cleaning one ton of fabric typically requires 150–250 m^3 of water. However, the use of biosurfactants offers an economical and environmentally friendly alternative. The textile industry is expanding rapidly, offering an evolving range of products. Simultaneously, rising energy and raw material costs, coupled with increasing demand for eco-friendly consumer products and textile cleaning services, necessitate advancements in cleaning technologies for fabrics in aqueous solutions (Fracchia et al. 2014).

Textile cleaning methods include washing, extraction, dyeing, and ironing (Fig. 4.5). Among these, washing is the most critical technological process for restoring the hygienic and physical properties of fabric. Modern technology provides consumers with a variety of services, such as washing, cleaning and preserving fur garments, stain removal, and ironing (Gayathiri et al. 2022).

Synthetic detergents are challenging to biodegrade in aqueous environments, leading to the formation of scum on water surfaces, which accelerates eutrophication. This process harms aquatic life and transforms freshwater ecosystems into

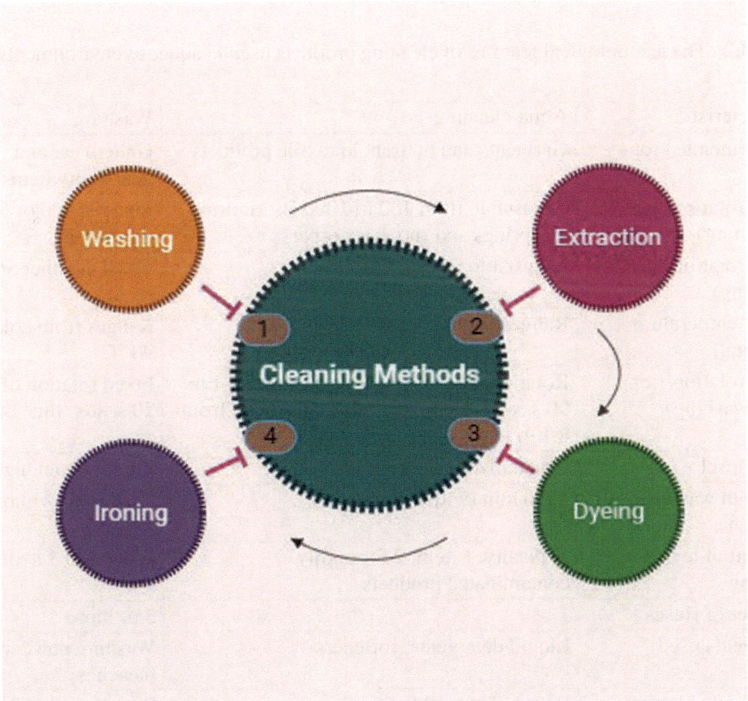

Fig. 4.5 Cleaning methods of textiles

polluted marshes. In contrast, biosurfactants are easily biodegradable and environmentally friendly. Their versatile properties make them effective in various laundry processes, including the formulation of washing detergents and domestic cleaners (Paraska et al. 2023).

The technological features of cleaning products used in aqueous environments are outlined in Table 4.2.

Fabric cleaning operations are inherently complex and often face challenges such as potential damage to textiles, chemical pollution of wastewater, high service costs, and excessive consumption of electricity and water. Consequently, the development of resource-saving technologies for cleaning fabrics is a critical task for the textile industry. Addressing this challenge can help improve the quality of modern cleaning processes, extend the lifespan of textiles, reduce environmental impacts, and lower service costs.

Resource-saving technologies focus on innovations and practices aimed at minimizing the consumption of energy, water, raw materials, and time, while maximizing efficiency and reducing waste. These technologies play a crucial role in promoting sustainable development and environmental preservation. Implementing resource-saving strategies is essential for achieving sustainable growth, minimizing ecological footprints, and creating a more resilient and efficient global infrastructure (Fracchia et al. 2014).

Table 4.2 The technological features of cleaning products used in aqueous environments (Paraska et al. 2023)

Characteristic	Aqua-cleaning	Washing
Recommended for	Garments and household textile products	Underwear and hospitality items
Drum rotation speed (extracting)	Adjustable from 400 to 1000 RPM, drum on springs and shock absorbers	Fixed at 400 or 500 RPM, rigid drum
Drum rotation speed (washing)	Adjustable from 5 to 40 RPM	Fixed at either 36 or 40 RPM
Water temperature (washing)	Ranges from 27 to 40 °C	Ranges from cold to 90 °C
Drum rotation/stop time (washing)	Rotation time can be adjusted from 0.1 to 90 s, with a stop time also adjustable from 0.1 to 90 s	Fixed rotation of 12 or 20 s, stop time of 3 or 4 s
Water level	Gradually varies from low to high	Ranges from high to low
One bath washing duration	4–10 min (with heating)	5–20 min (without heating)
Amount of textile washing	Typically, 1 item, 2 for highly contaminated products	From 1 to 3 items
Number of rinses	1	3 or more
Detergents used	Liquid detergents, softeners	Washing powders, spans, bleaches
Drying temperature	Ranges from 40 to 60 °C	Ranges from 50 to 110 °C

4.8 Applications of Biosurfactants in Textile Finishing

Pre-washing fabrics or clothes is crucial for producing high-quality end products during the textile finishing process. Specialized lipophilic materials, such as emollients, are used to achieve optimal friction behavior during fabric manufacturing. Biotechnology presents innovative solutions to address challenges in this process. Microorganisms metabolize lipophilic substrates like hydrocarbons, producing diverse surface-active substances known as biosurfactants.

Textile finishing consumes substantial amounts of water i.e. approximately 100–250 m^3 for every 1000 kg of textiles. Key manufacturing steps, including pretreatment, finishing, and dyeing, are essential to ensure top-quality production. Processes like desizing, bleaching, and washing are critical for removing natural polymers, fiber admixtures, and other contaminants from the fiber surface to prepare textiles for dyeing and finishing. Secondary compounds, such as mineral oils, fatty acid esters, and natural oils, are commonly used as emollients to enhance friction properties. However, their removal often causes issues like incomplete emulsification or oil redeposition on fibers.

Traditional washing methods negatively impact on the environment due to the presence of ecologically incompatible detergent residues in wastewater. Biotechnology offers sustainable alternatives to address these challenges, reducing the environmental impact while maintaining efficiency (Shah et al. 2016).

A variety of surface-active substances, known as biosurfactants, are produced by certain microorganisms that grow on lipophilic substrates such as hydrocarbons. These natural compounds have the potential to ecologically clean textiles, where fiber structures, such as emollients, can be contaminated by microbes and mineralized into carbon dioxide and water. The primary role of biosurfactants is to enhance the availability of water-insoluble substrates. By producing a broad range of natural surfactants, they facilitate the passive diffusion of hydrocarbons into microbial cells.

These substrates are incorporated into micelles and accumulate over the cell wall in the presence of natural liquid lipophilic surfactants. The biosurfactants in the cell wall effectively reduce the interfacial tension between oil droplets and the cell, leading to the formation of a monomolecular oil layer around the cell. This process aids in the efficient diffusion of oil into the cell's interior. This demonstrates the critical role of biosurfactants in enhancing the interaction between microbial cells and oil mixtures (Kosaric and Sukan 2014).

Biosurfactants are categorized based on their chemical structure and the microbial species that produce them. The primary class of biosurfactants is glycolipids, which are composed of carbohydrates and hydroxy or long-chain aliphatic acids. They can be further divided into ionic and non-ionic types, which may exist in extracellular or cell wall-bound forms. The chemical composition of these biosurfactants can vary depending on various factors such as culture conditions, including temperature, dilution, pH, and agitation. Biosurfactants are known to offer promising eco-friendly applications in textile processing. Their key functions in textile finishing include detergency, dispersing, wetting, emulsification, and solubilization (Sivaramakrishnan 2013).

4.9 Benefits of Biosurfactants

Biosurfactants come with several benefits that make them highly valuable in various industries. They are biodegradable, breaking down easily through microbial activity in both aquatic environments and soil, without causing any harm to the environment. These surfactants also possess high durability, able to withstand the high temperatures and pH levels commonly encountered in textile processes. Furthermore, biosurfactants remain effective even at low concentrations, due to their low critical micelle concentration (CMC). Their versatility allows them to offer a wide range of useful properties for diverse applications. Additionally, they contribute to waste treatment by serving as a food source for emerging microbes, thus addressing waste disposal problems (Sangeeta Patil 2023).

4.10 Uses in Textile Processing

In textile processing, biosurfactants have numerous practical applications. In the desizing process, they provide rapid wetting and enhance the effects of enzymatic desizing, leading to the efficient removal of size ingredients. During bio-scouring, they help achieve optimal absorbency and effectively remove soil under both natural and alkaline pH conditions. In dyeing, biosurfactants improve dye dissolution and dispersion, ensuring optimal and uniform dye exhaustion. They also play a role in dirty linen removal by enhancing the removal of hydrolyzed, unreacted dye from the fabric surface and preventing re-staining. Finally, in finishing, biosurfactants contribute to the stability of chemical suspensions used in the finishing process and help prevent the migration of silicone oils (Akbari et al. 2018).

4.11 Conclusions

Surfactants play an essential role in the textile industry, particularly in textile refining and fabric treatment processes. Biosurfactants, which are naturally produced by various microorganisms, present an eco-friendly alternative to traditional chemical surfactants. The use of conventional harsh and toxic chemical surfactants in the textile industry significantly contributes to environmental pollution and entails a high production cost. In contrast, biosurfactants are sustainable and economically viable, offering a natural and non-toxic alternative. These surfactants eliminate the need for harmful chemicals, thus promoting a greener approach to textile manufacturing. Although microorganisms naturally produce biosurfactants in small quantities, advancements in genetic engineering and genome editing tools can facilitate large-scale industrial production. Biosurfactants demonstrate a wide range of applications in textile treatment, including oil, soil, and stain removal, as well as in textile cleaning, processing, and finishing. Ongoing research continues to explore microorganisms capable of synthesizing these beneficial biosurfactants, paving the way for their expanded use in the textile industry.

References

Abdel-Mawgoud AM, Lépine F, Déziel E (2010) Rhamnolipids: diversity of structures, microbial origins and roles. Appl Microbiol Biotechnol 86:1323–1336

Adu SA, Naughton PJ, Marchant R, Banat IM (2020) Microbial biosurfactants in cosmetic and personal skincare pharmaceutical formulations. Pharmaceutics 12:1099

Akbari S, Abdurahman NH, Yunus RM, Fayaz F, Alara OR (2018) Biosurfactants—a new frontier for social and environmental safety: a mini review. Biotechnol Res Innov 2:81–90

Ambaye TG, Vaccari M, Prasad S, Rtimi S (2021) Preparation, characterization and application of biosurfactant in various industries: a critical review on progress, challenges and perspectives. Environ Technol Innov 24:102090

Asmer H-J, Lang S, Wagner F, Wray V (1988) Microbial production, structure elucidation and bioconversion of sophorose lipids. J Am Oil Chem Soc 65:1460–1466

Ayed HB, Jemil N, Maalej H, Bayoudh A, Hmidet N, Nasri M (2015) Enhancement of solubilization and biodegradation of diesel oil by biosurfactant from Bacillus amyloliquefaciens An6. Int Biodeterior Biodegrad 99:8–14

Bages-Estopa S, White D, Winterburn J, Webb C, Martin P (2018) Production and separation of a trehalolipid biosurfactant. Biochem Eng J 139:85–94

Banat IM, Franzetti A, Gandolfi I, Bestetti G, Martinotti MG, Fracchia L et al (2010) Microbial biosurfactants production, applications and future potential. Appl Microbiol Biotechnol 87:427–444

Benincasa M, Abalos A, Oliveira I, Manresa A (2004) Chemical structure, surface properties and biological activities of the biosurfactant produced by Pseudomonas aeruginosa LBI from soapstock. Antonie Van Leeuwenhoek 85:1–8

Bergström LM (2015) Explaining the growth behavior of surfactant micelles. J Colloid Interface Sci 440:109–118

Borah D, Agarwal K, Khataniar A, Konwar D, Gogoi SB, Kallel M (2019) A newly isolated strain of Serratia sp. from an oil spillage site of Assam shows excellent bioremediation potential. 3 Biotech 9:1–12

Broze G (1999) Handbook of detergents, part A. CRC Press, Boca Raton, FL, pp 1–6

Camargo FP, Menezes AJD, Tonello PS, Dos Santos ACA, Duarte ICS (2018) Characterization of biosurfactant from yeast using residual soybean oil under acidic conditions and their use in metal removal processes. FEMS Microbiol Lett 365:fny098

Cameron DR, Cooper DG, Neufeld R (1988) The mannoprotein of Saccharomyces cerevisiae is an effective bioemulsifier. Appl Environ Microbiol 54:1420–1425

Carrillo C, Teruel JA, Aranda FJ, Ortiz A (2003) Molecular mechanism of membrane permeabilization by the peptide antibiotic surfactin. BBA-Biomembranes 1611:91–97

Cazals F, Huguenot D, Crampon M, Colombano S, Betelu S, Galopin N et al (2020) Production of biosurfactant using the endemic bacterial community of a PAHs contaminated soil, and its potential use for PAHs remobilization. Sci Total Environ 709:136143

Chaprão MJ, Da Silva RDCFS, Rufino RD, Luna JM, Santos VA, Sarubbo LA (2018) Production of a biosurfactant from Bacillus methylotrophicus UCP1616 for use in the bioremediation of oil-contaminated environments. Ecotoxicology 27:1310–1322

Christova N, Lang S, Wray V, Kaloyanov K, Konstantinov S, Stoineva I (2015) Production, structural elucidation, and in vitro antitumor activity of trehalose lipid biosurfactant from Nocardia farcinica strain. J Microbiol Biotechnol 25:439–447

Da Rocha Junior RB, Meira HM, Almeida DG, Rufino RD, Luna JM, Santos VA et al (2019) Application of a low-cost biosurfactant in heavy metal remediation processes. Biodegradation 30:215–233

Das AJ, Kumar R (2018) Utilization of agro-industrial waste for biosurfactant production under submerged fermentation and its application in oil recovery from sand matrix. Bioresour Technol 260:233–240

Datta P, Tiwari P, Pandey LM (2018) Isolation and characterization of biosurfactant producing and oil degrading Bacillus subtilis MG495086 from formation water of Assam oil reservoir and its suitability for enhanced oil recovery. Bioresour Technol 270:439–448

De Araújo Padilha CE, Da Costa NC, De Santana Souza DF, De Oliveira JA, Dos Santos ES (2019) Valorization of green coconut fibre: use of the black liquor of organolsolv pretreatment for ethanol production and the washing water for production of rhamnolipids by Pseudomonas aeruginosa ATCC 27583. Ind Crop Prod 140:111604

De Sousa JR, Da Costa Correia JA, De Almeida JGL, Rodrigues S, Pessoa ODL, Melo VMM et al (2011) Evaluation of a co-product of biodiesel production as carbon source in the production of biosurfactant by P. aeruginosa MSIC02. Process Biochem 46:1831–1839

Dolman BM, Kaisermann C, Martin PJ, Winterburn JB (2017) Integrated sophorolipid production and gravity separation. Process Biochem 54:162–171

Drakontis CE, Amin S (2020) Biosurfactants: formulations, properties, and applications. Curr Opin Colloid Interface Sci 48:77–90

Durval IJB, Mendonça AHR, Rocha IV, Luna JM, Rufino RD, Converti A et al (2020) Production, characterization, evaluation and toxicity assessment of a Bacillus cereus UCP 1615 biosurfactant for marine oil spills bioremediation. Mar Pollut Bull 157:111357

Ejike Ogbonna K, Victor Agu C, Okonkwo CC, Tochukwu Ughamba K, Akor J, Njoku OU (2021) Use of Spondias Mombin fruit pulp as a substrate for biosurfactant production. Bioengineered 12:1–12

Elazzazy AM, Abdelmoneim T, Almaghrabi O (2015) Isolation and characterization of biosurfactant production under extreme environmental conditions by alkali-halo-thermophilic bacteria from Saudi Arabia. Saudi J Biol Sci 22:466–475

Fait ME, Bakas L, Garrote GL, Morcelle SR, Saparrat MC (2019) Cationic surfactants as antifungal agents. Appl Microbiol Biotechnol 103:97–112

Farias CBB, Almeida FCG, Silva IA, Souza TC, Meira HM, Soares Da Silva RDCF et al (2021) Production of green surfactants: market prospects. Electron J Biotechnol 51:28–39

Fracchia L, Ceresa C, Franzetti A, Cavallo M, Gandolfi I, Van Hamme J et al (2014) Biosurfactants: production and utilization—processes, technologies, and economics, vol 159. Routledge, London, pp 245–260

Fredell DL (2019) Cationic surfactants. CRC Press, Boca Raton, FL, pp 31–60

Gaur VK, Bajaj A, Regar RK, Kamthan M, Jha RR, Srivastava JK et al (2019a) Rhamnolipid from a Lysinibacillus sphaericus strain IITR51 and its potential application for dissolution of hydrophobic pesticides. Bioresour Technol 272:19–25

Gaur VK, Regar RK, Dhiman N, Gautam K, Srivastava JK, Patnaik S et al (2019b) Biosynthesis and characterization of sophorolipid biosurfactant by Candida spp.: application as food emulsifier and antibacterial agent. Bioresour Technol 285:121314

Gayathiri E, Prakash P, Karmegam N, Varjani S, Awasthi MK, Ravindran B (2022) Biosurfactants: potential and eco-friendly material for sustainable agriculture and environmental safety—a review. Agronomy 12:662

Gürkök S, Özdal M (2021) Microbial biosurfactants: properties, types, and production. Anatol J Biol 2:7–12

Hibbs J (2006) Chemistry and technology of surfactants, vol 91. Wiley, Hoboken, NJ

Ingham B, Winterburn J (2022) Developing an understanding of sophorolipid synthesis through application of a central composite design model. Microb Biotechnol 15:1744–1761

Ivanković T, Hrenović J (2010) Surfaktanti u okolišu. Arh Hig Rada Toksikol 61:95–109

Jakinala P, Lingampally N, Kyama A, Hameeda B (2019) Enhancement of atrazine biodegradation by marine isolate bacillus velezensis MHNK1 in presence of surfactin lipopeptide. Ecotoxicol Environ Saf 182:109372

Jamal P, Alam MZ, Zainuddin E, Nawawi W (2011) Production of biosurfactant in 2L bioreactor using sludge palm oil as a substrate. IIUM Eng J 12. https://doi.org/10.31436/iiumej.v12i4.212

Jiménez-Peñalver P, Gea T, Sánchez A, Font X (2016) Production of sophorolipids from winterization oil cake by solid-state fermentation: optimization, monitoring and effect of mixing. Biochem Eng J 115:93–100

Jiménez-Peñalver P, Rodríguez A, Daverey A, Font X, Gea T (2019) Use of wastes for sophorolipids production as a transition to circular economy: state of the art and perspectives. Rev Environ Sci Biotechnol 18:413–435

Jiménez-Peñalver P, Koh A, Gross R, Gea T, Font X (2020) Biosurfactants from waste: structures and interfacial properties of sophorolipids produced from a residual oil cake. J Surfactant Deterg 23:481–486

Johnson P, Trybala A, Starov V, Pinfield VJ (2021) Effect of synthetic surfactants on the environment and the potential for substitution by biosurfactants. Adv Colloid Interf Sci 288:102340

Joshi S, Yadav S, Desai AJ (2008) Application of response-surface methodology to evaluate the optimum medium components for the enhanced production of lichenysin by bacillus licheniformis R2. Biochem Eng J 41:122–127

Kandasamy R, Rajasekaran M, Venkatesan SK, Uddin M (2019) Next generation biomanufacturing technologies. ACS Publications, Washington, DC), pp 243–260

Kesting W, Tummuscheit M, Schacht H, Schollmeyer E (1996) Interfaces, surfactants and colloids in engineering. Cham, Springer, pp 125–130

Kosaric N, Sukan FV (Eds.) (2014) Biosurfactants: production and utilization—processes, technologies, and economics (Vol. 159). CRC press.

Kumar A, Singh SK, Kant C, Verma H, Kumar D, Singh PP et al (2021) Microbial biosurfactant: a new frontier for sustainable agriculture and pharmaceutical industries. Antioxidants 10:1472

Kurtzman CP, Price NP, Ray KJ, Kuo T-M (2010) Production of sophorolipid biosurfactants by multiple species of the Starmerella (Candida) bombicola yeast clade. FEMS Microbiol Lett 311:140–146

Kuyukina MS, Ivshina IB, Baeva TA, Kochina OA, Gein SV, Chereshnev VA (2015) Trehalolipid biosurfactants from nonpathogenic Rhodococcus actinobacteria with diverse immunomodulatory activities. New Biotechnol 32:559–568

Li J, Deng M, Wang Y, Chen W (2016) Production and characteristics of biosurfactant produced by bacillus pseudomycoides BS6 utilizing soybean oil waste. Int Biodeterior Biodegrad 112:72–79

Liu Z, Tian X, Chen Y, Lin Y, Mohsin A, Chu J (2019) Efficient sophorolipids production via a novel in situ separation technology by Starmerella bombicola. Process Biochem 81:1–10

Luna JM, Rufino RD, Albuquerque CD, Sarubbo LA, Campos-Takaki GM (2011) Economic optimized medium for tensio-active agent production by Candida sphaerica UCP0995 and application in the removal of hydrophobic contaminant from sand. Int J Mol Sci 12:2463–2476

Luna JM, Rufino RD, Jara AMA, Brasileiro PP, Sarubbo LA (2015) Environmental applications of the biosurfactant produced by Candida sphaerica cultivated in low-cost substrates. Colloids Surf A Physicochem Eng Asp 480:413–418

Marcelino PRF, Peres GFD, Terán-Hilares R, Pagnocca FC, Rosa CA, Lacerda TM et al (2019) Biosurfactants production by yeasts using sugarcane bagasse hemicellulosic hydrolysate as new sustainable alternative for lignocellulosic biorefineries. Ind Crop Prod 129:212–223

Marchut-Mikolajczyk O, Drożdżyński P, Pietrzyk D, Antczak T (2018) Biosurfactant production and hydrocarbon degradation activity of endophytic bacteria isolated from Chelidonium majus L. Microb Cell Factories 17:1–9

Martins PC, Martins VG (2018) Biosurfactant production from industrial wastes with potential remove of insoluble paint. Int Biodeterior Biodegrad 127:10–16

Mcclements DJ, Gumus CE (2016) Natural emulsifiers—biosurfactants, phospholipids, biopolymers, and colloidal particles: molecular and physicochemical basis of functional performance. Adv Colloid Interf Sci 234:3–26

Mireles JR, Toguchi A, Harshey RM (2001) Salmonella enterica serovar typhimurium swarming mutants with altered biofilm-forming abilities: surfactin inhibits biofilm formation. J Bacteriol 183:5848–5854

Mittal KL, Bahners, T (2017) Textile finishing: Recent developments and future trends. Wiley.

Moore SL (1997) The mechanisms of antibacterial action of some nonionic surfactants. PhD thesis. University of Brighton, Brighton

Nair AS, Al-Bahry S, Sivakumar N (2020) Co-production of microbial lipids and biosurfactant from waste office paper hydrolysate using a novel strain Bacillus velezensis ASN1. Biomass Convers Biorefinery 10:383–391

Nalini S, Parthasarathi R (2018) Optimization of rhamnolipid biosurfactant production from Serratia rubidaea SNAU02 under solid-state fermentation and its biocontrol efficacy against fusarium wilt of eggplant. Ann Agrar Sci 16:108–115

Naughton PJ, Marchant R, Naughton V, Banat IM (2019) Microbial biosurfactants: current trends and applications in agricultural and biomedical industries. J Appl Microbiol 127:12–28

Navon-Venezia S, Zosim Z, Gottlieb A, Legmann R, Carmeli S, Ron EZ et al (1995) Alasan, a new bioemulsifier from Acinetobacter radioresistens. Appl Environ Microbiol 61:3240–3244

Niu Y, Wu J, Wang W, Chen Q (2019) Production and characterization of a new glycolipid, mannosylerythritol lipid, from waste cooking oil biotransformation by Pseudozyma aphidis ZJUDM34. Food Sci Nutr 7:937–948

Paraska O, Synyuk O, Radek N, Zolotenko E, Mykhaylovskiy Y (2023) Usage of biosurfactants as environmental friendly detergents for textile products cleaning. Fibre Text 30:42–51

Paściak M, Sanchez-Carballo P, Duda-Madej A, Lindner B, Gamian A, Holst O (2010) Structural characterization of the major glycolipids from Arthrobacter globiformis and Arthrobacter scleromae. Carbohydr Res 345:1497–1503

Phulpoto IA, Yu Z, Hu B, Wang Y, Ndayisenga F, Li J et al (2020) Production and characterization of surfactin-like biosurfactant produced by novel strain Bacillus nealsonii S2MT and it's potential for oil contaminated soil remediation. Microb Cell Factories 19:1–12

Pinto M, Ribeiro B, Guerra J, Rufino R, Sarubbo L, Luna J (2018) Production in bioreactor, toxicity and stability of a low-cost biosurfactant. Chem Eng Trans 64:595–600

Quagliotto P, Montoneri E, Tambone F, Adani F, Gobetto R, Viscardi G (2006) Chemicals from wastes: compost-derived humic acid-like matter as surfactant. Environ Sci Technol 40:1686–1692

Rahman K, Rahman TJ, Mcclean S, Marchant R, Banat IM (2002) Rhamnolipid biosurfactant production by strains of Pseudomonas aeruginosa using low-cost raw materials. Biotechnol Prog 18:1277–1281

Rehm H, Reiff I (2005) Reactors and reactions. Springer, Cham, pp 175–215

Ribeiro BG, De Veras BO, Dos Santos AJ, Guerra JMC, Sarubbo LA (2020a) Biosurfactant produced by Candida utilis UFPEDA1009 with potential application in cookie formulation. Electron J Biotechnol 46:14–21

Ribeiro BG, Guerra JM, Sarubbo LA (2020b) Biosurfactants: production and application prospects in the food industry. Biotechnol Prog 36:e3030

Ribeiro BG, Guerra JMC, Sarubbo LA (2020c) Potential food application of a biosurfactant produced by Saccharomyces cerevisiae URM 6670. Front Bioeng Biotechnol 8:434

Rocha E, Silva NMP, Meira HM, Almeida FCG, Soares Da Silva RDCF, Almeida DG, Luna JM et al (2019) Natural surfactants and their applications for heavy oil removal in industry. Sep Purif Rev 48:267–281

Rodríguez A, Gea T, Font X (2021) Sophorolipids production from oil cake by solid-state fermentation. Inventory for economic and environmental assessment. Front Chem Eng 3:632752

Rodríguez-López L, Rincón-Fontán M, Vecino X, Moldes AB, Cruz JM (2020) Biodegradability study of the biosurfactant contained in a crude extract from corn steep water. J Surfactant Deterg 23:79–90
Rosenberg E, Ron EZ (1999) High-and low-molecular-mass microbial surfactants. Appl Microbiol Biotechnol 52:154–162
Rosenberg E, Zuckerberg A, Rubinovitz C, Gutnick D (1979) Emulsifier of Arthrobacter RAG-1: isolation and emulsifying properties. Appl Environ Microbiol 37:402–408
Rufino R, Luna J, Marinho P, Farias C, Ferreira S, Sarubbo L (2013) Removal of petroleum derivative adsorbed to soil by biosurfactant Rufisan produced by Candida lipolytica. J Pet Sci Eng 109:117–122
Sahebnazar Z, Mowla D, Karimi G, Yazdian F (2018) Zero-valent iron nanoparticles assisted purification of rhamnolipid for oil recovery improvement from oily sludge. J Environ Chem Eng 6:917–922
Salazar-Bryam AM, Yoshimura I, Santos LP, Moura CC, Santos CC, Silva VL et al (2021) Silver nanoparticles stabilized by ramnolipids: effect of pH. Colloids Surf B Biointerfaces 205:111883
Salerno-Kochan R, Kowalski M (2020) Safety management of textile products in the European Union and estimation of its efficiency. Part 2. Fibres Text East Eur 28:12–17
Sánchez A, Artola A, Gea T, Barrena R, Font X (2015) A new paradigm for waste management of organic materials. Waste Manag 42:1–2
Sangeeta Patil AA (2023) Biosurfactants in textiles: sustainable future. Trends Text Fash Des 5. https://doi.org/10.32474/LTTFD.2023.05.000214
Santos DKF, Rufino RD, Luna JM, Santos VA, Sarubbo LA (2016) Biosurfactants: multifunctional biomolecules of the 21st century. Int J Mol Sci 17:401
Santos DK, Resende AH, De Almeida DG, Soares Da Silva RDCF, Rufino RD, Luna JM et al (2017) Candida lipolytica UCP0988 biosurfactant: potential as a bioremediation agent and in formulating a commercial related product. Front Microbiol 8:767
Santos E, Teixeira M, Converti A, Porto A, Sarubbo L (2019) Production of a new lipoprotein biosurfactant by Streptomyces sp. DPUA1566 isolated from lichens collected in the Brazilian Amazon using agroindustry wastes. Biocatal Agric Biotechnol 17:142–150
Sarubbo LA, Lunaa JM, Rufinoa RD (2015) Application of a biosurfactant produced in low-cost substrates in the removal of hydrophobic contaminants. Chem Eng 43:295–300
Sarubbo LA, Maria Da Gloria CS, Durval IJB, Bezerra KGO, Ribeiro BG, Silva IA et al (2022) Biosurfactants: production, properties, applications, trends, and general perspectives. Biochem Eng J 181:108377
Satpute SK, Kulkarni GR, Banpurkar AG, Banat IM, Mone NS, Patil RH et al (2016) Biosurfactant/s from Lactobacilli species: properties, challenges and potential biomedical applications. J Basic Microbiol 56:1140–1158
Sekhon K (2012) Biosurfactant production and potential correlation with esterase activity. J Pet Environ Biotechnol 03. https://doi.org/10.4172/2157-7463.1000133
Seredyuk V, Alami E, Nydén M, Holmberg K, Peresypkin A, Menger F (2001) Micellization and adsorption properties of novel zwitterionic surfactants. Langmuir 17:5160–5165
Shah N, Nikam R, Gaikwad S, Sapre V, Kaur J (2016) Biosurfactant: types, detection methods, importance and applications. Indian J Microbiol Res 3:5–10
Shakeri F, Babavalian H, Amoozegar MA, Ahmadzadeh Z, Zuhuriyanizadi S, Afsharian MP (2021) Production and application of biosurfactants in biotechnology. Biointerface Res Appl Chem 11:10446–10460
Shen L, Zhu J, Lu J, Gong Q, Jin M, Long X (2019) Isolation and purification of biosurfactant mannosylerythritol lipids from fermentation broth with methanol/water/n-hexane. Sep Purif Technol 219:1–8
Silva RDCF, Almeida DG, Rufino RD, Luna JM, Santos VA, Sarubbo LA (2014) Applications of biosurfactants in the petroleum industry and the remediation of oil spills. Int J Mol Sci 15:12523–12542

Silva EJ, Correa PF, Almeida DG, Luna JM, Rufino RD, Sarubbo LA (2018) Recovery of contaminated marine environments by biosurfactant-enhanced bioremediation. Colloids Surf B Biointerfaces 172:127–135
Sivaramakrishnan C (2013) Advances in the dyeing and finishing of technical textiles. Elsevier, Amsterdam, pp 199–235
Soetaert W, Van Bogaert I, Roelants S (2017) Methods to produce bolaamphiphilic glycolipids
Souza K, Gudiña EJ, Schwan RF, Rodrigues LR, Dias DR, Teixeira JA (2018) Biosurfactant production by Wickerhamomyces anomalus CCMA 0358. J Hazard Mater 346:152–158
Twigg MS, Baccile N, Banat IM, Déziel E, Marchant R, Roelants S et al (2021) Microbial biosurfactant research: time to improve the rigour in the reporting of synthesis, functional characterization and process development. Microb Biotechnol 14:147–170
Van Renterghem L, Roelants SL, Baccile N, Uyttersprot K, Taelman MC, Everaert B et al (2018) From lab to market: an integrated bioprocess design approach for new-to-nature biosurfactants produced by Starmerella bombicola. Biotechnol Bioeng 115:1195–1206
Vandana P, Singh D (2018) Review on biosurfactant production and its application. Int J Curr Microbiol Appl Sci 7:4228–4241
Wang S-H, Tang TW-H, Wu E, Wang D-W, Wang C-F, Liao Y-D (2020) Inhibition of bacterial adherence to biomaterials by coating antimicrobial peptides with anionic surfactant. Colloids Surf B Biointerfaces 196:111364
Wittgens A, Rosenau F (2018) On the road towards tailor-made rhamnolipids: current state and perspectives. Appl Microbiol Biotechnol 102:8175–8185
Wongsirichot P, Ingham B, Winterburn J (2021) A review of sophorolipid production from alternative feedstocks for the development of a localized selection strategy. J Clean Prod 319:128727
Wu Y, Xu M, Xue J, Shi K, Gu M (2019) Characterization and enhanced degradation potentials of biosurfactant-producing bacteria isolated from a marine environment. ACS Omega 4:1645–1651
Xu C, Wang H, Wang D, Zhu Y, Zhu X, Yu H (2021) Study on the mechanism of polyethylene oxide groups improving the foamability of anionic surfactants in hard water. Colloids Surf A Physicochem Eng Asp 613:126046
Yang Z, Zu Y, Zhu J, Jin M, Cui T, Long X (2020) Application of biosurfactant surfactin as a pH-switchable biodemulsifier for efficient oil recovery from waste crude oil. Chemosphere 240:124946
Yuan C, Xu Z, Fan M, Liu H, Xie Y, Zhu T (2014) Estudio sobre la s caracteristicas y el dano de los tensioactivos. J Chem Pharm Res 6:2233–2237
Zenati B, Chebbi A, Badis A, Eddouaouda K, Boutoumi H, El Hattab M et al (2018) A non-toxic microbial surfactant from Marinobacter hydrocarbonoclasticus SdK644 for crude oil solubilization enhancement. Ecotoxicol Environ Saf 154:100–107

Chapter 5
Sustainable Microbial Biopolymers and Biocomposites for Textile Innovation

Azalfah Ibrar, Swebba Waheed, Fatima Mujahid, Muhammad Asif Zahoor, Sabeeka Shahwar, Ahsen Taqveem, and Mohsin Khurshid

Abstract Microbial biopolymers and biocomposites emerge as sustainable alternatives to synthetic petroleum-based polymers in the textile industry. These biopolymers offer inherent biodegradability and biocompatibility, addressing growing environmental concerns about persistent plastic waste and ecological pollution. Microbial biopolymers possess unique properties such as film-forming and gelling capabilities, which make them suitable for diverse textile applications. This chapter examines two prominent examples: bacterial cellulose (BC) and polyhydroxyalkanoates (PHAs). BC, produced by bacteria such as *Komagataeibacter xylinus*, boasts a nanofibrillar structure, high water retention, and potential as an eco-friendly vegan leather alternative. PHAs, synthesized by bacterial fermentation, are carbon-neutral, biodegradable polyesters with tunable mechanical and thermal properties. The chapter explores their biosynthesis pathways, emphasizing key enzymes and regulatory mechanisms. Despite their promise, challenges such as limited mechanical strength and high production costs hinder widespread adoption. To address these limitations, microbial biocomposites i.e. advanced materials integrating microbial biopolymers with natural or synthetic textile substrates are introduced. These biocomposites demonstrate enhanced thermal stability, durability, and functionality, thereby broadening their applicability in smart textiles, wearable technologies, and functional apparel. By bridging the gap between biotechnology and material science, this work underscores the transformative potential of microbial-based systems in fostering sustainable textile innovation.

Keywords Sustainable textiles. Biodegradable · Renewable resources · Eco-friendly textiles · Microbial polyesters · Smart textiles

A. Ibrar · S. Waheed · F. Mujahid · M. A. Zahoor · S. Shahwar · M. Khurshid (✉)
Institute of Microbiology, Government College University Faisalabad, Faisalabad, Pakistan
e-mail: drasifzahoor@gcuf.edu.pk; mohsinkhurshid@gcuf.edu.pk

A. Taqveem
School of Allied Health Sciences, Faculty of Sciences, Multan University of Science and Technology, Multan, Pakistan

M. Khurshid et al. (eds.), *Microbial Biotechnology in Textile Innovation*, SDGs and Textiles, https://doi.org/10.1007/978-981-95-0347-6_5

5.1 Introduction

Biopolymers are a distinct class of polymers naturally synthesized by living organisms through biological processes, in contrast to synthetic, petroleum-based polymers. Various enzymes facilitate these processes by linking monomeric subunits such as sugars, amino acids, and hydroxyl fatty acids to form long polymer chains with high molecular weight (Moradali et al. 2017). Microbial-based biopolymers refer to biopolymers produced by microorganisms, which serve as efficient cell factories capable of utilizing diverse nitrogen and carbon sources to generate a wide range of extracellular and intracellular biopolymers (Bhatia et al. 2021). The unique properties of microbial biopolymers, such as biodegradability, biocompatibility, and their ability to form films or gels, make them highly suitable for various textile applications. Derived from renewable resources, these biopolymers naturally decompose, unlike conventional synthetic fibers, thereby contributing to the reduction of global plastic waste and environmental pollution.

Microbial biopolymers are classified into three main groups based on their biological sources: algae, fungi, and bacteria (Table 5.1). They are further categorized by their chemical composition, which includes polysaccharides, polyphosphates, polyesters, and polyamides (Zhao et al. 2023). As photosynthetic microorganisms, algae can grow in wastewater, reducing dependency on freshwater resources. Fungi produce a diverse range of biopolymers with applications in enzyme immobilization, soil carbon sequestration, and biosurfactants (Schreiner et al. 2014; Ebinesar et al. 2021). Additionally, bacteria synthesize both intracellular and extracellular biopolymers through metabolic processes. Among microbial biopolymers, polyesters and polysaccharides are the most prevalent, produced via distinct metabolic pathways (Moradali and Rehm 2020; Ghosh et al. 2021). The production of microbial biopolymers involves various fermentation strategies tailored to specific microbial species and polymer types. The choice of production method depends on multiple factors, including the growth characteristics of the microorganisms, the properties of the desired polymers, and their intended applications. Recent advancements in metabolic engineering and bioprocess optimization have significantly enhanced the sustainability, cost-effectiveness, and yield of microbial biopolymers for industrial applications.

Many microbial biopolymers also exhibit functional properties that enhance their performance in textiles, such as antimicrobial activity and improved moisture regulation. These characteristics make them particularly suitable for medical textiles, technical fabrics, and antibacterial finishes, where durability and environmental resistance are essential (Li et al. 2017; Kiran et al. 2024). However, some microbial biopolymers have limitations, such as poor mechanical strength, limited functionality, and chemical sensitivity. To overcome these challenges, they can be combined with other materials to form composites known as "microbial biocomposites." These advanced materials integrate microbial components such as bacterial cellulose and fungal mycelium into textile substrates, offering eco-friendly alternatives to conventional fabrics.

Table 5.1 Fabrication of microbial biopolymers from different microbes

Microorganism	Nature of biopolymers	Biopolymers	Species	References
Algae	Polyesters	Polyhydroxyalkanoates	*Chlorella minutissima* *Spirulina spp.* *Synechococcus subsalsus*	Costa et al. (2019), Sharma et al. (2024)
		Polyhydroxybutyrate	*Chlorella spp.* *Stigeoclonium spp.*	Kumari et al. (2022), Sharma et al. (2024)
		Polylactide	*Scenedesmus abundans* *Corallina elongata*	Cheng and Wang (2022), Sharma et al. (2024)
	Polysaccharide	Alginate	*Nizimuddinia zanardini*	Torabi et al. (2022)
Fungi	Polysaccharides	β-glucan	*Aureobasidium thailandenese NRRL 58543* *Saccharomyces cerevisiae* *Aspergillus spp.*	Kayanna et al. (2022), Sharma et al. (2024)
		Chitosan	*Roccella montagnei* *Rhizopus stolonifera* *Mucorales*	Logesh et al. (2012), Sharma et al. (2024)
		Pullulan	*Aureobasidium spp.* *Aureobasidium pullulans ATCC 15233*	Hernandez-Tenorio and Giraldo-Estrada (2022), Sharma et al. (2024)
Bacteria	Polysaccharides	Bacterial cellulose	*Komagataeibacter spp.*	Fatima et al. (2023)
		Xanthan gum	*Xanthomonas campestris WXLB-006*	Wang et al. (2017a, b)
		Alginate	*Pseudomonas spp.* *Laminaria digitata*	Sharma et al. (2024)
		Hyaluronic acid	*Mytilus galloprovincialis* *Streptococcus spp.*	Kogan et al. (2007), Sharma et al. (2024)
	Polyamide	Poly(γ-glutamic acid)	*Bacillus spp.*	Parati et al. (2022)
	Polyester	Polyhydroxyalkanoates	*Bacillus megaterium*	Sharma et al. (2024)

Microbial bio-composites exhibit distinct thermal and mechanical properties, expanding their applications in smart textiles. For example, bacterial cellulose-based biocomposites reinforced with nanoparticles have gained significant attention due to their potential in textile manufacturing and tissue engineering (Liu et al. 2020). Similarly, fungi such as *Pleurotus ostreatus* can utilize textile waste to produce lightweight mycelium-based biocomposites, presenting a sustainable alternative to polystyrene. Additionally, photosynthetic algae can be integrated into rigid living materials that capture CO_2, contributing to sustainable infrastructure development. These advancements in polymer science highlight the versatility and environmental benefits of microbial-based biopolymers and biocomposites, positioning them as transformative materials for the textile industry.

5.2 Bacterial Cellulose

Cellulose, traditionally recognized as a plant-based product, is one of the most sustainable and abundant renewable polymers on Earth. However, recent advancements in fermentation techniques have revealed that several microorganisms, including fungi, algae, and a variety of aerobic bacteria, are also capable of producing cellulose (Ullah et al. 2016). The primary objective of this approach is to promote eco-friendly and sustainable production methods while conserving natural resources (Akinsemolu 2018).

Bacterial cellulose, a natural polysaccharide-based biopolymer with the general formula $(C_6H_{10}O_5)_n$, is widely regarded as a highly versatile biomaterial with significant potential in textile manufacturing (Jayabalan et al. 2014; Wang et al. 2017a, b; Raut et al. 2023). Unlike plant-derived cellulose, bacterial cellulose offers an environmentally friendly alternative, as it can be produced without reliance on forests or agricultural land (Felgueiras et al. 2021).

The bacterium *Komagataeibacter xylinus* (formerly classified under *Gluconacetobacter* and *Acetobacter*) is the most well-studied model organism for bacterial cellulose production due to its non-pathogenic nature and efficiency in large-scale fermentation (Blanco Parte et al. 2020). Additionally, several other bacterial genera, including *Acetobacter, Azotobacter, Rhizobium, Pseudomonas,* and *Achromobacter*, as well as the Gram-positive bacterium *Sarcina ventriculi*, are also known to produce cellulose (Vadanan et al. 2022).

5.2.1 *Biosynthesis and Structure of Bacterial Cellulose*

Komagataeibacter xylinus has been extensively studied to gain a comprehensive understanding of the biosynthetic pathway of cellulose production and its regulation. During bacterial cellulose biosynthesis, cellulose nanofibrils are secreted as an exopolysaccharide at the liquid-air interface. This intricate process involves polymerization and translocation of microfibrils, orchestrated by a membrane-associated multi-protein complex. This complex comprises various regulatory proteins, enzymes, and catalytic subunits that function in a highly coordinated manner. The latest framework for bacterial cellulose biosynthesis outlines a series of enzymatic reactions catalyzed by four key biocatalysts: (a) glucokinase, which phosphorylates glucose into glucose-6-phosphate (Glc-6-P); (b) phosphoglucomutase, which converts Glc-6-P into glucose-1-phosphate (Glc-1-P); (c) UDP-glucose pyrophosphorylase (UGPase), which produces UDP-glucose (UDPGlc); and (d) the cellulose synthase complex (BcsA, BcsB, BcsC, and BcsD), which polymerizes UDPGlc monomers into cellulose chains (Buldum and Mantalaris 2021). Among these enzymes, UGPase is considered the key catalyst in the biosynthetic process, playing a pivotal role in maintaining an adequate and continuous supply of UDPGlc for polymerization (Valla et al. 1989). Additionally, cyclic-di-GMP (C-di-GMP), a secondary messenger associated with biofilm regulation, has been identified as a crucial modulator of cellulose biosynthesis. It directly interacts with the cellulose synthase complex, specifically targeting the BcsA and BcsB subunits via the PilZ domain in the periplasm, thereby regulating bacterial cellulose production at the molecular level (Fujiwara et al. 2013; Römling et al. 2013).

Interestingly, a study reported that C-di-GMP serves a dual function, also regulating the expression of cellulose-degrading enzymes in anaerobic biofilm-producing microorganisms. This regulation ensures a balance between cellulose synthesis and degradation, which is essential for maintaining biofilm structure and function. The outer membrane subunits BcsC and BcsD further contribute to the biosynthetic pathway by assembling, crystallizing, and exporting cellulose chains (Raut et al. 2019). Beyond the core cellulose synthase complex, additional proteins such as CMCax, BlgAx, and CcpAx play vital roles in preserving the structural integrity of bacterial cellulose. Although these proteins are not directly involved in cellulose synthesis, they contribute to the mechanical strength and stability of the cellulose matrix (Gullo et al. 2018).

Bacterial cellulose (BC) possesses a distinctive chemical structure characterized by a three-dimensional ultra-fine network of fibers composed of glucose monomers linked by β-(1→4) glycosidic bonds. These monomers form glucan chains that aggregate into protofibrils, which are held together by hydrogen bonds. The protofibrils are secreted through the bacterial cell wall and subsequently combine into nanofibrils and microfibril ribbons (Aditya et al. 2022). This web-like 3D network, rich in hydroxyl groups, imparts unique properties to BC, including porosity, hydrophilicity, biodegradability, and the potential for chemical modification (Gregory et al. 2021).

The nano-porous structure, hydrophilic nature, and high surface area of BC contribute to its excellent water retention capacity (Lupaşcu et al. 2022). In the textile industry, bacterial cellulose is emerging as a sustainable alternative to traditional leather. Notably, ScobyTec, a German startup, has developed a vegan leather substitute using bacterial cellulose derived from kombucha. This material, known as ScobyTec BNC, has garnered significant attention due to its non-flammability and high mechanical strength (Nayak et al. 2024).

5.2.2 Production of Bacterial Cellulose

The production of bacterial cellulose (BC) involves several key steps, beginning with the selection of an appropriate bacterial strain. The efficiency and yield of BC production depend largely on the bacterial species used and the composition of the growth medium. Various species, such as *Komagataeibacter xylinus*, *Acetobacter hansenii*, and *Acetobacter pasteurianus*, have been widely studied for their high cellulose synthesis capabilities and adaptability to different substrates, making them suitable for industrial applications (Öz and Kalender 2023). Among these, *Komagataeibacter xylinus* is the most extensively utilized. The commonly used Hestrin and Schramm (HS) medium, which contains yeast extract, Difco Bactopeptone, glucose, citric acid, and disodium phosphate, plays a significant role in determining the quality and yield of BC (Hestrin and Schramm 1954). However, due to the high cost of yeast extract, HS medium has become less viable for commercial applications. To address this issue, the use of corn steep liquor (CSL) was proposed as an alternative nitrogen source, which significantly enhanced BC production (Toyosaki et al. 1995).

Following strain selection, bacterial cellulose is produced through fermentation, which can be conducted under static or agitated conditions. Static fermentation is the preferred method for textile applications due to its ability to produce a uniform and thick cellulose pellicle. In this process, the prepared culture medium is inoculated with a pre-culture of the selected bacterial strain and incubated at 25–30 °C for 1–14 days under sterile conditions to prevent contamination (Çakar et al. 2014). Maintaining static conditions allows the formation of a thick, gelatinous pellicle of BC at the air-liquid interface. In contrast, agitated fermentation is often used for large-scale production, but it results in irregular cellulose particles rather than a uniform sheet, requiring additional processing to meet textile industry standards (Öz and Kalender 2023).

Downstream processing involves the harvesting and purification of bacterial cellulose. The cellulose pellicles are carefully removed from the culture medium to prevent tearing and thoroughly washed with distilled water to eliminate residual medium components. They are then treated with 0.3 N NaOH and heated at 80 °C for 15–20 min to remove bacterial cells and impurities. After this step, multiple washes with distilled water ensure the complete removal of excess NaOH, bringing the pH to a neutral level (Shezad et al. 2009). The final drying process can be carried out using different methods such as air drying, freeze-drying, or hot pressing, depending on the desired application of the bacterial cellulose.

5.3 Polyhydroxyalkanoates

Polyhydroxyalkanoates (PHAs) are carbon-neutral, bio-based polymers that belong to the family of naturally occurring polyesters. They are produced by various microorganisms, particularly bacteria, as intracellular granules using renewable carbon resources (Tan et al. 2014). These granules are water-insoluble and typically range in diameter from 0.2 to 0.5 μm. PHAs were first discovered in 1926 by Lemogine (Anderson and Dawes 1990) and have since gained significant research and commercial interest due to their unique properties, including biocompatibility, chemical diversity, sourcing from renewable carbon resources, and biodegradability (Shah et al. 2008).

Microorganisms synthesize PHAs through the fermentation of organic compounds such as glucose and other carbon-containing substrates. Under nutrient-limiting conditions, these microorganisms store PHAs as intracellular carbon and energy reserves (Acharjee et al. 2023). urrently, over 90 genera of Gram-positive and Gram-negative bacteria have been identified as PHA producers in both anaerobic and aerobic environments. Among them, *Pseudomonas aeruginosa, P. stutzeri, Bacillus megaterium, Cupriavidus necator,* and *P. oleovorans* are the most extensively studied species due to their significant potential for large-scale biopolymer production (Muneer et al. 2020). PHAs constitute a diverse group of polymers with nearly one hundred different types of repeating units (Jain et al. 2010). These units primarily consist of hydrogen and carbon atoms, leading scientists and researchers to refer to them as "carbonosomes" (Koller 2018).

5.3.1 Biosynthesis and Structure of PHAs

Depending on the type or strain of microorganism and the growth conditions, polyhydroxyalkanoates (PHAs) can be produced as either homopolymers or copolymers, composed of 3-, 4-, 5-, and 6-hydroxyalkanoic acids. Among all PHAs, poly(3-hydroxybutyrate) (PHB) was the first to be identified and remains the most extensively studied. The primary enzyme involved in PHA synthesis is PHA synthase. A cluster of genes, including β-ketothiolase (PhaA), NADPH-acetoacetyl-CoA reductase, and PHA synthase (PhaC), is organized into a PHA biosynthetic operon (Rehm 2003).

The biosynthetic pathway for PHB, a short-chain-length PHA, begins with the condensation of two acetyl-CoA molecules to form acetoacetyl-CoA, a reaction catalyzed by β-ketothiolase. This compound is then reduced to (R)-3-hydroxybutyryl-CoA by the action of (R)-specific acetoacetyl-CoA reductase. Finally, PHA synthase catalyzes the polymerization of (R)-3-hydroxybutyryl-CoA into PHB (Urtuvia et al. 2018).

A typical PHA molecule consists of 600–35,000 repeating units of (R)-hydroxy fatty acid (Khanna and Srivastava 2005). Each repeating unit contains a side-chain R group, which is predominantly a saturated alkyl group but can also be an unsaturated, substituted, or branched alkyl group (Lu et al. 2009). Poly(3-hydroxybutyrate) is the most common and simplest form of PHA synthesized by prokaryotic organisms, particularly bacteria. This PHA consists of (R)-3-hydroxybutyrate (3HB) monomers, which polymerize to form a polymeric chain (Sudesh et al. 2000).

5.3.2 Production of PHAs

The key steps involved in the synthesis of these polyhydroxyalkanoates (PHAs) are as follows:

The synthesis of polyhydroxyalkanoates (PHAs) begins with the selection and preparation of suitable substrates, with carbon sources playing a crucial role in the process (Raza et al. 2018). Various organic wastes from industries such as food, paper, and sugar processing serve as rich carbohydrate sources for PHA production. Agricultural byproducts, including molasses and whey, also provide an abundant supply of carbon. Simple sugars like sucrose and glucose are the primary substrates used in PHA synthesis. Additionally, starch-based, cellulosic, and hemicellulosic materials can be utilized; however, these complex carbohydrates must first be hydrolyzed into simpler sugars, as microbes cannot directly metabolize them. Furthermore, fatty acid-based substrates such as oils and glycerol can also be employed, expanding the range of available carbon sources for PHA production.

Optimizing fermentation techniques and operational parameters is crucial for enhancing PHA production. Fed-batch fermentation has proven effective in increasing PHA accumulation by controlling the supply of nitrogen and carbon sources. A study reported that using glycerol as a carbon source resulted in the production of 81.2 g/L bacterial biomass and 54.3 g/L PHB when supplemented with 20 g/L sodium chloride (NaCl) under optimized ammonia and glycerol supply. PHB is one of the most well-known types of polyhydroxyalkanoates (Ibrahim and SteinbüChel 2009). Similarly, another study investigated large-scale fed-batch fermentation in a 60 L reactor using *A. eutrophus*, achieving 281 g/L cell biomass and 232 g/L PHB at a production rate of 3.14 g/L/h by limiting phosphate concentration and maintaining pH during the fermentation process (Ryu et al. 1997).

Another promising approach is the feast-famine strategy, in which microorganisms store PHA under nutrient-rich conditions and utilize these reserves when nutrients become scarce. Hafuka et al. (2011) proposed that this strategy could enhance PHA yield, while Fradinho et al. (2016) successfully implemented it in photosynthetic mixed cultures under an illumination-rich environment with a steady acetate supply, resulting in a 60% accumulation of PHB along with concurrent phosphate removal.

Additionally, solid-state fermentation (SSF) using waste biomass, such as polyurethane foam (PUF) and soy cake, has been explored for sustainable PHA production. Ramadas et al. (2013) studied SSF using *B. sphaericus* NII 0838 on PUF, achieving a PHB yield of 0.169 g/g and a biomass production of 0.4 g/g PUF. Likewise, Oliveira et al. (2007) developed an SSF process using *R. eutropha* and reported a 39% accumulation of PHB after 36 h of fermentation.

Continuous fermentation has also emerged as a potential strategy to reduce production costs and improve sustainability for large-scale PHA production. Lopar et al. (2013) demonstrated its effectiveness by maximizing the growth rate of selected microbial strains. Despite these advancements, further research is required to optimize fermentation techniques for the economically viable synthesis of PHAs.

The recovery and purification of PHA from bacterial biomass is a complex and resource-intensive process that often requires harsh chemicals, significantly increasing production costs (Table 5.2). In fact, downstream processing is estimated to account for 50% or more of the total production cost, making it a critical factor in determining the economic feasibility of PHA biomanufacturing (Li and Wilkins 2020). The recovery and purification process typically begins with the separation of bacterial cells after fermentation, followed by pretreatment methods such as freezing, heating, treatment with salt and liquid nitrogen, or exposure to warm water.

One of the most widely used methods for PHA recovery is solvent extraction, in which solvents enhance membrane permeability, facilitating PHA release.

Table 5.2 Comparative analysis of mechanical and chemical purification methods

Purification methods	Examples	Principle	Advantages	Limitations	References
Mechanical methods	Ultrasonic extraction method	Uses ultrasonic waves for the lysis of cells	Quick and energy-efficient	Expensive equipment	Hafuka et al. (2011), Kosseva and Rusbandi (2018)
	Mechanical disruption	Physical crushing and grinding	Scalable and simple	Can disrupt the structure of PHA	
	High-pressure homogenization	Mechanical disruption at high-pressure	No chemical usage	High energy requirements	
Chemical methods	Solvent extraction (e.g. DMSO)	Organic solvent dissolves non-PHA materials	High efficiency, high PHA purity	Expensive solvent, leaves toxic compounds behind	Saratale and Oh (2015), Murugan et al. (2016), Kosseva and Rusbandi (2018)
	Sodium hypochlorite extraction	Chemically degrades the non-PHA materials	Simple and effective	If overused, it can also degrade PHA molecules	
	SDS sonication method	Surfactants disrupt the cells	Preserves the polymeric properties	Time-consuming	
	Enzymatic digestion	Enzymatic degradation of non-PHA materials	Eco-friendly	Expensive	
	Flotation	PHA granules are separated by air bubbles	Cost-effective	Not used widely	
	Aqueous two-phase system	Phase separation of PHA and non-PHA materials	Biocompatible	Process optimization required	

Traditionally, halogenated solvents like chloroform, often combined with sodium hypochlorite, have been used to achieve high recovery rates. However, recent innovations have focused on "green solvents" under pressurized conditions, ionic liquids, supercritical solvents, switchable anionic surfactants, hypotonic cell disintegration, and even animal digestion of non-PHA biomass (Koller 2018).

An alternative method, aqueous two-phase extraction, has also been explored for PHA recovery and purification. This technique utilizes different phase systems, including polymer/polymer, ionic liquid/salt, polymer/salt, and mixtures of ionic and non-ionic surfactants (Saratale and Oh 2015; Yau et al. 2015). More recently, researchers have investigated biological tools for PHA recovery, such as using mealworms and lytic organisms, which have shown promising results in achieving high-purity PHA (Murugan et al. 2016). Advancements in extraction techniques are crucial for enhancing the efficiency of PHA recovery while reducing costs and minimizing environmental impact. Continued research in this area will help drive the development of more sustainable and economically viable PHA biomanufacturing processes.

5.4 Other Microbial Polysaccharides

Alginate is a naturally abundant polyanionic carbohydrate polymer composed of α-L-guluronic acid (G) and β-D-mannuronic acid (M) blocks, arranged in both homogeneous and heterogeneous patterns (Zhang et al. 2023). It is primarily derived from brown seaweed (Phaeophyceae) (Hay et al. 2013). Additionally, certain bacterial species, including Pseudomonas and Acetobacter, produce alginates to enhance their resistance to cytotoxic compounds (Remminghorst and Rehm 2006).

Due to its strong gelling potential, exceptional biocompatibility, biodegradability, and ease of modification, alginate has found extensive applications in biomedical and industrial fields (Kibungu et al. 2021; Wang et al. 2021; Zhang et al. 2023). t has also emerged as a promising antibacterial finishing agent for textiles (Li et al. 2017). However, the high production cost of microbial alginates remains a major challenge. To mitigate this, industrial waste from dairy, biodiesel, and sugar industries, which are rich in carbon sources, can replace pure sugars in microbial alginate production, reducing substrate costs. Additionally, genetic modifications and optimized fermentation conditions can further enhance production efficiency (Wang et al. 2023).

Dextran is another microbial exopolysaccharide, consisting of D-glucose units linked by α-(1 → 6) bonds (Díaz-Montes 2021). It is synthesized by lactic acid bacteria from various genera, including Enterococcus, Lactococcus, Alloiococcus, Lactobacillus, and Aerococcus (Mozzi et al. 2013). In the textile industry, dextran is used for fabric coatings, enhancing water resistance and printing applications.

Similarly, pullulan is a natural fibrous extracellular polysaccharide with the molecular formula $(C_6H_{10}O_5)_n$. It consists of linear, unbranched maltotriose units connected by α-1,6-glucosidic bonds and is primarily produced by the fungi

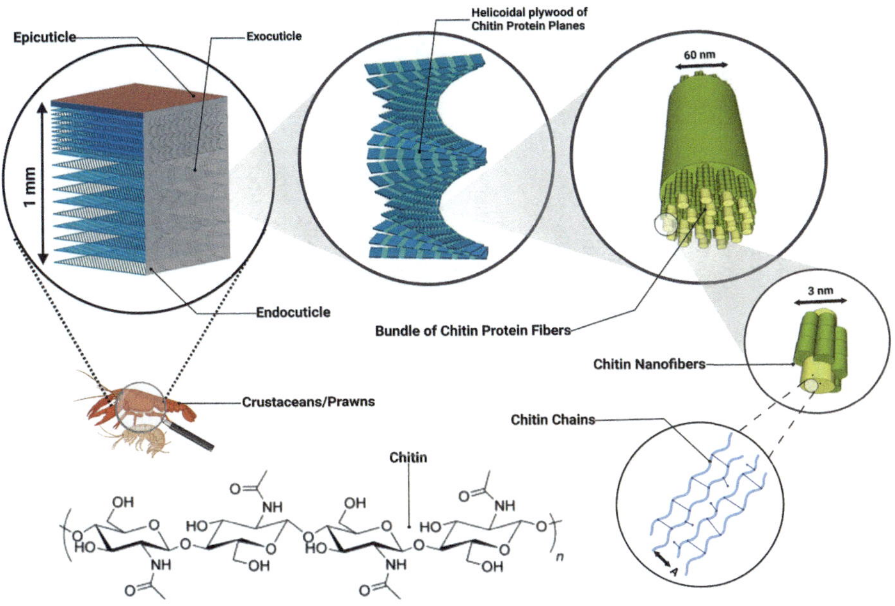

Fig. 5.1 Hierarchical structure of chitin, a model biopolymer for sustainable material development. (The image was created by authors using BioRender licensed account)

Aureobasidium pullulans and *Eurotium chevalieri* (Gaur et al. 2010; Shukla et al. 2019). The molecular weight of pullulan varies widely, ranging from 4.5×10^4 to 6×10^5 Da. Due to its film-forming ability and moisture resistance, pullulan is widely used in textile coatings (Cheng et al. 2011).

Another important polysaccharide, chitosan, is the second most abundant polysaccharide on Earth. It is derived from chitin, the primary component of crustacean shells and some fungi (Fig. 5.1). Due to its antibacterial properties, chitosan is widely used in antibacterial textile production, a rapidly growing sector (Lim and Hudson 2003).

Xanthan gum, with the molecular formula $(C_{35}H_{49}O_{29})n$, is a microbial exopolysaccharide produced by *Xanthomonas campestris* (Bhat et al. 2022). Due to its light yellow or white powdery appearance, it is also referred to as corn sugar gum. Structurally, xanthan gum features a β-(1→4) linked glucan backbone and contains D-glucosyl, D-mannosyl, and D-glucuronyl residues in a molar ratio of 2:2:1, along with varying degrees of pyruvyl and O-acetyl substitutions (Sanderson 1982; Becker et al. 1998). While widely used in the food industry, xanthan gum is increasingly being incorporated into textiles to improve fabric texture, dyeing properties, and durability (Sandford and Baird 1983).

The various types of microbial biopolymers in textile production are summarized in Table 5.3. This outlines key microorganisms involved in biosynthesis, along with fermentation methods used to produce these polymers. With their distinct structural and functional properties, these polysaccharides play a crucial role in the textile

Table 5.3 Microbial biopolymers in textiles: Overview of production and applications

Types of biopolymers	Microorganisms Used	Substrates utilized	Type of fermentation	Parameters of fermentation	Duration of fermentation	Yield	Textile applications	References
Alginate	*Azotobacter vinelandii, Acetobacter, Pseudomonas aeruginosa* (bacteria)	Sucrose, glucose, glycerol	Static	pH = 7–7.2 Temp = 30 °C	20–72 h	3–5.4 g/L	Antimicrobial textiles, wound dressings, bio-based coatings	Li et al. (2017), Kibungu et al. (2021)
Bacterial cellulose (BC)	*Gluconacetobacter hansenii, Komagataeibacter xylinus* (bacteria)	Glycerol, glucose, agricultural waste	Static	pH = 4.8–6.4 Temp = 30 °C	1–14 days	6.13 g/L	Eco-friendly leather substitutes, moisture-absorbing fabrics, sustainable fabrics	Ross et al. (1991), Gregory et al. (2021)
Dextran	*Streptococcus mutans, lactic-acid bacteria, Leuconostoc mesenteroides* (bacteria)	Sucrose, glucose	–	–	–	–	Biodegradable fabric printing and finishing, textile coating for extra softness	Naessens et al. (2005)
Polyhydroxyalkanoates (PHAs)	*Cupriavidus necator, Ralstonia eutropha, bacillus megaterium* (bacteria)	Sugarcane molasses, whey, waste fats, plant oils	Fed batch	pH = 7–10 Temp = 60 °C	18–62 h	4.94 g/L	Biodegradable synthetic fiber alternatives, as coating agent	Anderson and Dawes (1990)

Chitosan	*Mucor rouxii, aspergillus Niger* (fungus)	Crab or shrimp shell waste, fungal cell wall	Submerged	pH = 7.5–8.5 Temp = 25 °C	70 h	1.53 g/L	Medical textiles, odor-resistant fabrics, antimicrobial fabrics or sportswear	Lim and Hudson (2003), Rinaudo (2006)
Xanthan gum	*Xanthomonas campestris* (bacteria)	Glucose, molasses, corn steep liquor, starch hydrolysates	Batch	pH = 7 Temp = 28 °C	72 h	15.5 g/L	Textile coating	Sutherland (2001), Bhat et al. (2022)
Pullulan	*Aureobasidium pullulans* (fungus)	Sucrose, starch hydrolysates, agro-industrial waste	Batch	pH = 6 Temp = 26 °C	5–7 days	23–50 g/L	Bio-based fabric finishing	Shukla et al. (2019)

industry, contributing to sustainability, quality, and performance. Continued advancements in microbial production and modification techniques are expected to further expand their industrial applications.

5.5 Properties and Characteristics of Microbial Biopolymers

Microbial biopolymers play a crucial role in various biological and industrial applications due to their multifunctional nature and adaptability. These biopolymers can be modified to enhance their functionality in response to changing environmental conditions (Awasthi et al. 2022). One of their key roles is acting as catalysts in biological reactions, particularly those involving polypeptides or proteins. This catalytic function reduces activation energy and enables microbes to efficiently break down substrates and synthesize essential products required for proper cellular functioning (Schnepp 2013). Additionally, microbial biopolymers serve as energy storage reservoirs, allowing microorganisms to accumulate excess energy for future metabolic demands. This ability is vital for their adaptation and survival in fluctuating environments (Van Loosdrecht et al. 1997). Beyond energy storage, these biopolymers also function as protective agents, safeguarding microbial cells from harsh environmental conditions. By forming biofilms and adhering to cell surfaces, they create a defensive barrier against chemical and physical stresses, enhancing microbial resilience in challenging circumstances. Furthermore, microbial biopolymers facilitate cellular communication, enabling microorganisms to effectively interact with their surroundings (Vu et al. 2009; Yin et al. 2019).

In addition to these fundamental characteristics, microbial biopolymers possess several unique properties that make them highly valuable in various industries, including textiles. One significant attribute is their antibacterial activity. Microbial biopolymers play a crucial role in developing resistance against microbes, which is particularly important in textiles since moisture absorption and high surface area of fabrics promote microbial growth (Pinho et al. 2011). Chitosan, for instance, is well known for its positively charged amino groups, which interact with negatively charged bacterial cell membranes, leading to structural damage and cell death. Similarly, alginate exhibits antimicrobial properties and is widely used in medical textiles, particularly in hydrogel dressings that not only prevent microbial infections but also promote wound healing (Shirvan et al. 2019). Another critical advantage of microbial biopolymers is their biodegradability. Unlike synthetic polymers, microbial biopolymers degrade into non-toxic by-products through microbial activity, making them environmentally friendly and crucial for sustainability in textile production (Egan and Salmon 2022).

Moreover, microbial biopolymers exhibit excellent biocompatibility, making them suitable for use in surgical textiles. Their non-toxic nature significantly reduces the risk of allergic reactions, with chitosan being a notable example (Rivera-Briso and Serrano-Aroca 2018). Beyond their biocompatibility, microbial biopolymers can be engineered to provide specific functional properties such as flame resistance, hydrophobicity, and UV protection. These properties enable innovative applications in textile industries, enhancing the performance and durability of fabrics (El-Sayed and Hassabo 2021).

5.6 Microbial Bio-composites in Textile

Microbial biocomposites are advanced materials that integrate biological components, such as bacteria and fungi, with conventional textile substrates to create sustainable and functional textile products. These biocomposites leverage the natural properties of microorganisms to enhance textile performance while reducing environmental impact, often resulting in biodegradable materials with unique functionalities. A key application of microbial biocomposites is the incorporation of microbial polymers to impart specific characteristics to textiles. For instance, microbial cellulose exhibits superior tensile strength, high water-holding capacity, and crystallinity, making it an ideal material for wound dressings and advanced clothing for burn patients due to its high porosity. Additionally, microbial polymers can be modified to develop water-repellent fabrics suitable for raincoats and vehicle covers (Retegi et al. 2010).

Alginate, a linear polysaccharide derived from soil bacteria, is another valuable microbial polymer widely used in sports apparel for its excellent moisture absorption properties, which help keep athletes dry by wicking away sweat. Furthermore, alginate serves as a flame retardant in various textile applications, enhancing safety and durability (Gowri et al. 2010). Microbial biocomposites also enable the incorporation of nanoparticles into fabrics, significantly enhancing properties such as antimicrobial activity, flame resistance, water and dust repellency, UV protection, and improved electrical conductivity. These advancements make microbial biocomposites particularly valuable in medical textiles, where fibers like aubasidan and rodexman provide protective and therapeutic benefits in wound dressings, promoting healing and tissue repair without adverse effects (Shamolina 1997).

Moreover, bacterial cellulose can be combined with hydrophobic polymers to create water-impermeable yet breathable nanocomposites, which are highly suitable for applications in the textile and footwear industries. These innovative materials hold significant potential for the future of sustainable and high-performance textiles, offering eco-friendly solutions without compromising functionality or durability (Fernandes et al. 2019).

5.7 Components of Microbial Biocomposites

Bacterial cellulose (BC), produced by bacteria such as *Komagataeibacter*, is a biomaterial with a wide range of distinct properties, making it an excellent industrial alternative to conventional plant cellulose. BC demonstrates exceptional mechanical strength compared to plant-derived cellulose, and its high purity due to the absence of lignin and hemicellulose. These properties makes it ideal for various textile applications (Wu et al. 2016). One innovative approach in sustainable textiles involves integrating living microalgae into textile substrates, enabling these materials to perform photosynthesis, effectively capturing carbon dioxide and potentially improving air quality. This integration can be achieved using advanced techniques such as 3D printing (Whitton 2012).

Plant-derived natural fibers, including cotton, hemp, and flax, serve as excellent sustainable substrates for microbial growth. These fibers provide a nutrient-rich environment for microorganisms while also being biodegradable, contributing to environmental sustainability. BC is also emerging as a promising alternative to leather due to its unique material properties, minimal infrastructure requirements, and biodegradability. Its ease of cultivation has made BC an attractive prototype biomaterial for designers and fashion innovators exploring sustainable textile production (Gupta and Bhaumik 2007; Gao and Cranston 2008).

Additionally, biopolymers such as chitosan (extracted from crustacean shells) and kappa-carrageenan (sourced from red seaweeds) are commonly incorporated into microbial biocomposites to enhance structural integrity and antimicrobial properties (Stefanova et al. 2021). Many microbial biocomposites also contain antimicrobial compounds derived from natural sources, preventing the growth of harmful microorganisms—a particularly beneficial feature for healthcare and hygiene-related textiles (Bonaldi 2018).

Another emerging material in sustainable textiles is mycelium, the root structure of fungi, which can be cultivated into strong, biodegradable fabrics. Mycelium thrives on organic substrates, effectively utilizing agricultural waste, and can be engineered for specific properties, expanding its applications in the textile industry (Sharma et al. 2024).

With ongoing research and development, these microbial biocomposites offer promising solutions for sustainable, high-performance textiles, balancing functionality with environmental responsibility.

5.8 Types of Microbial Biocomposites

5.8.1 Cellulose Based Biocomposites

As a biopolymer, cellulose is biodegradable and exhibits exceptional mechanical properties (Gibson 2012). Cellulose-fiber-reinforced polymer composites have gained significant interest due to their low density, non-abrasive nature, combustibility, non-toxicity, affordability, and biodegradability (Kalia et al. 2009). The concept of ACC (All-cellulose composites) was initially introduced by Nishino et al. (2004). ACCs are composed entirely of cellulose, where the matrix consists of dissolved and regenerated cellulose, while the reinforcement comprises undissolved or partially dissolved cellulose (Nishino et al. 2004). At the end of their service life, ACCs can be disposed of or composted without harming the environment, or they can be recycled for energy recovery (Baillie and Jayasinghe 2004). Their complete biodegradability makes them environmentally friendly and sustainable, which are key factors in their growing appeal.

Additionally, foam-formed biocomposites that incorporate cellulose fibers and additives such as lignin produce lightweight materials ideal for packaging and insulation applications (Reichler et al. 2021). Nanocellulose is primarily used in paper and packaging products; however, it is also being explored for applications in construction, automotive, furniture, electronics, pharmaceuticals, and cosmetics (Fig. 5.2). Furthermore, nanocellulose is employed as a membrane to enhance sound quality in electroacoustic devices (Norizan et al. 2022).

Cellulose-based composites have shown great potential in various biomedical fields due to their biocompatibility and mechanical properties, which closely resemble those of natural tissues. Their biomedical applications include tissue engineering, wound healing, and drug delivery. These composites can be engineered to provide structural support and facilitate healing processes, thereby improving patient outcomes. On the other hand, foam-formed cellulose biocomposites leverage their lightweight properties to create sustainable packaging solutions that are biodegradable and offer excellent insulation capabilities. This dual functionality is crucial for reducing plastic waste while ensuring effective product protection. The integration of cellulose in both biomedical applications and eco-friendly packaging highlights its potential to address contemporary challenges in health and environmental sustainability (Pattnaik and Swain 2022).

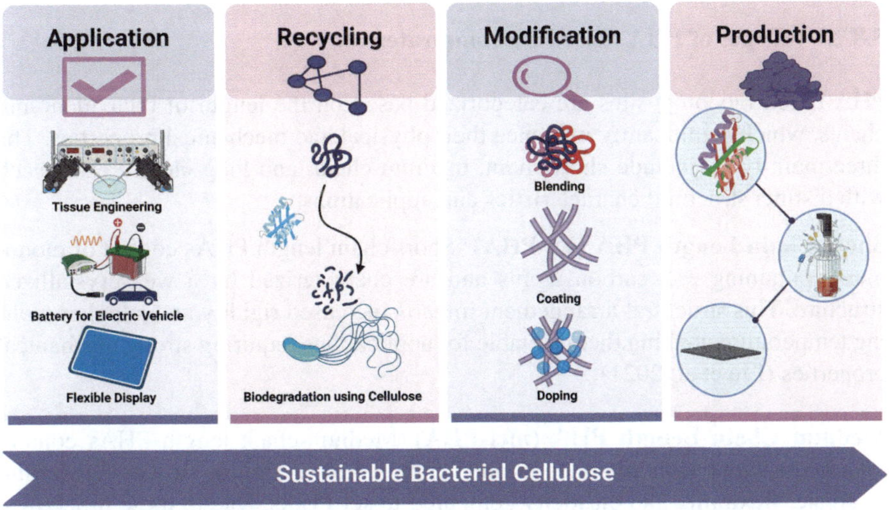

Fig. 5.2 Bacterial cellulose: a renewable biopolymer for advanced applications and Green Technologies. (The image was created by authors using BioRender licensed account)

5.8.2 *PHA Based Biocomposites*

Polyhydroxyalkanoates (PHAs) are a group of biodegradable and biocompatible polyesters produced through fermentation processes, making them an environmentally friendly and sustainable alternative to conventional plastics. Unlike other bio-based polymers such as polylactide (PLA), polybutylene succinate (PBS), and bio-polyethylene (bio-PE), PHAs are entirely biosynthesized and biopolymerized, featuring over 150 different monomer variations. This extensive variety allows PHAs to exhibit a broad range of material properties, making them highly versatile for applications in industries such as packaging, medicine, and agriculture (Wang et al. 2014).

The PHA family consists of various monomers, each contributing unique properties to the resulting polymer. Commonly encountered monomers include 3-hydroxybutyrate (3HB or C4), 3-hydroxyvalerate (3HV or C5), 3-hydroxydecanoate (3HD or C10), and 3-hydroxydodecanoate (3HDD or C12), as well as 4-hydroxybutyrate (4HB) (Steinbüchel and Valentin 1995). In recent years, several companies have commercialized PHAs as biodegradable plastics, promoting them as eco-friendly alternatives characterized by lower CO_2 emissions, sustainability, and reduced dependence on petroleum resources (Chen and Patel 2012).

Despite their potential, large-scale commercialization of PHAs has faced challenges, including high production costs, inconsistent material properties, and processing difficulties compared to conventional petroleum-based plastics. To enhance the competitiveness of PHAs, greater efforts are needed to optimize production processes and reduce costs associated with raw materials and fermentation (Sudesh et al. 2000).

5.8.2.1 Types of PHA-Based Biocomposites

PHA-based biocomposites are categorized based on the length of their monomer chains, which significantly influence their physical and mechanical properties. The three main types include short-chain, medium-chain, and long-chain PHAs, each with distinct structural characteristics and applications.

Short Chain Length PHA (scl-PHA) Short-chain length PHAs consist of monomers containing 3–5 carbon atoms and are characterized by a well-crystallized structure. This structural arrangement imparts increased rigidity and a higher melting temperature, making them suitable for applications requiring strong mechanical properties (Liu et al. 2021).

Medium Chain Length PHA (mcl-PHA) Medium-chain length PHAs contain monomers with 6–14 carbon atoms and exhibit a less crystalline structure, resulting in greater flexibility and elasticity compared to scl-PHAs. Due to these properties, mcl-PHAs are commonly used in applications where flexibility is essential, such as in packaging materials (Reddy et al. 2003).

Long Chain Length PHA (lcl-PHA) Long-chain length PHAs consist of monomers with 15 or more carbon atoms and exhibit flexibility similar to mcl-PHAs. However, their extended chain structure imparts unique properties, making them suitable for specialized applications in biomedical fields and high-performance materials (Zhang et al. 2024).

5.8.2.2 Applications for PHA Based Biocomposites

PHAs are increasingly recognized as sustainable alternatives to petroleum-based plastics due to their biodegradability and suitability for food packaging (Khatami et al. 2022). In particular, PHBV is widely used in tissue engineering, drug delivery systems, and various medical applications due to its biocompatibility and controlled degradation rates (Bidone et al. 2009). The environmentally friendly properties of PHAs make them ideal for agricultural films and other disposable products that decompose naturally after use (Mehrpouya et al. 2021).

The potential of PHAs as a drug delivery system for the sustained release of therapeutics into the bloodstream has garnered significant interest (Bansal et al. 2011). Polyhydroxyesters, such as PEG-b-PCL block copolymers, are being developed into micelles and nanoparticles for the parenteral delivery of taxanes (Gaucher et al. 2010). Additionally, nanoparticles composed of poly-3-hydroxybutyrate-co-3-hydroxyhexanoate (PHBHXx) have been utilized as a matrix for delivering antineoplastic agents directly to cancer cells (Kılıçay et al. 2011).

To enhance targeted drug delivery, researchers have explored PHB nanoparticles modified with tumor-specific ligands for selective breast cancer treatment. Furthermore, the non-steroidal anti-inflammatory drug ibuprofen has been conjugated with non-toxic oligo-3-hydroxybutyrate to potentially improve drug delivery; however, this novel formulation requires further evaluation (Zawidlak-Węgrzyńska et al. 2010).

5.8.3 Hybrid Biocomposite

Hybrid biocomposites are innovative materials that integrate bio-based fibers or matrices with synthetic components to enhance performance while addressing environmental concerns.

Biofiber-Synthetic Fiber Composites combine natural fibers, such as jute and hemp, with synthetic fibers like glass and carbon, all within a polymer matrix. This integration improves mechanical properties while retaining a degree of biodegradability, making these composites suitable for applications requiring both strength and durability (Singh et al. 1995).

Natural Fiber-Natural Fiber Composites involve the combination of two or more types of natural fibers, such as bast fibers derived from plant stems and leaf fibers. This hybridization enhances stiffness and toughness by balancing the properties of different fibers. Interply or Tow-by-Tow Composites are formed by arranging layers or tows of different fiber types in either an alternating or random pattern. This configuration allows customization of mechanical properties depending on the fiber arrangement and composition. Sandwich Hybrids (Core-Shell Structures) consist of a core material enclosed between two layers of another material. This design improves impact resistance and structural integrity, making these hybrids valuable for applications such as protective gear and structural components. Intimately Mixed Hybrids blend various fibers at a microscopic level to form a homogeneous material. This method optimizes stress distribution and enhances the overall performance of the composite. Laminated Composites are created by stacking alternating layers of different materials in a regular pattern. This structure enhances strength and stiffness while maintaining a certain degree of flexibility, depending on the composition of the layers (John et al. 2009).

5.9 Fabrication Techniques for Biocomposites

Biocomposites are experiencing significant growth across various sectors and research domains. The increasing demand for these materials is driven by rising concerns over environmental pollution and the implementation of stricter environmental regulations by both local and global authorities. To achieve the desired properties and enhance the cost-effectiveness of biocomposites, various chemical modification treatments and advanced processing techniques are essential (Zwawi 2021).

Despite advancements, the available information and methodologies for the effective design and manufacturing of biocomposites remain limited, particularly regarding the diversity of raw materials such as textiles, fibers, fabrics, and wood (Hasan et al. 2021). The selection of an appropriate processing technique should prioritize maintaining a high fiber aspect ratio while ensuring uniform dispersion and orientation of the fibers. A combination of a high aspect ratio and consistent fiber distribution contributes to the development of homogeneous biocomposites with superior mechanical properties (Chaitanya and Singh 2017).

Currently, biocomposite fabrication is conducted using established methods such as hand lay-up, injection molding, compression molding, solution casting, and extrusion. Among these, compression and injection molding are recognized as two of the leading technologies for mass production due to their high precision and rapid cycle times (Leong et al. 2014).

5.9.1 Injection Molding Technique

For nearly 150 years, injection molding technology has continued to advance, driven by the development of new application sectors, particularly in the medical, hardware, packaging, and automotive industries. This process involves melting the material using a screw mechanism, combined with an external heating source, and subsequently injecting the molten material into a mold. The product takes shape as the mold cools and the material solidifies (Fu et al. 2020).

Due to the complexity of the process, a deeper understanding of material behavior during key stages is essential, including the physical phenomena that occur and their impact on the properties and performance of the final molded product (Rabbi et al. 2021). Injection molding is generally limited to producing components with short to intermediate-length fibers, as fiber length is influenced by the size of the pellets formed during the compounding stage.

Parts manufactured through injection molding require minimal post-processing, reducing labor costs and lowering overall product expenses. However, conventional injection molding tends to produce components with inferior mechanical properties compared to compression-molded products. This limitation arises because injection molding is restricted to short reinforcement fibers with low aspect ratios, whereas compression molding allows for the incorporation of long or continuous fibers. Although injection molding offers high precision and rapid dimensional accuracy, it is incapable of producing continuous fiber-reinforced composites. To address this limitation, the injection-compression molding technique has been employed for natural fiber textile insert molding, achieving some degree of success (Leong et al. 2014).

5.9.2 Compression Molding Technique

Compression molding is widely recognized as an efficient manufacturing technique due to its high production efficiency and short processing time. This method encompasses both compression and flow compression molding. Bio-composites produced using this technique exhibit sufficient mechanical properties, making them suitable for various applications (Akampumuza et al. 2017).

However, compression molding can lead to a reduction in fiber strength, which is influenced by the initial fiber length and several processing parameters, including mold design, screw speed, and melt viscosity. Additionally, the incompatibility between natural fibers and polymer matrices further weakens both the fibers and the resulting bio-composite (Ho et al. 2012). Compression molding is categorized into two primary methods: autoclave and hot pressing. The starting materials, such as sheet and bulk molding compounds, typically occupy 30–70% of the mold cavity. Once the material is placed, the mold is securely closed, and heat and pressure are applied. It is important to note that natural fibers are susceptible to degradation or breakage under high temperature and pressure conditions. To mitigate shrinkage and enhance the strength of the final product, short fibers are sometimes blended before the compression molding process (Mallick 2007).

5.9.3 *Extrusion Technique*

One of the most common methods for producing large-scale products involves incorporating filler materials, available in either chopped or powdered forms, along with polymers supplied in pellet form. Numerous studies on natural fiber-based bio-composites have explored both single-screw and twin-screw extrusion techniques, with a preference for twin-screw extrusion due to its superior mixing capabilities, which enhance the mechanical properties of the final composite (Ruz-Cruz et al. 2022).

This approach is particularly suitable for fabricating polymer blends and hybrid composites, as it enables intensive material mixing through the application of heat (Gallos et al. 2017). Additionally, it is often used in combination with other manufacturing methods, such as injection molding and additive manufacturing, to ensure proper integration of bio-composite components before final shaping (Rath et al. 2023).

5.9.4 *Hand-Lay-Up Method*

This is one of the oldest, most economical, straightforward, and widely used methods for producing composites. The process begins with the application of an anti-adhesive agent to the mold, facilitating the easy removal of the finished composite. Next, fibers are either chopped, ground into powder, or cut to a predetermined length and then positioned in the mold. The biopolymer resin is applied to the fibers using rollers, a brush, or by pouring, forming the first layer of the composite. Additional layers can be added in a similar manner to build up the final structure (Raji et al. 2019).

5.9.5 *Solution Casting*

This methodology involves dissolving a biopolymer in a suitable solvent, followed by the incorporation of a filler into the polymer solution under continuous stirring. Once a uniform mixture is achieved, it is cast into a mold of the desired shape and allowed to dry gradually, either at ambient conditions or in an air oven. This approach is particularly suited for producing biocomposite films containing short fibers or powdered fillers (Anbukarasu et al. 2021).

To achieve high-strength composites, proper mixing and a controlled drying process are essential. The drying temperature must remain below the solvent's boiling point during both the mixing and drying phases; otherwise, bubble formation may occur. These trapped bubbles can compromise the composite's structural integrity, leading to diminished mechanical properties (Kong et al. 2015).

5.10 Applications in Textiles

5.10.1 *Medical Textiles*

Alginate is widely used in medical applications, particularly in wound dressings and drug delivery. Alginate-based wound dressings can absorb significant amounts of exudate while maintaining a moist environment that facilitates healing (Winter 1963). hese non-adherent dressings can remain in place for up to 7 days without irritating the skin. Available in various sizes to accommodate different wound types, they are also relatively inexpensive, making them an accessible option for wound care management (Varaprasad et al. 2020). Additionally, these dressings provide antibacterial protection, helping to prevent infections and promote faster recovery.

Similarly, chitosan-based wound dressings are highly valued for their excellent hydrophilicity, allowing them to absorb exudate while retaining moisture (Zhang et al. 2020). Their low adhesion force minimizes repeated damage during dressing changes, ensuring greater patient comfort (Khorasani et al. 2018). Another promising material for wound dressings is bacterial cellulose, known for its high-water retention capacity and biodegradability. Its ability to maintain a hydrated wound environment is crucial for tissue regeneration. However, despite these advantages, bacterial cellulose lacks inherent antibacterial activity, which limits its use in medical textiles (Wei et al. 2011). Additionally, the polysaccharide pullulan exhibits excellent film-forming properties, making it valuable in medical textiles for the synthesis of coated wound dressings. By enhancing healing efficacy, infection control, and patient comfort, the integration of these diverse polymers into medical textiles holds great promise for revolutionizing wound care.

5.10.2 *Functional Textiles*

In functional textiles, alginate can be used to produce hydrogels that respond to temperature and pH fluctuations, making them ideal for wound care and drug delivery systems. Similarly, polyhydroxyalkanoates (PHAs) can be utilized to create eco-friendly fabrics. Their properties, such as mechanical resilience and water repellency, make them valuable for protective gear and outdoor clothing (Abourehab et al. 2022).

5.10.3 *Sustainable Fashion*

The sustainable use of microbial biopolymers marks a promising shift toward eco-friendly textile production. Bacterial cellulose and polyhydroxyalkanoates (PHAs) serve as excellent alternatives to synthetic fibers due to their durability and biodegradability. These materials not only mitigate environmental pollution but also provide a sustainable solution for high-performance textiles (Dias et al. 2025). Additionally, biopolymers such as alginate and chitosan enhance fabric functionality by imparting antibacterial properties, making them ideal for applications in medical textiles and sportswear. By inhibiting bacterial growth, these biopolymers contribute to improved hygiene and user comfort. Integrating microbial biopolymers into sustainable fashion presents exciting opportunities to reduce ecological footprints while enhancing textile performance (Emam 2019). As manufacturing technologies advance, these biopolymers could play a pivotal role in transforming the fashion industry toward more sustainable practices.

5.11 Advantages of Microbial Based Biopolymers and Biocomposites

Microbial-based biopolymers and biocomposites offer significant environmental and material advantages over traditional synthetic polymers. Their biodegradability, sustainable production, and diverse applications make them highly valuable across various industries. One of the key benefits is environmental sustainability, as microbial biopolymers naturally decompose, reducing the accumulation of non-biodegradable waste in landfills and oceans (Angelina 2015). Furthermore, they can be produced from low-cost waste materials, such as agricultural residues and industrial by-products, contributing to pollution reduction and sustainable waste management (Sharma et al. 2024). Their biocompatibility also makes them highly suitable for medical applications. For instance, polyhydroxyalkanoates (PHAs) are known for their compatibility with human cells, making them ideal for drug delivery systems and implantable devices (Yadav et al. 2015).

In addition to their environmental and biomedical benefits, microbial biopolymers exhibit enhanced mechanical properties. Materials such as bacterial cellulose (BC) and PHAs possess remarkable tensile strength and elasticity. BC, for example, has mechanical properties comparable to human tissues, making it an excellent choice for biomedical applications like tissue scaffolding (Moradali and Rehm 2020). Moreover, microbial biocomposites offer improved barrier properties, providing superior moisture resistance compared to traditional fabrics, which enhances their performance and durability in textile applications (Abbineni and Mao 2010). Another notable advantage is their customizability, as these biopolymers can be functionalized with additives such as essential oils or nanoparticles to enhance their properties. The incorporation of essential oils can improve flexibility and elongation at break while maintaining acceptable tensile strength, whereas the addition of nanoparticles has been shown to enhance tensile properties through better dispersion and interaction within the polymer matrix (Mittal et al. 2023).

5.12 Current Challenges and Limitations for Bio-composites

Currently, bio-composites face significant challenges due to the inherent variability in natural fiber properties. These variations arise from geographical factors, seasonal changes, plant-specific traits, and environmental conditions such as soil mineral content. Consequently, producing bio-composites with consistent properties is challenging, limiting their industrial scalability and broad adoption (Raj et al. 2023).

In response to environmental and sustainability demands, bio-composites have emerged as pivotal green materials in recent decades, aiming to replace synthetic petroleum-based composites with bio-based alternatives. Petroleum-derived materials are not only hazardous and costly to produce but also nonrenewable. In contrast, bio-composites offer biodegradability, recyclability, and mechanical properties comparable to synthetic fibers, while being lightweight, cost-effective, and eco-friendly. Despite their growing adoption across industries, manufacturing processes for natural fiber-based bio-composites still face persistent challenges (Akter et al. 2022).

PLA-based materials represent a "green" evolution of conventional eco-composites, which typically combine synthetic polymers with natural fibers or micro/nanofillers. However, performance limitations and the higher cost of biodegradable polymers compared to traditional ones hinder commercial scalability. While biodegradable polymers currently have higher market prices than conventional options, their reduced disposal costs and projected price declines as driven by innovations in manufacturing, which could make them viable replacements for traditional composites. Their compostability addresses polymer waste challenges, positioning bio-composites as key twenty-first-century materials for novel applications (Avella et al. 2009).

Nevertheless, designing green bio-composites faces significant hurdles. A major challenge is the lack of design data stemming from the wide variety of fibers, matrices, and production methods. Designers must address key factors like processing parameters, additive selection, polymer compatibility, and part optimization. Complexities in predicting material behavior often necessitate high safety factors, resulting in over-engineered components. To overcome these issues, rigorous testing and optimized manufacturing processes are critical for developing functional bio-composites at both macro and nano scales (AL-Oqla et al. 2017).

5.13 Future Scope for Bio-composites

Despite growing interest in bio-composites, a significant challenge persists in substituting traditional composites with alternatives that demonstrate comparable structural and functional stability during storage, usage, and environmental degradation post-disposal. Consequently, the market for bio-composites is projected to expand significantly, driven by rising social and environmental awareness among consumers and producers, the discovery of novel applications, and advancements in processing technologies (Gurunathan et al. 2015). Recent studies highlight that a major barrier to commercializing natural fiber-reinforced bio-composites has been the lack of recognition for research and development efforts in developing nations, where these fibers are abundant. In contrast, many industrialized countries particularly in Europe have overcome this challenge and emerged as leaders in the field (Faruk et al. 2012).

Materials scientists face key challenges regarding the acceptance criteria for bio-composites, especially concerning performance and cost. While the renewability and recyclability of both matrix and reinforcement materials are advantageous, establishing product standards to ensure material efficacy remains urgent. Bio-composites intended for structural applications must comply with regulations governing large-scale waste management. Additionally, next-generation bio-composites are anticipated to serve diverse applications, from mass-produced, short-term consumer products to durable solutions for long-term indoor use (Summerscales et al. 2010).

The weakened mechanical properties of bio-composites are primarily attributed to poor interfacial bonding between biopolymers and most natural fibers. Further research is therefore critical to deepen understanding of individual fiber morphology, surface modification techniques, and mechanical behavior. Such insights are vital for improving interfacial bonding and fiber distribution. Although extensive studies exist on fiber surface modification via various treatments, research gaps persist in single-fiber processing, which is crucial for enhancing mechanical performance (Raj et al. 2023).

Advancements in surface modification techniques, the use of coupling agents, and optimized compounding processes are urgently needed to advance bio-composite manufacturing. Additionally, exploring unconventional or underutilized fibers could broaden applications and directly benefit rural communities where these materials are readily available (Rajan et al. 2020).

5.14 Conclusions

Microbial biopolymers and biocomposites represent a groundbreaking shift toward sustainable textile innovation, offering eco-friendly alternatives to petroleum-based materials. As these are derived from renewable resources like bacteria, fungi, and algae, the biodegradability, biocompatibility, and functional versatility of these materials can be helpful in addressing critical environmental challenges including the reduction of plastic waste, lowering carbon footprints. However, despite their promise, several challenges such as high production costs, scalability issues, and material limitations persist. Advances in metabolic engineering, fermentation optimization, and green processing techniques are key to overcoming these hurdles. Future progress hinges on interdisciplinary collaboration, policy support, and consumer adoption to mainstream these solutions. By integrating microbial biopolymers into textile ecosystems, the industry can achieve a balance between performance and planetary health, paving the way for a circular economy where innovation aligns with ecological stewardship.

References

Abbineni G, Mao C (2010) Design and applications of genetically engineered nanocomposites. Wiley-VCH, Hobken, NJ

Abourehab MAS, Rajendran RR, Singh A, Pramanik S, Shrivastav P, Ansari MJ et al (2022) Alginate as a promising biopolymer in drug delivery and wound healing: a review of the state-of-the-art. Int J Mol Sci 23:9035

Acharjee SA, Bharali P, Gogoi B, Sorhie V, Walling B, Alemtoshi (2023) PHA-based bioplastic: a potential alternative to address microplastic pollution. Water Air Soil Pollut 234:21

Aditya T, Allain JP, Jaramillo C, Restrepo AM (2022) Surface modification of bacterial cellulose for biomedical applications. Int J Mol Sci 23:610

Akampumuza O, Wambua PM, Ahmed A, Li W, Qin XH (2017) Review of the applications of biocomposites in the automotive industry. Polym Compos 38:2553–2569

Akinsemolu AA (2018) The role of microorganisms in achieving the sustainable development goals. J Clean Prod 182:139–155

Akter M, Uddin MH, Tania IS (2022) Biocomposites based on natural fibers and polymers: a review on properties and potential applications. J Reinf Plast Compos 41:705–742

Al-Oqla FM, Almagableh A, Omari MA (2017) Green biocomposites: design and applications. Springer, Cham, pp 45–67

Anbukarasu P, Sauvageau D, Elias A (2021) The effects of solvent casting temperature and physical aging on polyhydroxybutyrate-graphene nanoplatelet composites. Polym Compos 42:1451–1461

Anderson AJ, Dawes E (1990) Occurrence, metabolism, metabolic role, and industrial uses of bacterial polyhydroxyalkanoates. Microbiol Rev 54:450–472

Angelina VS (2015) Microbial factories: biodiversity, biopolymers, bioactive molecules, vol 2. Springer, Cham, pp 113–125

Avella M, Buzarovska A, Errico ME, Gentile G, Grozdanov A (2009) Eco-challenges of bio-based polymer composites. Materials 2:911–925

Awasthi SK, Kumar M, Kumar V, Sarsaiya S, Anerao P, Ghosh P et al (2022) A comprehensive review on recent advancements in biodegradation and sustainable management of biopolymers. Environ Pollut 307:119600

Baillie C, Jayasinghe R (2004) Green composites: polymer composites and the environment. Elsevier, Amsterdam

Bansal SS, Goel M, Aqil F, Vadhanam MV, Gupta RC (2011) Advanced drug delivery systems of curcumin for cancer chemoprevention. Cancer Prev Res 4:1158–1171

Becker A, Katzen F, Pühler A, Ielpi L (1998) Xanthan gum biosynthesis and application: a biochemical/genetic perspective. Appl Microbiol Biotechnol 50:145–152

Bhat IM, Wani SM, Mir SA, Masoodi F (2022) Advances in xanthan gum production, modifications and its applications. Biocatal Agric Biotechnol 42:102328

Bhatia SK, Gurav R, Choi Y-K, Choi T-R, Kim H-J, Song H-S et al (2021) Bioprospecting of exopolysaccharide from marine Sphingobium yanoikuyae BBL01: production, characterization, and metal chelation activity. Bioresour Technol 324:124674

Bidone J, Melo APP, Bazzo GC, Carmignan F, Soldi MS, Pires AT et al (2009) Preparation and characterization of ibuprofen-loaded microspheres consisting of poly (3-hydroxybutyrate) and methoxy poly (ethylene glycol)-b-poly (D, L-lactide) blends or poly (3-hydroxybutyrate) and gelatin composites for controlled drug release. Mater Sci Eng C 29:588–593

Blanco Parte FG, Santoso SP, Chou C-C, Verma V, Wang H-T, Ismadji S et al (2020) Current progress on the production, modification, and applications of bacterial cellulose. Crit Rev Biotechnol 40:397–414
Bonaldi R (2018) High-performance apparel. Amsterdam, Elsevier, pp 129–156
Buldum G, Mantalaris A (2021) Systematic understanding of recent developments in bacterial cellulose biosynthesis at genetic, bioprocess and product levels. Int J Mol Sci 22:7192
Çakar F, Özer I, Aytekin AÖ, Şahin F (2014) Improvement production of bacterial cellulose by semi-continuous process in molasses medium. Carbohydr Polym 106:7–13
Chaitanya S, Singh I (2017) Processing of PLA/sisal fiber biocomposites using direct-and extrusion-injection molding. Mater Manuf Process 32:468–474
Chen G-Q, Patel MK (2012) Plastics derived from biological sources: present and future: a technical and environmental review. Chem Rev 112:2082–2099
Cheng Y-R, Wang H-Y (2022) Highly effective removal of microplastics by microalgae Scenedesmus abundans. Chem Eng J 435:135079
Cheng K-C, Demirci A, Catchmark JM (2011) Pullulan: biosynthesis, production, and applications. Appl Microbiol Biotechnol 92:29–44
Costa SS, Miranda AL, De Morais MG, JaV C, Druzian JI (2019) Microalgae as source of polyhydroxyalkanoates (PHAs)—a review. Int J Biol Macromol 131:536–547
Dias JC, Marques S, Branco PC, Rodrigues T, Torres CAV, Freitas F et al (2025) Biopolymers derived from forest biomass for the sustainable textile industry. Forests 16:163
Díaz-Montes E (2021) Dextran: sources, structures, and properties. Polysaccharides 2:554–565
Ebinesar A, More VS, Ramya D, Amrutha G, More SS (2021) Microbial polymers: applications and ecological perspectives. Springer, Cham, pp 311–340
Egan J, Salmon S (2022) Strategies and progress in synthetic textile fiber biodegradability. SN Appl Sci 4:1–36
El-Sayed E, Hassabo AG (2021) Recent advances in the application of plasma in textile finishing (a review). J Text Color Polym Sci 18:33–43
Emam HE (2019) Antimicrobial cellulosic textiles based on organic compounds. 3 Biotech 9:29
Faruk O, Bledzki AK, Fink H-P, Sain M (2012) Biocomposites reinforced with natural fibers: 2000–2010. Prog Polym Sci 37:1552–1596
Fatima A, Ortiz-Albo P, Neves LA, Nascimento FX, Crespo JG (2023) Biosynthesis and characterization of bacterial cellulose membranes presenting relevant characteristics for air/gas filtration. J Membr Sci 674:121509
Felgueiras C, Azoia NG, Gonçalves C, Gama M, Dourado F (2021) Trends on the cellulose-based textiles: raw materials and technologies. Front Bioeng Biotechnol 9:608826
Fernandes M, Gama M, Dourado F, Souto AP (2019) Development of novel bacterial cellulose composites for the textile and shoe industry. Microb Biotechnol 12:650–661
Fradinho J, Reis M, Oehmen A (2016) Beyond feast and famine: selecting a PHA accumulating photosynthetic mixed culture in a permanent feast regime. Water Res 105:421–428
Fu H, Xu H, Liu Y, Yang Z, Kormakov S, Wu D et al (2020) Overview of injection molding technology for processing polymers and their composites. ES Mater Manuf 8:3–23
Fujiwara T, Komoda K, Sakurai N, Tajima K, Tanaka I, Yao M (2013) The c-di-GMP recognition mechanism of the PilZ domain of bacterial cellulose synthase subunit a. Biochem Biophys Res Commun 431:802–807
Gallos A, Paës G, Allais F, Beaugrand J (2017) Lignocellulosic fibers: a critical review of the extrusion process for enhancement of the properties of natural fiber composites. RSC Adv 7:34638–34654
Gao Y, Cranston R (2008) Recent advances in antimicrobial treatments of textiles. Text Res J 78:60–72
Gaucher G, Marchessault RH, Leroux J-C (2010) Polyester-based micelles and nanoparticles for the parenteral delivery of taxanes. J Control Release 143:2–12

Gaur R, Singh R, Tiwari S, Yadav S, Daramwal N (2010) Optimization of physico-chemical and nutritional parameters for a novel pullulan-producing fungus, Eurotium chevalieri. J Appl Microbiol 109:1035–1043
Ghosh S, Lahiri D, Nag M, Dey A, Sarkar T, Pathak SK et al (2021) Bacterial biopolymer: its role in pathogenesis to effective biomaterials. Polymers 13:1242
Gibson LJ (2012) The hierarchical structure and mechanics of plant materials. J R Soc Interface 9:2749–2766
Gowri S, Almeida L, Amorim T, Carneiro N, Pedro Souto A, Fátima Esteves M (2010) Polymer nanocomposites for multifunctional finishing of textiles-a review. Text Res J 80:1290–1306
Gregory DA, Tripathi L, Fricker AT, Asare E, Orlando I, Raghavendran V et al (2021) Bacterial cellulose: a smart biomaterial with diverse applications. Mater Sci Eng R Rep 145:100623
Gullo M, La China S, Falcone PM, Giudici P (2018) Biotechnological production of cellulose by acetic acid bacteria: current state and perspectives. Appl Microbiol Biotechnol 102:6885–6898
Gupta D, Bhaumik S (2007) Antimicrobial treatments for textiles. Indian J Fibre Text 32:254–263
Gurunathan T, Mohanty S, Nayak SK (2015) A review of the recent developments in biocomposites based on natural fibres and their application perspectives. Compos A Appl Sci Manuf 77:1–25
Hafuka A, Sakaida K, Satoh H, Takahashi M, Watanabe Y, Okabe S (2011) Effect of feeding regimens on polyhydroxybutyrate production from food wastes by Cupriavidus necator. Bioresour Technol 102:3551–3553
Hasan KF, Horváth PG, Zsolt K, Alpár T (2021) Value-added biocomposites. Boca Raton, FL, CRC Press, pp 157–188
Hay ID, Rehman ZU, Moradali MF, Wang Y, Rehm BH (2013) Microbial alginate production, modification and its applications. Microb Biotechnol 6:637–650
Hernandez-Tenorio F, Giraldo-Estrada C (2022) Characterization and chemical modification of pullulan produced from a submerged culture of Aureobasidium pullulans ATCC 15233. Polym Test 114:107686
Hestrin S, Schramm M (1954) Synthesis of cellulose by Acetobacter xylinum. 2. Preparation of freeze-dried cells capable of polymerizing glucose to cellulose. Biochem J 58:345–352
Ho M-P, Wang H, Lee J-H, Ho C-K, Lau K-T, Leng J et al (2012) Critical factors on manufacturing processes of natural fibre composites. Compos Part B Eng 43:3549–3562
Ibrahim MH, SteinbüChel A (2009) Poly (3-hydroxybutyrate) production from glycerol by Zobellella denitrificans MW1 via high-cell-density fed-batch fermentation and simplified solvent extraction. Appl Environ Microbiol 75:6222–6231
Jain R, Kosta S, Tiwari A (2010) Polyhydroxyalkanoates: a way to sustainable development of bioplastics. Chron Young Sci 1:10–10
Jayabalan R, Malbaša RV, Lončar ES, Vitas JS, Sathishkumar M (2014) A review on kombucha tea—microbiology, composition, fermentation, beneficial effects, toxicity, and tea fungus. Compr Rev Food Sci Food Saf 13:538–550
John MJ, Anandjiwala RD, Thomas S (2009) Old City Publishing
Kalia S, Kaith B, Kaur I (2009) Pretreatments of natural fibers and their application as reinforcing material in polymer composites—a review. Polym Eng Sci 49:1253–1272
Kayanna N, Suppavorasatit I, Bankeeree W, Lotrakul P, Punnapayak H, Prasongsuk S (2022) Production of prebiotic aubasidan-like β-glucan from Aureobasidium thailandense NRRL 58543 and its potential as a functional food additive in gummy jelly. LWT 163:113617
Khanna S, Srivastava AK (2005) Recent advances in microbial polyhydroxyalkanoates. Process Biochem 40:607–619
Khatami K, Perez-Zabaleta M, Cetecioglu Z (2022) Pure cultures for synthetic culture development: next level municipal waste treatment for polyhydroxyalkanoates production. J Environ Manag 305:114337
Khorasani MT, Joorabloo A, Moghaddam A, Shamsi H, Mansoorimoghadam Z (2018) Incorporation of ZnO nanoparticles into heparinised polyvinyl alcohol/chitosan hydrogels for wound dressing application. Int J Biol Macromol 114:1203–1215

Kibungu C, Kondiah PP, Kumar P, Choonara YE (2021) This review recent advances in chitosan and alginate-based hydrogels for wound healing application. Front Mater 8:681960

Kılıçay E, Demirbilek M, Türk M, Güven E, Hazer B, Denkbas EB (2011) Preparation and characterization of poly (3-hydroxybutyrate-co-3-hydroxyhexanoate)(PHBHHX) based nanoparticles for targeted cancer therapy. Eur J Pharm Sci 44:310–320

Kiran S, Shahzaib IS, Munir B, Hafiz I (2024) In biopolymers in the textile industry: opportunities and limitations. Springer, Cham, pp 1–28

Kogan G, Šoltés L, Stern R, Gemeiner P (2007) Hyaluronic acid: a natural biopolymer with a broad range of biomedical and industrial applications. Biotechnol Lett 29:17–25

Koller M (2018) Biodegradable and biocompatible polyhydroxy-alkanoates (PHA): auspicious microbial macromolecules for pharmaceutical and therapeutic applications. Molecules 23:362

Kong I, Tshai K, Hoque ME (2015) in Manufacturing of natural fibre reinforced polymer composites. Springer, Cham, pp 331–349

Kosseva MR, Rusbandi E (2018) Trends in the biomanufacture of polyhydroxyalkanoates with focus on downstream processing. Int J Biol Macromol 107:762–778

Kumari P, Kiran BR, Mohan SV (2022) Polyhydroxybutyrate production by Chlorella sorokiniana SVMIICT8 under nutrient-deprived mixotrophy. Bioresour Technol 354:127135

Leong Y, Thitithanasarn S, Yamada K, Hamada H (2014) Natural fibre composites. Elsevier, Amsterdam, pp 216–232

Li M, Wilkins MR (2020) Recent advances in polyhydroxyalkanoate production: feedstocks, strains and process developments. Int J Biol Macromol 156:691–703

Li J, He J, Huang Y (2017) Role of alginate in antibacterial finishing of textiles. Int J Biol Macromol 94:466–473

Lim S-H, Hudson SM (2003) Review of chitosan and its derivatives as antimicrobial agents and their uses as textile chemicals. J Macromol Sci C 43:223–269

Liu W, Du H, Zhang M, Liu K, Liu H, Xie H et al (2020) Bacterial cellulose-based composite scaffolds for biomedical applications: a review. ACS Sustain Chem Eng 8:7536–7562

Liu W, Zhang J, Liu H, Guo X, Zhang X, Yao X et al (2021) A review of the removal of microplastics in global wastewater treatment plants: characteristics and mechanisms. Environ Int 146:106277

Logesh A, Thillaimaharani K, Sharmila K, Kalaiselvam M, Raffi S (2012) Production of chitosan from endolichenic fungi isolated from mangrove environment and its antagonistic activity. Asian Pac J Trop Biomed 2:140–143

Lopar M, Špoljarić IV, Atlić A, Koller M, Braunegg G, Horvat P (2013) Five-step continuous production of PHB analyzed by elementary flux, modes, yield space analysis and high structured metabolic model. Biochem Eng J 79:57–70

Lu J, Tappel RC, Nomura CT (2009) Mini-review: biosynthesis of poly (hydroxyalkanoates). J Macromol Sci C 49:226–248

Lupaşcu RE, Ghica MV, Dinu-Pîrvu C-E, Popa L, Velescu BŞ, Arsene AL (2022) An overview regarding microbial aspects of production and applications of bacterial cellulose. Materials 15:676

Mallick PK (2007) Fiber-reinforced composites: materials, manufacturing, and design, vol 83. CRC Press, Boca Raton, FL

Mehrpouya M, Vahabi H, Barletta M, Laheurte P, Langlois V (2021) Additive manufacturing of polyhydroxyalkanoates (PHAs) biopolymers: materials, printing techniques, and applications. Mater Sci Eng C 127:112216

Mittal M, Ahuja S, Yadav A, Aggarwal NK (2023) Development of poly (hydroxybutyrate) film incorporated with nano silica and clove essential oil intended for active packaging of brown bread. Int J Biol Macromol 233:123512

Moradali MF, Rehm BH (2020) Bacterial biopolymers: from pathogenesis to advanced materials. Nat Rev Microbiol 18:195–210

Moradali MF, Ghods S, Rehm BH (2017) Pseudomonas aeruginosa lifestyle: a paradigm for adaptation, survival, and persistence. Front Cell Infect Microbiol 7:39

Mozzi F, Ortiz ME, Bleckwedel J, De Vuyst L, Pescuma M (2013) Metabolomics as a tool for the comprehensive understanding of fermented and functional foods with lactic acid bacteria. Food Res Int 54:1152–1161

Muneer F, Rasul I, Azeem F, Siddique MH, Zubair M, Nadeem H (2020) Microbial polyhydroxyalkanoates (PHAs): efficient replacement of synthetic polymers. J Polym Environ 28:2301–2323

Murugan P, Han L, Gan C-Y, Maurer FH, Sudesh K (2016) A new biological recovery approach for PHA using mealworm, Tenebrio molitor. J Biotechnol 239:98–105

Naessens M, Cerdobbel A, Soetaert W, Vandamme EJ (2005) Leuconostoc dextransucrase and dextran: production, properties and applications. J Chem Technol Biotechnol 80:845–860

Nayak R, Cleveland D, Tran G, Joseph F (2024) Potential of bacterial cellulose for sustainable fashion and textile applications: a review. J Mater Sci 59:1–26

Nishino T, Matsuda I, Hirao K (2004) All-cellulose composite. Macromolecules 37:7683–7687

Norizan MN, Shazleen SS, Alias AH, Sabaruddin FA, Asyraf MRM, Zainudin ES et al (2022) Nanocellulose-based nanocomposites for sustainable applications: a review. Nanomaterials 12:3483

Oliveira FC, Dias ML, Castilho LR, Freire DM (2007) Characterization of poly (3-hydroxybutyrate) produced by Cupriavidus necator in solid-state fermentation. Bioresour Technol 98:633–638

Öz YE, Kalender M (2023) A novel static cultivation of bacterial cellulose production from sugar beet molasses: series static culture (SSC) system. Int J Biol Macromol 225:1306–1314

Parati M, Khalil I, Tchuenbou-Magaia F, Adamus G, Mendrek B, Hill R et al (2022) Building a circular economy around poly (D/L-γ-glutamic acid)-a smart microbial biopolymer. Biotechnol Adv 61:108049

Pattnaik S, Swain K (2022) Cellulose-based composites and their biomedical applications. Cellul Chem Technol 56:115–122

Pinho E, Magalhães L, Henriques M, Oliveira R (2011) Antimicrobial activity assessment of textiles: standard methods comparison. Ann Microbiol 61:493–498

Rabbi M, Islam T, Islam GS (2021) Injection-molded natural fiber-reinforced polymer composites–a review. Int J Mech Mater Eng 16:1–21

Raj H, Tripathi S, Bauri S, Choudhury AM, Mandal SS, Maiti P (2023) Green composites using naturally occurring fibers: a comprehensive review. Sustain Polym Energy 1:10010

Rajan VS, Govindaraju M, Ramu M, Satheeshkumar V (2020) Influence of metal foam properties on performance of polymer composite spur gear. Mater Today Proc 24:1244–1250

Raji M, Abdellaoui H, Essabir H, Kakou C-A, Bouhfid R (2019) Durability and life prediction in biocomposites, fibre-reinforced composites and hybrid composites. Elsevier, Amsterdam, pp 27–62

Ramadas NV, Sindhu R, Binod P, Pandey A (2013) Development of a novel solid-state fermentation strategy for the production of poly-3-hydroxybutyrate using polyurethane foams by Bacillus sphaericus NII 0838. Ann Microbiol 63:1265–1274

Rath A, Grisin B, Pallicity TD, Glaser L, Guhathakurta J, Oehlsen N et al (2023) Fabrication of chitosan-flax composites with differing molecular weights and its effect on mechanical properties. Compos Sci Technol 235:109952

Raut MP, Couto N, Karunakaran E, Biggs CA, Wright PC (2019) Deciphering the unique cellulose degradation mechanism of the ruminal bacterium Fibrobacter succinogenes S85. Sci Rep 9:16542

Raut MP, Asare E, Syed Mohamed SMD, Amadi EN, Roy I (2023) Bacterial cellulose-based blends and composites: versatile biomaterials for tissue engineering applications. Int J Mol Sci 24:986

Raza ZA, Abid S, Banat IM (2018) Polyhydroxyalkanoates: characteristics, production, recent developments and applications. Int Biodeterior Biodegradation 126:45–56

Reddy C, Ghai R, Kalia VC (2003) Polyhydroxyalkanoates: an overview. Bioresour Technol 87:137–146

Rehm BH (2003) Polyester synthases: natural catalysts for plastics. Biochem J 376:15–33

Reichler M, Rabensteiner S, Törnblom L, Coffeng S, Viitanen L, Jannuzzi L et al (2021) Scalable method for bio-based solid foams that mimic wood. Sci Rep 11:24306

Remminghorst U, Rehm BH (2006) Bacterial alginates: from biosynthesis to applications. Biotechnol Lett 28:1701–1712

Retegi A, Gabilondo N, Peña C, Zuluaga R, Castro C, Gañán P et al (2010) Bacterial cellulose films with controlled microstructure–mechanical property relationships. Cellulose 17:661–669

Rinaudo M (2006) Chitin and chitosan: properties and applications. Prog Polym Sci 31:603–632

Rivera-Briso AL, Serrano-Aroca Á (2018) Poly (3-Hydroxybutyrate-co-3-Hydroxyvalerate): enhancement strategies for advanced applications. Polymers 10:732

Römling U, Galperin MY, Gomelsky M (2013) Cyclic di-GMP: the first 25 years of a universal bacterial second messenger. Microbiol Mol Biol Rev 77:1–52

Ross P, Mayer R, Benziman M (1991) Cellulose biosynthesis and function in bacteria. Microbiol Rev 55:35–58

Ruz-Cruz M, Herrera-Franco P, Flores-Johnson E, Moreno-Chulim M, Galera-Manzano L, Valadez-González A (2022) Thermal and mechanical properties of PLA-based multiscale cellulosic biocomposites. J Mater Res Technol 18:485–495

Ryu HW, Hahn SK, Chang YK, Chang HN (1997) Production of poly (3-hydroxybutyrate) by high cell density fed-batch culture of Alcaligenes eutrophus with phospate limitation. Biotechnol Bioeng 55:28–32

Sanderson G (1982) Interactions of xanthan gum in food systems. Prog Food Nutr Sci 6:77–87

Sandford PA, Baird J (1983) The polysaccharides. Elsevier, Amsterdam, pp 411–490

Saratale GD, Oh M-K (2015) Characterization of poly-3-hydroxybutyrate (PHB) produced from Ralstonia eutropha using an alkali-pretreated biomass feedstock. Int J Biol Macromol 80:627–635

Schnepp Z (2013) Biopolymers as a flexible resource for nanochemistry. Angew Chem Int Ed 52:1096–1108

Schreiner KM, Blair NE, Levinson W, Egerton-Warburton LM (2014). Soil carbon. p 155–161

Shah AA, Hasan F, Hameed A, Ahmed S (2008) Biological degradation of plastics: a comprehensive review. Biotechnol Adv 26:246–265

Shamolina I (1997) Prospects for use of microbial raw material for fabrication of fibre and film materials. Review. Fibre Chem 29:1–8

Sharma M, Tellili N, Kacem I, Rouissi T (2024) Microbial biopolymers: from production to environmental applications—a review. Appl Sci 14:5081

Shezad O, Khan S, Khan T, Park JK (2009) Production of bacterial cellulose in static conditions by a simple fed-batch cultivation strategy. Korean J Chem Eng 26:1689–1692

Shirvan AR, Shakeri M, Bashari A (2019) The impact and prospects of green chemistry for textile technology. Elsevier, Amsterdam, pp 107–133

Shukla A, Mehta K, Parmar J, Pandya J, Saraf M (2019) Depicting the exemplary knowledge of microbial exopolysaccharides in a nutshell. Eur Polym J 119:298–310

Singh B, Gupta M, Verma A (1995) Mechanical behaviour of particulate hybrid composite laminates as potential building materials. Constr Build Mater 9:39–44

Stefanova A, In-Na P, Caldwell GS, Bridgens B, Armstrong R (2021) Photosynthetic textile biocomposites: using laboratory testing and digital fabrication to develop flexible living building materials. Sci Eng Compos Mater 28:223–236

Steinbüchel A, Valentin HE (1995) Diversity of bacterial polyhydroxyalkanoic acids. FEMS Microbiol Lett 128:219–228

Sudesh K, Abe H, Doi Y (2000) Synthesis, structure and properties of polyhydroxyalkanoates: biological polyesters. Prog Polym Sci 25:1503–1555

Summerscales J, Dissanayake NP, Virk AS, Hall W (2010) A review of bast fibres and their composites. Part 1–fibres as reinforcements. Compos A Appl Sci Manuf 41:1329–1335

Sutherland IW (2001) Biofilm exopolysaccharides: a strong and sticky framework. Microbiology 147:3–9

Tan G-YA, Chen C-L, Li L, Ge L, Wang L, Razaad IMN et al (2014) Start a research on biopolymer polyhydroxyalkanoate (PHA): a review. Polymers 6:706–754

Torabi P, Hamdami N, Keramat J (2022) Microwave-assisted extraction of sodium alginate from brown macroalgae Nizimuddinia zanardini, optimization and physicochemical properties. Sep Sci Technol 57:872–885

Toyosaki H, Naritomi T, Seto A, Matsuoka M, Tsuchida T, Yoshinaga F (1995) Screening of bacterial cellulose-producing Acetobacter strains suitable for agitated culture. Biosci Biotechnol Biochem 59:1498–1502

Ullah H, Wahid F, Santos HA, Khan T (2016) Advances in biomedical and pharmaceutical applications of functional bacterial cellulose-based nanocomposites. Carbohydr Polym 150:330–352

Urtuvia V, Villegas P, Fuentes S, González M, Seeger M (2018) Burkholderia xenovorans LB400 possesses a functional polyhydroxyalkanoate anabolic pathway encoded by the pha genes and synthesizes poly (3-hydroxybutyrate) under nitrogen-limiting conditions. Int Microbiol 21:47–57

Vadanan SV, Basu A, Lim S (2022) Bacterial cellulose production, functionalization, and development of hybrid materials using synthetic biology. Polym J 54:481–492

Valla S, Coucheron DH, Fjærvik E, Kjosbakken J, Weinhouse H, Ross P et al (1989) Cloning of a gene involved in cellulose biosynthesis in Acetobacter xylinum: complementation of cellulose-negative mutants by the UDPG pyrophosphorylase structural gene. Mol Gen Genet 217:26–30

Van Loosdrecht M, Pot M, Heijnen J (1997) Importance of bacterial storage polymers in bioprocesses. Water Sci Technol 35:41–47

Varaprasad K, Jayaramudu T, Kanikireddy V, Toro C, Sadiku ER (2020) Alginate-based composite materials for wound dressing application: a mini review. Carbohydr Polym 236:116025

Vu B, Chen M, Crawford RJ, Ivanova EP (2009) Bacterial extracellular polysaccharides involved in biofilm formation. Molecules 14:2535–2554

Wang Y, Yin J, Chen G-Q (2014) Polyhydroxyalkanoates, challenges and opportunities. Curr Opin Biotechnol 30:59–65

Wang S-S, Han Y-H, Ye Y-X, Shi X-X, Xiang P, Chen D-L et al (2017a) Physicochemical characterization of high-quality bacterial cellulose produced by Komagataeibacter sp. strain W1 and identification of the associated genes in bacterial cellulose production. RSC Adv 7:45145–45155

Wang Z, Wu J, Gao M-J, Zhu L, Zhan X-B (2017b) High production of xanthan gum by a glycerol-tolerant strain Xanthomonas campestris WXLB-006. Prep Biochem Biotech 47:468–472

Wang J, Guo H, Huang J, Jiang S, Hou S, Chen X et al (2021) L-lactic acid production from fructose by chitosan film-coated sodium alginate-polyvinyl alcohol immobilized lactobacillus pentosus cells and its kinetic analysis. Bioresour Bioprocess 8:1–17

Wang J, Liu S, Huang J, Ren K, Zhu Y, Yang S (2023) Alginate: microbial production, functionalization, and biomedical applications. Int J Biol Macromol 242:125048

Wei B, Yang G, Hong F (2011) Preparation and evaluation of a kind of bacterial cellulose dry films with antibacterial properties. Carbohydr Polym 84:533–538

Whitton BA (2012) Ecology of cyanobacteria II: their diversity in space and time. Springer, Cham

Winter GD (1963) Effect of air exposure and occlusion on experimental human skin wounds. Nature 200:378–379

Wu Z-Y, Liang H-W, Chen L-F, Hu B-C, Yu S-H (2016) Bacterial cellulose: a robust platform for design of three dimensional carbon-based functional nanomaterials. Acc Chem Res 49:96–105

Yadav P, Yadav H, Shah VG, Shah G, Dhaka G (2015) Biomedical biopolymers, their origin and evolution in biomedical sciences: a systematic review. J Clin Diagn Res 9:ZE21

Yau YK, Ooi CW, Ng E-P, Lan JC-W, Ling TC, Show PL (2015) Current applications of different type of aqueous two-phase systems. Bioresour Bioprocess 2:1–13

Yin W, Wang Y, Liu L, He J (2019) Biofilms: the microbial "protective clothing" in extreme environments. Int J Mol Sci 20:3423

Zawidlak-Węgrzyńska B, Kawalec M, Bosek I, Łuczyk-Juzwa M, Adamus G, Rusin A et al (2010) Synthesis and antiproliferative properties of ibuprofen–oligo (3-hydroxybutyrate) conjugates. Eur J Med Chem 45:1833–1842

Zhang B, He J, Shi M, Liang Y, Guo B (2020) Injectable self-healing supramolecular hydrogels with conductivity and photo-thermal antibacterial activity to enhance complete skin regeneration. Chem Eng J 400:125994
Zhang X, Qu Q, Yang A, Wang J, Cheng W, Deng Y et al (2023) Chitosan enhanced the stability and antibiofilm activity of self-propelled Prussian blue micromotor. Carbohydr Polym 299:120134
Zhang Z, Wang Y, Wang X, Zhang Y, Zhu T, Peng L et al (2024) Towards scaling-up implementation of polyhydroxyalkanoate (PHA) production from activated sludge: Progress and challenges. J Clean Prod. 447:141542
Zhao C, Liu G, Tan Q, Gao M, Chen G, Huang X et al (2023) Polysaccharide-based biopolymer hydrogels for heavy metal detection and adsorption. J Adv Res 44:53–70
Zwawi M (2021) A review on natural fiber bio-composites, surface modifications and applications. Molecules 26:404

Chapter 6
Application of Green Nanomaterials in Textile Biotechnology

Faluk Shair, Khalil Ahmad, Muhammad Kashif, Muhammad Ashfaq, and Muhammad Arif Khan

Abstract Textile dye-contaminated wastewater poses significant threats to the environment, biological diversity, and human health. While physical and chemical methods have been widely employed to mitigate dye pollution, these approaches suffer from inherent drawbacks and limitations, including high operational costs, generation of secondary hazardous by-products, and substantial energy requirements. Biological methods, such as those utilizing microorganisms and enzymes, present eco-friendly alternatives; however, challenges related to low efficiency, sensitivity to environmental variability, and slow degradation rates hinder their large-scale application. Recent advances in green nanotechnology offer promising solutions using eco-friendly nanoparticles for dye detoxification. Nevertheless, scalability, stability, and economic feasibility issues continue to impede their widespread adoption. Future advancements in nanoparticle synthesis, particularly through the integration of transgenic microorganisms, are anticipated to enhance the efficiency and sustainability of dye elimination processes.

Keywords Textile dyes · Bioremediation · Green nanomaterial (GNMs) · Sustainable detoxification · Biological diversity

6.1 Introduction

The textile industry, encompassing processes such as knitting, crocheting, and weaving, relies on three core principles: design, production, and distribution of finished and semi-finished goods, including yarn, fabrics, and clothing. As a

F. Shair · K. Ahmad (✉) · M. Kashif · M. A. Khan
Department of Chemistry, Emerson University Multan, Multan, Pakistan
e-mail: muhammad.kashif@eum.edu.pk

M. Ashfaq
Institute of Chemistry, The Islamia University of Bahawalpur, Bahawalpur, Punjab, Pakistan

M. Khurshid et al. (eds.), *Microbial Biotechnology in Textile Innovation*, SDGs and Textiles, https://doi.org/10.1007/978-981-95-0347-6_6

cornerstone of the global economy, the textile sector contributed USD 1837.27 billion to the global market in 2023, with a projected annual growth rate of 7.3% until 2030 (Karahan et al. 2023). However, its rapid expansion has come at significant environmental cost. Industry remains one of the largest sources of pollution worldwide, releasing toxic chemicals such as Sulphur or nitrogen-based oxides, volatile organic compounds (VOCs), and airborne dust particles. Additionally, it generates substantial solid waste including yarn scraps, fabric remnants, and packaging materials as well as hazardous byproducts like textile sludge, organic matter, pathogenic microorganisms, and heavy metals (Khan 2025).

6.2 Textile Industry and Environmental Pollution

Excessive water consumption in the textile industry is a primary driver of water contamination, as untreated, non-degradable effluents containing synthetic dyes are discharged into ecosystems. Globally, 80% of textile-related pollution enters water bodies, posing severe environmental risks (Bhatia et al. 2017; Lellis et al. 2019). A key issue lies in the weak binding affinity of dyes to fabrics, wool, nylon, and leather, which leads to their release often without prior treatment into aquatic environments such as lakes and rivers (Tkaczyk et al. 2020; Al-Tohamy et al. 2022; Babuji et al. 2023; Fida et al. 2023; Vaiano et al. 2023). These dyes, which include derivatives, modified complexes, and polymers, are highly persistent pollutants, with over 1,000,000 tons released annually into the environment from textile and other industries. During the dyeing process alone, approximately 200 billion liters of dye effluents are discharged yearly at rates of 5–50%, depending on the fabric material (e.g., leather, wool, or nylon) (Ali 2010; Kant 2012; Tkaczyk et al. 2020; Dutta et al. 2024; Rathod et al. 2024).

More than 100,000 synthetic organic dyes are utilized across industries, many of which are toxic even at low concentrations (ng/L to μg/L or 1 ppm) (Lavanya et al. 2014; Tan et al. 2015; Tkaczyk et al. 2020; Rane et al. 2021; Eltaboni et al. 2022a, b). When released, these dyes exhibit carcinogenic, mutagenic (e.g., chromosomal aberrations in *Allium cepa*), and allergenic effects (Kant 2012; Tan et al. 2015; Lellis et al. 2019; Tkaczyk et al. 2020; Roy et al. 2022; Babuji et al. 2023; Dutta et al. 2024). Beyond aesthetic harm, they disrupt aquatic and terrestrial ecosystems by absorbing light, altering microbial populations, and impairing biochemical oxygen demand (BOD) and chemical oxygen demand (COD) (Jinqi et al. 1992; Abadulla et al. 2000; Sen et al. 2016; Lellis et al. 2019; Aragaw et al. 2021; Al-Tohamy et al. 2022; Vaiano et al. 2023).

The use of untreated dye-contaminated wastewater for irrigation damages plant physiology (e.g., germination, photosynthesis, growth), degrades soil health, and harms aquatic flora (e.g., microbiome, fish), ultimately entering the human food chain directly or indirectly (Ali 2010; Aragaw et al. 2021; Fida et al. 2023; Razali et al. 2023; Vaiano et al. 2023). Additionally, synthetic dyes exhibit antimicrobial properties, disrupting bacteria, fungi, and molds, thereby threatening local biodiversity (Di Martino et al. 2022; Prasad Mishra et al. 2024). Paradoxically, certain dyes are repurposed to enhance fish growth in aquaculture or act as antifungal agents in poultry feed. Others serve as biological stains in laboratories due to their anionic, cationic, or fat-soluble properties (Roybal et al. 1989; Clark et al. 1993; Tkaczyk et al. 2020; Khan et al. 2023). Given their dual role as pollutants and functional agents, synthetic dyes present a complex regulatory and sustainability challenge. Researchers must address critical gaps in decontamination strategies to mitigate their environmental impact effectively.

6.3 Classification of Textile Dyes

Textile dyes can be broadly categorized as inorganic or organic. Organic dyes are further classified into synthetic and natural variants, depending on their production methods and sources of isolation (Mabuza et al. 2023; Alegbe et al. 2024; Ali 2024). Natural dyes are subdivided into three groups: vegetable dyes (e.g., indigo, Indian madder, turmeric, marigold, henna, tea, and onion), animal-derived dyes (e.g., lac dye, Tyrian purple, and cochineal), and mineral-based dyes (e.g., iron, Prussian blue, chrome yellow, and manganese brown). These natural dyes are celebrated for their eco-friendly properties, UV resistance, antimicrobial activity, flame retardancy, and ease of extraction, coupled with sustainable raw material availability. However, their practical application faces challenges such as storage instability, labor-intensive extraction processes, and inconsistent reproducibility of color quantity and shade due to impurities. Furthermore, their availability is often season-dependent, limiting year-round use (Sudarshan et al. 2023).

6.3.1 Based on Structure and Application

Synthetic organic dyes are classified based on physical properties (e.g., solubility, stability, and chromogenic activity), dyeing processes, chemical structure, application (e.g., textile, leather, paper, food, cosmetic, and inkjet dyes), and colorfastness/optical brighteners (e.g., lightfastness, wash fastness, and heat stability) (Tkaczyk et al. 2020; Eltaboni et al. 2022a, b; Roy et al. 2022; Khan et al. 2024). Key synthetic dye types include pigment dyes, azoic (naphthol) dyes, sulfur dyes, disperse dyes, vat dyes, basic dyes, acid dyes, reactive dyes, direct dyes, and mordant (chrome) dyes. Among these, chromogenic dyes with distinct conjugated structures (Fig. 6.1) are particularly significant (Stolz 2001; Ali 2010; Hutchings et al. 2010; Alegbe et al. 2024). These dyes are characterized by their chromophoric and conjugative properties and enhance product visualization across industries such as food and beverage, printing, paper, plastics, pharmaceuticals, cosmetics, construction, art and crafts, wood, electronics, automotive, agriculture, and medical diagnostics (Hutchings et al. 2010; Tkaczyk et al. 2020; Alegbe et al. 2024). Despite their

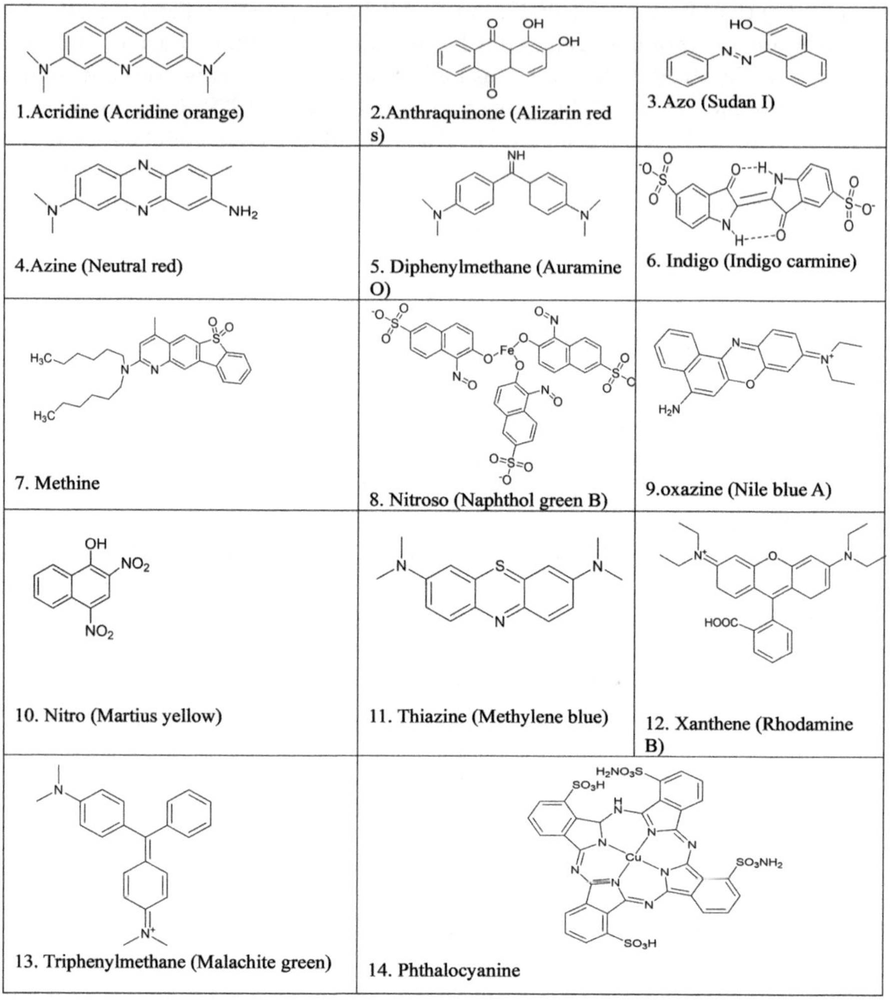

Fig. 6.1 Basic structure of some important chromogenic dyes

utility, textile dyes pose environmental challenges due to their resistance to conventional removal methods from water and soil, slow biodegradability, and long-term persistence (Ali 2010; Kant 2012; Farah Maria Drumond et al. 2013; Benkhaya et al. 2017; Tkaczyk et al. 2020). For instance, Reactive Blue 19 exhibits a hydrolysis half-life of approximately 46 years under natural environmental conditions, underscoring its recalcitrance (Razali et al. 2023).

6.3.2 *Biosynthetic and Bio-based Dyes*

Microbial pigments, such as those derived from *Monascus*, *Streptomyces*, and *Talaromyces* species, offer non-toxic and biodegradable alternatives to conventional dyes. Advances in genetic engineering have enhanced microbial production of natural dyes like indigo, violacein, prodigiosin, and anthraquinone-based compounds. Additionally, sustainable dyeing processes leveraging microbial fermentation significantly reduce chemical use and water consumption. However, challenges remain, including a limited color range with stability issues, restricted variability of microbial dyes, specialized equipment requirements for extraction, research gaps leading to inconsistent results, regulatory hurdles, cost competitiveness, consumer acceptance barriers, and advancements in synthetic dye technologies (e.g., green chemistry and biodegradability) (Rahman 2016). These limitations have spurred innovation, particularly in CRISPR-based techniques for identifying, isolating, cloning, and transferring genes to engineer microorganisms capable of efficiently degrading azo and anthraquinone dyes. Further, transgenic plants expressing heterologous laccase enzymes demonstrate promise in phytoremediation, effectively breaking down organic dyes and organophosphorus pesticides in polluted soils and hydroponic systems (Jadhav et al. 2024).

6.4 Conventional and Advanced Methods for Dye Degradation

Dye removal is achieved through chemical methods (e.g., electrolysis, oxidation, ozonation, ion exchange), physical methods (e.g., filtration, adsorption, osmosis, flocculation), and biological approaches (e.g., bacterial, fungal, algal, or enzymatic degradation) (Table 6.1). Adsorption, for instance, relies on highly porous materials with large surface areas, such as activated carbon (500–2000 m^2/g), to capture dye molecules (Rehman et al. 2023). However, the practical application of activated carbon is hindered by high costs linked to its generation and regeneration under specific pressure conditions (Panda et al. 2021). In contrast, biotechnology offers sustainable solutions by integrating enzymes, biosurfactants, microbial biopigments, and bio-based biocides into textile processing. These innovations are applied across critical stages, including sizing and desizing (where starch is removed via enzymatic or fermentation processes instead of harsh chemicals), scouring, bleaching (using H_2O_2), mercerizing, dyeing (with natural dyes), printing (via 3D methods), and finishing (using biodegradable polylactic acid, PLA) (Chung 2016; Mokif 2019).

Table 6.1 Methods for textile dye degradation: advantages and limitations

Methods	Technique	Type	Mechanism	Advantages	Limitations
Physical methods	Adsorption	Conventional	Dyes adhere to surface of adsorbent	Simple, effective, reusable materials (e.g., activated carbon)	Saturation limits, disposal of spent adsorbent, costly for large-scale (Alsukaibi 2022)
	Membrane filtration	Conventional	Size exclusion through membranes	High efficiency, no chemical additives	Membrane fouling, high cost, not suitable for all dye types
Chemical methods	Coagulation-flocculation	Conventional	Destabilizes dye particles to settle	Simple, fast, cost-effective	Sludge generation, secondary pollution, ineffective for soluble dyes
	Oxidation (e.g. O_3, H_2O_2)	Conventional	Break down dye molecules	Effective for many dyes, color removal	Requires precise control, high cost, incomplete degradation sometimes
	Advanced oxidation processes (AOPs)	Emerging	Produces hydroxyl radicals	Decomposes complex dyes, no sludge	High energy cost, specialized equipment
Biological methods	Aerobic biodegradation	Conventional	Microbial metabolism breaks down dyes	Environmentally friendly, low cost	Slow, ineffective for recalcitrant dyes, sensitive to toxic dyes
	Anaerobic biodegradation	Conventional	Reduction of azo bonds by anaerobes	Good for azo dyes, low energy	Often needs post-treatment, slow process
	Algal/bacterial treatment	Emerging	Uptake and degradation by algae/bacteria	Eco-friendly, potentially cost-effective	Sensitive to environment, slow, not fully efficient (Khan et al. 2023)
Other emerging methods	Electrochemical treatment	Emerging	Electrolysis to degrade dye compounds	No chemical additives, automation possible	High energy requirement, electrode fouling
	Photocatalysis (e.g. TiO_2)	Emerging	Light-activated catalysts degrade dyes	Renewable, low sludge	Needs UV light, slow process, expensive catalysts
	Nanotechnology-based adsorption	Emerging	Nanomaterials adsorb or degrade dyes	High surface area, efficient at low concentrations	Cost, regeneration/recovery issues, potential toxicity (Jadoun et al. 2022)

6.4.1 Biological Methods

Microbial biotechnology provides eco-friendly alternatives to conventional dyeing and wastewater treatment by leveraging biodegradation, biosorption, and biocatalysis to mitigate environmental harm. This approach fosters opportunities in the textile industry, including sustainable dye production, innovative dyeing techniques for brand differentiation, bioprocess optimization, functional textile enhancements, cross-sector collaborations, educational initiatives, and new market development (Mokif 2019). Microorganisms are employed as single pure cultures or co-cultures/consortia with fungi, algae, and molds. These microbes operate effectively under aerobic or anaerobic conditions, relying on extracellular enzymes such as reductases, hydrogenases, and laccases (Saroj et al. 2015; Lellis et al. 2019; Patel et al. 2022; Thulasisingh et al. 2023). For instance, *Bacillus* species and *Staphylococcus epidermidis* degrade Congo Red dye via laccase (an oxidoreductase) and catalase activity, as well as NADH-DCIP reductase and azoreductase enzymes (Lellis et al. 2019). Fungi such as *Trametes versicolor* synthesize cellulolytic laccases to break down dyes (Singh et al. 2020), while bacterial co-cultures like *Ochrobactrum* species, *Pseudomonas aeruginosa*, and *Providencia vermicola* degrade red-violet dyes by producing oxidoreductive enzymes that generate naphthalene intermediates (Dana et al. 2017; Lellis et al. 2019). Similarly, species such as *Shewanella aquimarina*, *Scheffersomyces spartinae*, *Pichia occidentalis*, *Aeromonas punctata*, *Shewanella putrefaciens*, and *Pseudomonas aeruginosa* degrade azo dyes (e.g., Acid Red 27, Reactive Black 5, Disperse Orange 3) via azo reduction, deamination, and desulfurization (Madhu et al. 2017). Enzymes also replace harsh chemicals in textile processing, such as cellulases for bio-polishing and proteases for wool finishing which are further advancing sustainability.

6.4.2 Enzymatic Pre-treatment

Microbial enzymes such as laccases, peroxidases, and azoreductases play a pivotal role in degrading synthetic dyes into non-toxic byproducts while enhancing dye uptake, thereby reducing dye wastage and effluent loads. To address persistent challenges in textile processing including environmental pollution and resource inefficiency, GNMs have emerged as sustainable alternatives. These materials offer eco-friendly solutions for dye removal, antimicrobial finishing, and improved fabric functionalities. This chapter examines the applications of GNMs in textile biotechnology, emphasizing their contributions to environmental sustainability, the development of advanced functional textiles, and their viability for industrial-scale adoption (Saroj et al. 2015; Lellis et al. 2019; Patel et al. 2022; Khan et al. 2023).

6.4.3 Microbial Dye Degradation and Bioremediation

The application of GNMs in textile biotechnology offers promising strategies for sustainable wastewater treatment. A diverse array of microorganisms, including bacteria and fungi, can be explored in combination with plants to decontaminate wastewater from hazardous dyes, providing cost-effective and eco-friendly solutions (Sen et al. 2016; Jamee et al. 2019). During bioremediation, halotolerant, halophilic, or extremophilic microorganisms along with their extremozymes exhibit resistance to plasmolysis, extreme pH, temperature fluctuations, and high salinity, even in the presence of metallic cations such as Mg^{2+}, Ca^2, and Pb^{2+}. These traits are particularly advantageous for degrading textile Azo dyes, which often contain high salt concentrations. For instance, *Aeromonas hydrophila* produces laccase and lignin peroxidase enzymes under sucrose and yeast extract substrates, enabling efficient degradation of crystal violet dye. Similarly, bacterial consortia comprising *Bacillus*, *Sedimentibacter*, *Pseudomonas*, *Clostridiales*, and *Streptomyces* have demonstrated efficacy in breaking down high concentrations of Reactive Brilliant Red X-3B, a common textile Azo dye. Additionally, *Escherichia coli* has been shown to acquire CotA laccase which is a degradative enzyme transferred from *Pseudomonas luteola* and *Bacillus subtilis*, thereby gaining the ability to biodegrade azo dyes. The effectiveness of microbial treatment for dye-laden wastewater encompassing decontamination, decolorization, mineralization, and detoxification depends on several factors. These include pH, temperature, dye type and concentration, and the ability of microorganisms to adapt to environmental stressors while sustaining its population (Stolz 2001; Pandey et al. 2007; Sadia Afrin et al. 2021). Such biological systems highlight the potential for integrating microbial resilience with GNMs to advance sustainable textile biotechnology.

6.5 Limitations of Conventional Methods

Organic dye effluents in water can be removed through physical and chemical methods such as oxidation, adsorption, electrochemical treatments, coagulation, flocculation, and filtration (Afrin et al. 2021; Roy et al. 2022; Vaiano et al. 2023). However, these approaches face significant limitations, including high material costs, the generation of hazardous byproducts, and energy-intensive operational requirements. Additionally, residual sludge often containing residual dyes must be extracted from treatment sites for further purification or decontamination, adding to logistical and environmental challenges (Tanaka et al. 2000; Cristóvão 2010; Afrin et al. 2021; Roy et al. 2022).

6.5.1 Microbial Decontamination and Limitation

The bioaccumulation of dyes within cell walls and dead plant tissues, as well as their degradation by resistant microbes, transforms these compounds into simpler, recalcitrant forms. Microbes act as biosorbents, forming complexes with nitrogen or carbon constituents of textile dyes inside their cellular structures, thereby facilitating bioremediation. A second mechanism involves the release of enzymes such as NADH-DCIP, laccase, azoreductase, peroxidase, MG reductase, and tyrosine enzymes, which degrade the (-N=N-) azo bonds in dyes (Jinqi et al. 1992; Abadulla et al. 2000; Sen et al. 2016; Lellis et al. 2019; Aragaw et al. 2021; Al-Tohamy et al. 2022; Vaiano et al. 2023). For example, *Penicillium oxalicum* (DS-2) and *Aspergillus tubingensis* (DS-4), along with consortium RH-2, produce laccase and manganese peroxidase enzymes capable of degrading 100 mg/L Congo red dye, depending on the specific growth media used (Thakor et al. 2022).

Microbial decolorization processes occur through three primary pathways. The first is structural modification, where minor alterations are made to the organic compound while retaining its core structure, often resulting in recalcitrant intermediates (Eltaboni et al. 2022a, b). The second pathway involves fragmentation, in which dye molecules are broken down into smaller branches; however, these fragments may reassemble into their original structure under specific environmental conditions (Aragaw 2024). The third and most effective mechanism is mineralization, where organic compounds are completely transformed into inorganic byproducts such as carbon dioxide, methane, and simple inorganic molecules, ensuring permanent detoxification (Sadia Afrin et al. 2021; Mishra et al. 2022).

Notably, microbial communities have achieved 90% decolorization and decontamination of Reactive Black 5 (a textile Azo dye) via biostimulation or bioaugmentation under anaerobic conditions. This process occurs in 4–5-L containers containing activated textile sludge, where aromatic amines are generated following azo bond reduction. Additionally, *Acinetobacter* sp. SW30, which biosynthesizes gold and silver nanoparticles under anaerobic conditions with sodium borohydride ($NaHB_4$), has demonstrated the ability to degrade Direct Black 22 and Reactive Yellow 186 dyes (Lellis et al. 2019).

6.5.2 Phytoremediation and Limitations

Phytoremediation of textile dyes is achieved through several key processes, including biotransformation, rhizofiltration, phytoextraction/phytoaccumulation, phytodegradation, and rhizodegradation. Decontamination of dyes is facilitated in natural or artificially constructed ponds or phytoreactors that combine plants and rhizosphere microbial communities. For instance, a consortium of the plant *Fimbristylis dichotoma* and the fungus *Saccharomyces cerevisiae* produces oxidoreductive enzymes that biodegrade the azo dye Rubine GFL (92%) and actual textile effluent (68%) within 4 days. Transgenic *Agrobacterium rhizogenes* (plasmid Ri) enhances phytochemical production, such as secondary metabolites, and secretes enzymes like lignin peroxidases, Mn peroxidases, tyrosinases, and azoreductases from the root hairs of *Tagetes patula* (Asteraceae), enabling the biotransformation of azo dyes such as Red 198. Similarly, *Tecoma stans* (Bignoniaceae) is particularly effective for phytoremediation of the textile dye Malachite Green in callus cultures when combined with calcium alginate and peroxidase enzymes. To optimize these processes, constructed wetlands can be established to enhance dye phytoremediation using specialized plant species (Lellis et al. 2019).

6.6 GNMs as a Sustainable Solution

GNMs are nanoscale materials (1–100 nm in size) synthesized through environmentally benign processes such as those derived from plants, microbes (fungi, bacteria, algae), or biopolymers. The preparation of biobased nanoparticles follows a multi-step process as shown in Fig. 6.2. These HNMs prioritize eco-friendliness, biocompatibility, biodegradability, and reduced toxicity. These methods minimize hazardous chemical use and energy consumption compared to conventional

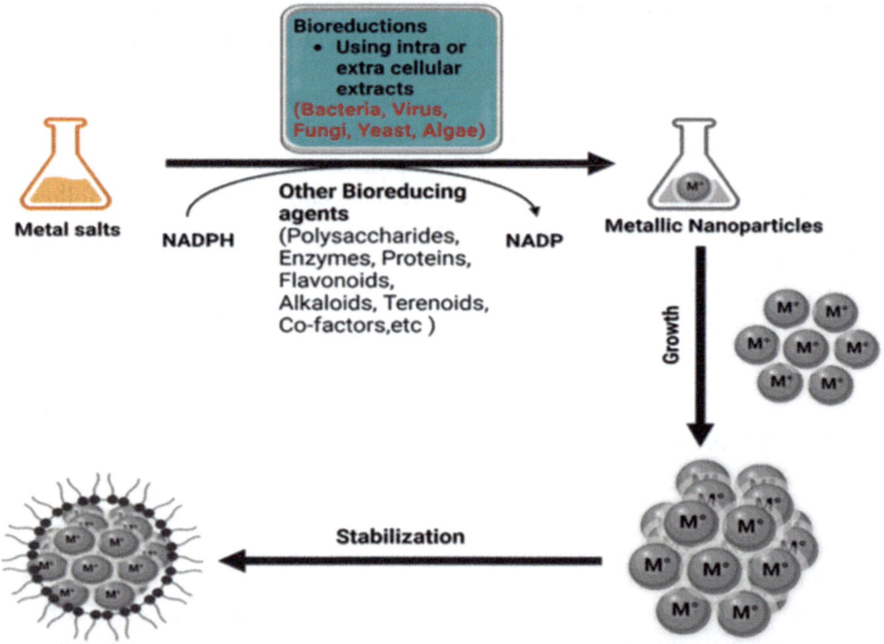

Fig. 6.2 A general mechanism to prepare the biobased nano particles (Ref. Velusamy et al. 2016)

physical or chemical synthesis approaches (Alves et al. 2024). GNMs, when engineered with precise size and morphology, have emerged as sustainable alternatives to traditional nanomaterials, offering cost-effectiveness, safety, and vast potential across industries. Their minimal ecological footprint makes them ideal for applications in biomedicine, agriculture, and textiles. Notably, GNMs play a pivotal role in integrating microbial biotechnology with textile applications, such as dye degradation and effluent treatment (Lin et al. 2022; Singh et al. 2023; Vijayaram et al. 2024).

Nanomaterials have garnered significant scientific attention due to their small size, functional diversity, and robust properties. These materials are broadly classified into organic and inorganic categories. Organic nanoparticles, such as carbon-based fullerenes and graphene, are distinct from their inorganic counterparts, which include magnetic nanoparticles (e.g., iron), metallic nanoparticles (e.g., gold, silver), and semiconductor nanoparticles (Velusamy et al. 2016; Alves et al. 2024; Vijayaram et al. 2024). While chemical and physical synthesis methods yield nanoparticles with uniform size, high purity, and tailored properties, they are often costly and rely on toxic chemicals, raising environmental and economic concerns (Ghosh et al. 2021).

Synthesis approaches are categorized into two primary strategies: top-down and bottom-up. The top-down approach involves breaking down bulk materials into nanoscale particles through physical or chemical reduction techniques, such as UV irradiation, sonochemistry, radiolysis, or laser ablation. In contrast, the bottom-up method builds nanoparticles from atomic or molecular components via chemical or biological assembly, ensuring precise control at the nanoscale. Notably, green or plant-mediated metallic nanoparticles have emerged as sustainable alternatives, leveraging biocompatible processes for applications such as environmental remediation, energy storage, photocatalytic activity, and reduction reactions. These eco-friendly nanoparticles, synthesized using plant extracts or microbial systems, minimize toxic byproducts and align with circular economy principles (Groeneveld et al. 2023; Rehman et al. 2023; Vijayaram et al. 2024).

6.6.1 Synthesis of GNMs

GNMs are classified based on their composition (metal-based, carbon-based, polymer-based, silica-based, hybrid), synthesis methods (biogenic, phyto-synthesized, microbial-synthesized, enzyme-assisted), and applications (energy, textile, biomedical, environmental, agricultural) (Selvaraj et al. 2021; Singh et al. 2023). These nanoparticles are synthesized using microorganisms such as bacteria (*Bacillus* and *Pseudomonas* spp.), fungi or plant-derived protocols, with enzymatic activity playing a pivotal role in their functionalization and stabilization (Patel et al. 2022). Enzymatic synthesis offers advantages like high specificity, efficiency, and eco-friendly operating conditions, making GNMs ideal for textile applications. For instance, microbial enzymes act as biocatalysts, enabling nanoparticle formation under mild conditions without harsh chemicals (Stolz 2001; Cristóvão 2010; Patel et al. 2022).

Key enzymes involved include nitrate reductase, found in fungi like *Fusarium oxysporum* and bacteria, which reduces metal salts to silver or gold nanoparticles (Alihosseini et al. 2023), and laccases (p-benzenediol: oxygen oxidoreductase), multicopper proteins present in plants (e.g., apples, potatoes), bacteria (*Streptomyces lavendulae*), and fungi (*Aspergillus niger*). These enzymes degrade lignin, catalyze phenolic compound oxidation, and aid in nanocellulose composite synthesis (Dana et al. 2017; Mehta et al. 2021). Hydrogenases from bacteria like *Desulfovibrio desulfuricans* reduce metal ions to produce biogenic platinum/palladium nanoparticles, while cysteine-rich proteins from *Lactobacillus* species stabilize nanoparticles for textile use (Dana et al. 2017; Mehta et al. 2021). Metabolite-assisted stabilization involves microbial exopolysaccharides (EPS), alginates, chitosan, and organic acids (citric acid, glutamic acid), which enhance nanoparticle stability in extracellular environments (Patel et al. 2022; Ying et al. 2022; Singh et al. 2023; Thulasisingh et al. 2023; Rathod et al. 2024).

Nanoparticle synthesis occurs intracellularly (within microorganisms or plants) via absorption and reduction of metal ions, yielding stable, size-controlled nanoparticles (Mishra et al. 2021; Patel et al. 2022), or phytochemically using plant-derived compounds like polyphenols, flavonoids, and alkaloids. These phytochemicals act as eco-friendly reducing, capping, and stabilizing agents, enabling cost-effective, large-scale synthesis (Singh et al. 2023). GNMs can be metallic (e.g., gold, silver), polymeric (chitosan, starch), or hybrid composites, tailored for diverse industrial applications (Velusamy et al. 2016; Patel et al. 2022).

Microorganisms stabilize metallic elements such as gold, titanium, silver, zinc, platinum, copper, palladium, and cadmium by reducing them into nanoparticles (<100 nm in size) through biological synthesis routes (Velusamy et al. 2016; Mishra et al. 2022; Singh et al. 2023). The eco-friendly nature of green nanoparticles extends to agricultural applications, where they minimize reliance on conventional fertilizers and pesticides. Their high nutrient-holding capacity and controlled release properties also aid in remediating residual dyes, pesticides, and excess fertilizers in contaminated soils (Roy et al. 2022; Singh et al. 2023). In contrast, nanoparticles synthesized via physical or chemical methods often involve hazardous chemicals, posing risks of secondary contamination at scale and direct or indirect harm to human health (Ahmad et al. 2019; Lin et al. 2022).

Bio-based nanoparticles, such as those derived from cellulose, chitosan, or lignin, offer superior environmental compatibility due to their natural origins, antimicrobial properties, catalytic activity, biodegradability, and stability under diverse conditions (Selvaraj et al. 2021). Green chemistry principles address these challenges by prioritizing eco-friendly synthesis methods with low energy inputs and minimal toxicity (Ghosh et al. 2021; Patel et al. 2022). Plant-derived phytochemicals further enhance sustainability, enabling cost-effective, large-scale extraction of natural reducing and stabilizing agents that reduce environmental toxicity (Tan et al. 2015; Patel et al. 2022).

6.7 GNMs in Textile Biotechnology

GNMs are synthesized using biological methods such as plant extracts, microbial processes, and biopolymers, offering sustainable alternatives to conventional nanomaterials. Nanoparticles like silver (AgNPs) and zinc oxide (ZnO NPs), produced via plant-mediated or microbial reduction, are widely employed in textiles for their antimicrobial properties and UV protection. Similarly, titanium dioxide nanoparticles (TiO_2 NPs) enhance self-cleaning and UV-resistant fabrics. Detoxification processes in textile biotechnology often rely on synergistic combinations of nanomaterials, including TiO_2, ZnO, Ag NPs, CuO, carbon-based nanoparticles (carbon dots, graphene), iron-based nanoparticles (Fe NPs, Fe_3O_4, Fe_2O_3), metal-organic frameworks (MOFs), and bio-nanocomposites, tailored to specific industrial conditions (Groeneveld et al. 2023; Tamjidi et al. 2024). Biopolymer-based nanomaterials further advance sustainability by leveraging renewable resources: chitosan nanoparticles provide antimicrobial coatings for medical textiles, starch-based nanomaterials enable biodegradable fabric finishes, and protein-based nanoparticles (e.g., silk fibroin, casein) enhance textile durability through functional coatings (Tanaka et al. 2000; Dutta et al. 2021; Rathod et al. 2024). These innovations underscore the versatility of GNMs in reducing environmental footprints while meeting diverse industrial demands.

6.7.1 GNMs for Dye Degradation

GNMs serve as efficient adsorbents and catalysts for removing toxic dyes from textile effluents. For instance, iron oxide nanoparticles, synthesized using plant extracts, effectively degrade azo dyes in wastewater treatment systems, while chitosan-based nanocomposites adsorb reactive dyes, significantly reducing environmental toxicity. These eco-friendly techniques are not only more effective for treating dye-contaminated wastewater but also improve human health by minimizing exposure to hazardous pollutants (Aragaw et al. 2021; Al-Tohamy et al. 2022). GNMs encompass a range of structures including nanoparticles, nanocomposites, nanofibers, and nano-coatings that are derived from biological or natural sources. Their versatility enables critical advancements in integrating microbial biotechnology with textile innovations, particularly in sustainable dye degradation, antimicrobial finishing, and the development of functionalized fabrics with enhanced performance.

6.7.2 Photocatalysis

Processes such as ozonation (O_3) and oxidation using hypochlorite ions, Fenton's reagent, or photocatalytic systems involving ultraviolet (UV) radiation combined with TiO_2, H_2O_2, and O_3 have proven highly effective in degrading organic dyes

(Tanaka et al. 2000; Rathod et al. 2024). Analytical studies of dye effluents reveal elevated concentrations of diverse dye classes, including Azo dyes (e.g., Methyl Orange, Congo Red), Reactive dyes (e.g., Reactive Black 5, Reactive Blue 19), disperse dyes (e.g., Disperse Blue 79, Disperse Red 60), Vat dyes (e.g., Indigo, Vat Blue 6), and Acid dyes (e.g., Acid Blue 25, Acid Red 88) (Omar et al. 2023; Tamjidi et al. 2024). These pollutants are detoxified through a combination of advanced oxidation processes (AOPs), photocatalytic degradation, adsorption, redox reactions, Fenton reactions, catalytic reduction, enzymatic biodegradation, and chemical oxidation (Groeneveld et al. 2023).

6.7.3 UV Protection and Photocatalytic Activity

GNMs have revolutionized the textile industry by integrating functional properties such as antimicrobial resistance, UV protection, water repellency, and self-cleaning capabilities. These innovations enhance fabric durability, sustainability, and performance, making them indispensable for modern textile technologies. For instance, zinc oxide (ZnO) and titanium dioxide (TiO_2) nanoparticles, synthesized via eco-friendly methods, significantly improve UV resistance in textiles, preventing fabric degradation and safeguarding human skin from harmful radiation. Carbon-based nanomaterials, derived from renewable sources like lignin, cellulose, and chitosan, further elevate textile functionality: graphene oxide (GO) enables the production of high-strength, conductive, and flame-retardant fabrics, while nanocellulose provides bio-compatible coatings that enhance mechanical properties such as flexibility and tensile strength. By leveraging these tailored attributes, GNMs address both performance and environmental demands, positioning green nanotechnology at the forefront of sustainable textile innovation (Dutta et al. 2021).

6.8 Challenges and Future Perspectives

The integration of GNMs into textiles presents transformative opportunities for sustainable innovation, though significant challenges must be addressed to realize their full potential. Current advancements highlight the efficacy of GNMs in dye decontamination, with biogenic iron oxide nanoparticles and graphene-based materials enabling efficient adsorption and catalytic breakdown of textile dyes in wastewater (Rowe 1999). Antimicrobial textiles leveraging silver and ZnO nanoparticles demonstrate promise in medical and sportswear applications, inhibiting pathogenic bacteria (Eweis et al. 2024), while UV-protective fabrics infused with TiO_2 and ZnO nanoparticles enhance user safety and reduce reliance on chemical detergents through self-cleaning and water-repellent properties. Innovations like graphene oxide coatings and carbon nanotubes further advance flame-retardant and conductive textiles for wearable technologies (Rowe 1999; Eweis et al. 2024).

Environmentally, GNMs align with circular economic goals by reducing chemical waste and energy consumption, yet scalability and cost remain critical barriers. Biotechnology-based synthesis, though promising for cost-effectiveness, demands specialized infrastructure and raises concerns about process reliability. Material stability, environmental safety, and ethical governance also require urgent attention. Looking ahead, CRISPR-edited microbes and AI-driven optimization offer pathways to enhance nanoparticle synthesis efficiency, while machine learning can refine size and functionality for targeted applications. Interdisciplinary collaboration will be pivotal in developing multifunctional smart textiles with integrated sensors and self-cleaning capabilities, alongside global policies to ensure ethical adoption. By addressing these challenges through innovation and regulation, GNMs can revolutionize textile manufacturing, balancing performance with planetary health (Eweis et al. 2024).

6.9 Conclusion

The application of GNMs in textile biotechnology offers a transformative and sustainable alternative to conventional textile processing. By integrating eco-friendly nanoparticles such as biogenic iron oxide, graphene-based composites, and plant-derived silver or zinc oxide, the textile industry can significantly reduce environmental pollution while enhancing fabric functionality. GNMs enable antimicrobial resistance, UV protection, self-cleaning capabilities, and flame retardancy, improving performance across sectors like healthcare, sportswear, and protective gear. Despite challenges in scalability and cost, advancements in microbial biotechnology and green synthesis methods underscore the potential for biodegradable, energy-efficient, and economically viable textile innovations. Continued progress in this field promises to align textile manufacturing with global sustainability goals, fostering eco-conscious consumer practices and reducing reliance on hazardous chemicals.

References

Abadulla E, Tzanov T, Costa S, Robra K-H, Cavaco-Paulo A, Gübitz Georg M (2000) Decolorization and detoxification of textile dyes with a laccase from *Trametes hirsuta*. Appl Environ Microbiol 66:3357–3362

Afrin S, Shuvo HR, Sultana B, Islam F, Rus'd AA, Begum S et al (2021) The degradation of textile industry dyes using the effective bacterial consortium. Heliyon 7:e08102

Ahmad S, Munir S, Zeb N, Ullah A, Khan B, Ali J et al (2019) Green nanotechnology: a review on green synthesis of silver nanoparticles—an ecofriendly approach. Int J Nanomedicine 14:5087–5107

Alegbe EO, Uthman TO (2024) A review of history, properties, classification, applications and challenges of natural and synthetic dyes. Heliyon 10:e33646

Ali H (2010) Biodegradation of synthetic dyes—a review. Water Air Soil Pollut 213:251–273

Ali AN (2024) A comprehensive study of natural and synthetic dyes: their properties, methods of preparation, and uses. SHIFAA 2024:1–17

Alihosseini F, Azimi SH, Mostajeran M, Feiz M (2023) Removal of different vat dyes by *Aspergillus niger*: a comparative study focusing on the molecular structure. J Microbiol Methods 208:106720

Alsukaibi AKD (2022) Various approaches for the detoxification of toxic dyes in wastewater. Processes 10:1968

Al-Tohamy R, Ali SS, Li F, Okasha KM, Mahmoud YAG, Elsamahy T et al (2022) A critical review on the treatment of dye-containing wastewater: Ecotoxicological and health concerns of textile dyes and possible remediation approaches for environmental safety. Ecotoxicol Environ Saf 231:113160

Alves MBN, Jorge AMS, Pereira JFB (2024) The biotechnology revolution in textile dyeing. Trends Biotechnol 42:1211–1214

Aragaw TA (2024) A review of dye biodegradation in textile wastewater, challenges due to wastewater characteristics, and the potential of alkaliphiles. J Hazard Mater Adv 16:100493

Aragaw TA, Bogale FM (2021) Biomass-based adsorbents for removal of dyes from wastewater: a review. Front Environ Sci 9. https://doi.org/10.3389/fenvs.2021.764958

Babuji P, Thirumalaisamy S, Duraisamy K, Periyasamy G (2023) Human health risks due to exposure to water pollution: a review. Water 15:2532

Benkhaya S, El Harfi S, El Harfi A (2017) Classifications, properties and applications of textile dyes: a review. Appl J Environ Eng Sci 3:311–320

Bhatia SC, Devraj S (2017) Pollution control in textile industry. 1st ed. WPI Publishing. p 342

Farah Maria Drumond Chequer, Gisele Augusto Rodrigues De Oliveira, Elisa Raquel Anastácio Ferraz, Juliano Carvalho Cardoso, Maria Valnice Boldrin Zanoni, Danielle Palma de Oliveira (2013) In: Melih G (ed) Eco-Friendly textile dyeing and finishing. IntechOpen, London. Ch. 6, 150–176

Chung K-T (2016) Azo dyes and human health: a review. J Environ Sci Health C 34:233–261

Clark FD, Frame DD, Jensen MM (1993) Occlusive laryngotracheitis in turkeys following drinking water administration of gentian violet. Avian Dis 37:226–228

Cristóvão RO (2010) Degradation of dye-containing textile effluents by enzymatic catalysis PhD in Chemical and Biological Engineering thesis, Universidade do Porto (Portugal)

Dana M, Bakhshi Khaniki G, Mokhtarieh AA, Davarpanah SJ (2017) Biotechnological and industrial applications of laccase: a review. J Appl Biotechnol Rep 4:675–679

Di Martino M, Sessa L, Di Matteo M, Panunzi B, Piotto S, Concilio S (2022) Azobenzene as antimicrobial molecules. Molecules 27:5643

Dutta D, Das BM (2021) Scope of green nanotechnology towards amalgamation of green chemistry for cleaner environment: a review on synthesis and applications of green nanoparticles. Environ Nanotechnol Monit Manag 15:100418

Dutta S, Adhikary S, Bhattacharya S, Roy D, Chatterjee S, Chakraborty A et al (2024) Contamination of textile dyes in aquatic environment: adverse impacts on aquatic ecosystem and human health, and its management using bioremediation. J Environ Manag 353:120103

Eltaboni F, Bader N, El-Kailany R, Elsharif N, Ahmida A (2022a) Dyes: a comprehensive review. J Chem Rev 4:313–330

Eltaboni F, Bader N, El-Kailany R, Elsharif N, Ahmida A (2022b) Chemistry and applications of azo dyes: a comprehensive review. J Chem Rev 4:313–330

Eweis AA, El-Raheem HA, Ahmad MS, Hozzein WN, Mahmoud R (2024) Green fabrication of nanomaterials using microorganisms as Nano-factories. J Clust Sci 35:2149–2176

Fida M, Li P, Wang Y, Alam SMK, Nsabimana A (2023) Water contamination and human health risks in Pakistan: a review. Expo Health 15:619–639

Ghosh S, Ahmad R, Zeyaullah M, Khare SK (2021) Microbial Nano-factories: synthesis and biomedical applications. Front Chem 9:626834

Groeneveld I, Kanelli M, Ariese F, Van Bommel MR (2023) Parameters that affect the photodegradation of dyes and pigments in solution and on substrate—an overview. Dyes Pigments 210:110999
Hutchings M, Bamfield P (2010) Chromic phenomena: technological applications of colour chemistry. The Royal Society of Chemistry, Cambridge
Jadhav RR, Tapase SR, Chandanshive VV, Gophane AD, Jadhav JP (2024) Plant and yeast consortium for efficient remediation of dyes and effluents: a biochemical and toxicological study. Int J Microbiol 27:1269–1283
Jadoun S, Chauhan NPS, Zarrintaj P, Barani M, Varma RS, Chinnam S et al (2022) Synthesis of nanoparticles using microorganisms and their applications: a review. Environ Chem Lett 20:3153–3197
Jamee R, Siddique R (2019) Biodegradation of synthetic dyes of textile effluent by microorganisms: an environmentally and economically sustainable approach. Eur J Microbiol Immunol 9:114–118
Jinqi L, Houtian L (1992) Degradation of azo dyes by algae. Environ Pollut 75:273–278
Kant R (2012) Textile dyeing industry an environmental hazard. Nat Sci 04:22–26
Karahan M, Ahrari M, Karahan N (2023) Technical textiles market research and added value analysis: a Regio-global case study. Rezult Cercet Noastre Teh 24:162–180
Khan MF (2025) Recent advances in microbial enzyme applications for sustainable textile processing and waste management. Sci 7:46
Khan WU, Ahmed S, Dhoble Y, Madhav S (2023) A critical review of hazardous waste generation from textile industries and associated ecological impacts. J Indian Chem Soc 100:100829
Khan A, Ju P, Han Z, Ni C (2024) A comprehensive review on adsorptive removal of azo dyes using functional materials. Aqua Water Infrastruct Ecosyst Soc 73:266–285
Lavanya C, Rajesh Dhankar RD, Sunil Chhikara SC, Sarita Sheoran SS (2014) Degradation of toxic dyes: a review. Int J Curr Microbiol Appl Sci 3:189–199
Lellis B, Fávaro-Polonio CZ, Pamphile JA, Polonio JC (2019) Effects of textile dyes on health and the environment and bioremediation potential of living organisms. Biotechnol Res Innov 3:275–290
Lin L, Yang H, Xu X (2022) Effects of water pollution on human health and disease heterogeneity: a review. Front. Environ Sci 10. https://doi.org/10.3389/fenvs.2022.880246
Mabuza L, Sonnenberg N, Marx-Pienaar N (2023) Natural versus synthetic dyes: consumers' understanding of apparel coloration and their willingness to adopt sustainable alternatives. Resour Conserv Recycl Adv 18:200146
Madhu A, Chakraborty JN (2017) Developments in application of enzymes for textile processing. J Clean Prod 145:114–133
Mehta M, Sharma M, Pathania K, Jena PK, Bhushan I (2021) Degradation of synthetic dyes using nanoparticles: a mini-review. Environ Sci Pollut Res 28:49434–49446
Mishra S, Cheng L, Maiti A (2021) The utilization of agro-biomass/byproducts for effective bio-removal of dyes from dyeing wastewater: a comprehensive review. J Environ Chem Eng 9:104901
Mishra A, Takkar S, Joshi NC, Shukla S, Shukla K, Singh A et al (2022) An integrative approach to study bacterial enzymatic degradation of toxic dyes. Front Microbiol. 12:802544
Mokif LA (2019) Removal methods of synthetic dyes from industrial wastewater: a review. Mesop Environ J 5:23–40
Omar AZ, El-Rahman MA, Hamed EA, El-Sadany SK, El-Atawy MA (2023) Synthesis, spectroscopic characterization and dyeing performance of novel bis azo dyes derived from benzidine. Sci Rep 13:7826
Panda SK, Aggarwal I, Kumar H, Prasad L, Kumar A, Sharma A et al (2021) Magnetite nanoparticles as sorbents for dye removal: a review. Environ Chem Lett 19:2487–2525
Pandey A, Singh P, Iyengar L (2007) Bacterial decolorization and degradation of azo dyes. Int Biodeterior Biodegradation 59:73–84

Patel H, Yadav VK, Yadav KK, Choudhary N, Kalasariya H, Alam MM et al (2022) A recent and systemic approach towards microbial biodegradation of dyes from textile industries. Water 14:3163

Prasad Mishra D, Kumar Sahu P, Acharya B, Prasad Mishra S, Bhati S (2024) A review of the synthesis and application of azo dyes and metal complexes for emerging antimicrobial therapies. Results Chem 10:101712

Rahman F (2016) The treatment of industrial effluents for the discharge of textile dyes using by techniques and adsorbents. J Text Sci Eng 6:1000242

Rane A, Joshi SJ (2021) Biodecolorization and biodegradation of dyes: a review. Open Biotechnol J 15:97–108

Rathod S, Preetam S, Pandey C, Bera SP (2024) Exploring synthesis and applications of green nanoparticles and the role of nanotechnology in wastewater treatment. Biotechnol Rep 41:e00830

Razali NA, Jamain Z (2023) Synthesis, chemical identification and biological application of azo-based molecules containing different terminal group: a review. J Mol Struct 1284:135329

Rehman MU, Taj MB, SaC C (2023) Biogenic adsorbents for removal of drugs and dyes: a comprehensive review on properties, modification and applications. Chemosphere 338:139477

Rowe HD (1999) Biotechnology in the textile/clothing industry—a review. J Con Stud Home Econ 23:53–61

Roy A, Murthy HA, Ahmed HM, Islam MN, Prasad R (2022) Phytogenic synthesis of metal/metal oxide nanoparticles for degradation of dyes. J Renew Mater 10:1911–1930

Roybal JE, Munns RK, Hurlbut JA, Shimoda W (1989) High-performance liquid chromatography of gentian violet, its demethylated metabolites, leucogentian violet and methylene blue with electrochemical detection. J Chromatogr A 467:259–266

Saroj S, Dubey S, Agarwal P, Prasad R, Singh RP (2015) Evaluation of the efficacy of a fungal consortium for degradation of azo dyes and simulated textile dye effluents. Sustain Water Resour Manag 1:233–243

Selvaraj V, Swarna Karthika T, Mansiya C, Alagar M (2021) An over review on recently developed techniques, mechanisms and intermediate involved in the advanced azo dye degradation for industrial applications. J Mol Struct 1224:129195

Sen SK, Raut S, Bandyopadhyay P, Raut S (2016) Fungal decolouration and degradation of azo dyes: a review. Fungal Biol Rev 30:112–133

Singh G, Dwivedi SK, Mishra J (2020) Role of Fungal Enzymes in the Removal of Azo Dyes. In: Arora NK, Mishra J, Mishra V (eds) Microbial enzymes: roles and applications in industries. Springer, Singapore, pp 231–257

Singh NA, Narang J, Garg D, Jain V, Payasi D, Suleman S et al (2023) Nanoparticles synthesis via microorganisms and their prospective applications in agriculture. Plant Nano Biol 5:100047

Stolz A (2001) Basic and applied aspects in the microbial degradation of azo dyes. Appl Microbiol Biotechnol 56:69–80

Sudarshan S, Harikrishnan S, Rathibhuvaneswari G, Alamelu V, Aanand S, Rajasekar A et al (2023) Impact of textile dyes on human health and bioremediation of textile industry effluent using microorganisms: current status and future prospects. J Appl Microbiol 134:lxac064

Tamjidi S, Kamyab Moghadas B, Abdollahi P, Parnian N, Farhadi A (2024) Industrial by-products as potentially economic and promising adsorbents for removing dyes from effluents: a review study. Phys Chem Res 12:361–382

Tan KB, Vakili M, Horri BA, Poh PE, Abdullah AZ, Salamatinia B (2015) Adsorption of dyes by nanomaterials: recent developments and adsorption mechanisms. Sep Purif Technol 150:229–242

Tanaka K, Padermpole K, Hisanaga T (2000) Photocatalytic degradation of commercial azo dyes. Water Res 34:327–333

Thakor R, Mistry H, Tapodhan K, Bariya H (2022) Efficient biodegradation of Congo red dye using fungal consortium incorporated with *Penicillium oxalicum* and *aspergillus tubingensis*. Folia Microbiol 67:33–43

Thulasisingh A, Ananthakrishnan K, Raja A, Kannaiyan S (2023) Bioprospecting of novel and industrially appropriate enzymes: a review. Water Air Soil Pollut 235:12

Tkaczyk A, Mitrowska K, Posyniak A (2020) Synthetic organic dyes as contaminants of the aquatic environment and their implications for ecosystems: a review. Sci Total Environ 717:137222

Vaiano V, De Marco I (2023) Removal of azo dyes from wastewater through heterogeneous Photocatalysis and supercritical water oxidation. Separations 10

Velusamy P, Kumar GV, Jeyanthi V, Das J, Pachaiappan R (2016) Bio-inspired green nanoparticles: synthesis, mechanism, and antibacterial application. Toxicol Res 32:95–102

Vijayaram S, Razafindralambo H, Sun Y-Z, Vasantharaj S, Ghafarifarsani H, Hoseinifar SH et al (2024) Applications of green synthesized metal nanoparticles — a review. Biol Trace Elem Res 202:360–386

Ying S, Guan Z, Ofoegbu PC, Clubb P, Rico C, He F et al (2022) Green synthesis of nanoparticles: current developments and limitations. Environ Technol Innov 26:102336

Chapter 7
Microbial Biofilm on Textile Surfaces

Faiqa Munir, Manal, Waseem Safdar, Saeed Ahmed, and Faiza Saleem

Abstract Biofilm formation on textile surfaces has garnered significant attention due to its implications across diverse sectors, including healthcare and industry. Textiles, as essential components of daily life, are particularly vulnerable to biofilm initiation due to sweat absorption during wear. Notably, biofilm persistence on healthcare textiles, such as hospital linens and uniforms, is a critical contributor to healthcare-associated infections (HAIs). Inadequate laundering practices allow pathogenic bacterial strains to adhere to fabrics, exacerbating infectious disease transmission while compromising material functionality and durability. Additionally, physicochemical properties of textiles such as surface charge, chemical composition, and hydrophilicity/hydrophobicity play pivotal roles in bacterial adhesion, the foundational step in biofilm formation. Once established, biofilm eradication becomes exceedingly challenging. Consequently, pretreatment strategies targeting both antimicrobial and antibiofilm activity have been developed to mitigate these issues. This chapter explores the mechanisms underlying biofilm development on various fabric materials (natural and synthetic), factors influencing microbial adhesion, associated challenges, detrimental consequences, and strategies for biofilm removal. Moreover, advancements in innovative approaches have been discussed which emphasize the urgent need for next-generation functional textiles engineered with anti-adhesion properties. Prospects for future research include refining fabric pretreatment protocols and developing textiles impregnated with antimicrobial and antibiofilm agents. Such innovations aim to foster healthier environments in daily life, healthcare settings, and industrial applications by sustainably addressing biofilm-related challenges.

F. Munir · Manal · S. Ahmed (✉) · F. Saleem
Department of Biological Sciences, National University of Medical Sciences, Rawalpindi, Pakistan
e-mail: saeed.ahmed@numspak.edu.pk

W. Safdar (✉)
Department of Chemical and Biomolecular Engineering, Whiting School of Engineering, Johns Hopkins University, Baltimore, MD, USA
e-mail: waseem.safdar@numspak.edu.pk

M. Khurshid et al. (eds.), *Microbial Biotechnology in Textile Innovation*, SDGs and Textiles, https://doi.org/10.1007/978-981-95-0347-6_7

Keywords Biofilm maturation · Bacterial adhesion · Physicochemical factors · Next generation textiles · Interaction · Self-cleaning textiles · Antimicrobial coating

7.1 Introduction

The formation of biofilms on unique conditioning films (composed of various fibers) with specific atomic combinations has garnered significant attention in recent years. Biofilms on textiles are not only relevant to daily clothing (e.g., odor resistance) (Møllebjerg et al. 2021) but also hold critical implications for healthcare (infection control) and industrial sectors (enhanced material performance). Textiles exhibit immense diversity, ranging from natural fibers (cotton, jute, silk) to synthetic fibers (nylon, polyester) and semi-synthetic fibers (rayon), each possessing distinct properties. The primary objectives of this chapter are to establish a comprehensive understanding of biofilms on fabrics, elucidate the factors influencing their formation, address the challenges and benefits associated with biofilms, and evaluate control strategies to manage these resilient microbial communities. A nuanced grasp of these topics could pave the way for developing next-generation textiles with anti-biofilm properties, anti-adhesive surfaces, or antimicrobial coatings to inhibit bacterial adhesion which is the critical first step in biofilm formation (Caschera et al. 2019).

The cascade of biofilm formation begins with the adherence of planktonic cells to a substratum. Microorganisms inherently rely on mutualistic interactions and environmental conditions for survival, whether through nutrient acquisition, surface attachment, or habitat colonization. Surface attachment, a prerequisite for microbial growth and proliferation, can be reversible or irreversible (Donlan 2001). The transition from planktonic to biofilm states is marked by multicellular cooperative behavior (Kostakioti et al. 2013). Key environmental factors such as oxygen availability, optimal temperature, nutrient richness, and surface hydrophobicity act as catalysts for biofilm initiation. Mature biofilms often adopt mushroom-like architecture (Sauer et al. 2022), though their structural diversity is vast.

Among the factors influencing biofilm formation, the attachment surface is paramount. Textiles, owing to their porosity and hydrophobicity, serve as efficient substrates for biofilm development. This characteristic underscores their role in nosocomial infections (hospital-acquired infections) (Dixit et al. 2023) and highlights their potential in sports and industrial applications. Factors affecting biofilm formation are broadly categorized as textile-based, environment-based (Mitchell et al. 2015), and microbe-based. A foundational focus on textile-based factors is critical, as these directly or indirectly govern the "interaction process" between microbes and surfaces. Since adhesion is the initial step of biofilm formation (Donlan 2001), strategies for biofilm removal must prioritize understanding physicochemical factors (e.g., surface charge, chemical composition) and constructional factors (e.g., porosity, roughness) (Ivankovic et al. 2022).

The subsequent section of the chapter explores the dual nature of biofilms on textiles, addressing their benefits and challenges across environmental, healthcare, and daily wear contexts. This is followed by an analysis of current anti-biofilm strategies aimed at producing germ-free, durable, and functional fabrics (Dixit et al. 2023). Emerging advancements such as small-scale innovations and methods to enhance antibacterial and anti-biofilm efficacy are also discussed. Next-generation textiles, including antimicrobial, self-cleaning, and biodegradable variants, are poised to revolutionize diverse sectors with their unique properties (Caschera et al. 2019).

7.2 Mechanism

To effectively design strategies for eradicating and preventing biofilms, a thorough understanding of bacterial behavior is critical. Fabric hydrophobicity plays a dominant role in governing the bacterial life cycle across different textile types (Møllebjerg et al. 2021). Textiles act as microbial vectors, significantly contributing to infection occurrence and transmission in diverse contexts from everyday items like towels to hospital-acquired infections linked to medical gowns and attire (Dixit et al. 2023). Biofilm formation is not limited to extreme environments or specific bacterial strains; fabric type itself is a pivotal determinant (Dixit et al. 2023). A robust theoretical framework for biofilm development is essential to mitigate pathogenic microorganisms while leveraging beneficial microbial communities.

Biofilm initiation and progression constitute a dynamic, multi-stage process (Fig. 7.1). The three-dimensional architecture of biofilms critically influences bacterial survival, transmission, resistance, and persistence. These microbial communities offer comprehensive protection to their inhabitants, enabling resilience across diverse ecological niches. The baseline strategy for biofilm formation on any surface can be distilled into five key stages (Yin et al. 2019):

1. Initial attachment: Planktonic cells adhere to the surface.
2. Colonization: Hydrophobic interactions stabilize microbial aggregation.
3. Development: Extracellular polymeric substance (EPS) matrix formation reinforces structural integrity.
4. Maturation: Nutrient acquisition and protective mechanisms are optimized.
5. Active dispersal: Critical for biofilm dissemination and ecological spread.

The persistent propagation and redistribution of biofilms render them resistant to extreme biotic and abiotic stressors. While bacterial strategies on textile surfaces may vary, the overarching mechanisms align with the principles above. Bacterial adhesion to textile fibers can induce surface modification, alteration, biofilm establishment, or biodegradation. Notably, clothing and textiles can create microenvironments that favor the proliferation of opportunistic skin-associated microorganisms (Catovic et al. 2022).

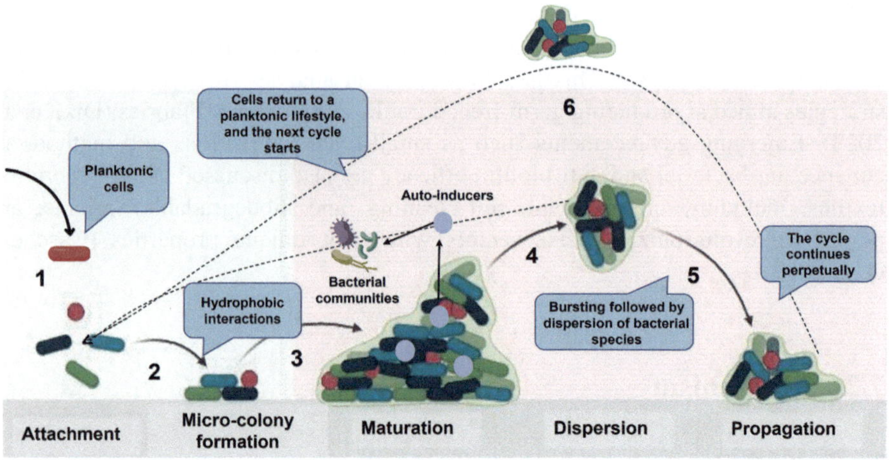

Fig. 7.1 Schematic representation of biofilm development stages on surfaces. (1) Reversible attachment: Planktonic cells adhere to surfaces via weak interactions (e.g., van der Waals forces). (2) Irreversible adhesion: Hydrophobic interactions stabilize microbial attachment, forming micro-colonies. (3) Maturation: Biofilms develop mushroom-shaped structures with extracellular polymeric substances (EPS) matrices. Autoinducers (quorum-sensing molecules) regulate EPS production, enhancing resistance to environmental stressors and antimicrobial agents. (4–6) Dispersion: Biofilm fragments detach and disperse, enabling colonization of new surfaces

Cotton and polyester are widely used fabrics exemplify contrasting hygroscopic properties due to their differing hydrophobicity levels. These materials often stimulate distinct biofilm types. For instance, biofilm formation on cellulose-based fabrics (e.g., cotton) follows a defined sequence: initial attachment, maturation, maintenance, and potential dispersal. Each phase is shaped by microbial species diversity and their metabolic activities, particularly cellulose degradation processes (Møllebjerg et al. 2021).

The process begins when microorganisms colonize the cellulose surface. Initial attachment is facilitated by specific adhesins, such as cellulase proteins associated with bacterial cells (Young et al. 2012). For example, *Clostridium thermocellum* exhibits a binding behavior where cells adhere to localized cellulose regions, forming a slackened monolayer (Dumitrache et al. 2013). This attachment and subsequent microcolony formation represent the foundational steps in biofilm progression.

Maturation is the stage following attachment that involves bacterial propagation and proliferation, culminating in a mushroom-like structure. This architecture not only shelters biofilm communities but also facilitates nutrient acquisition through the production and transport of autoinducers (Dumitrache et al. 2013). These signaling molecules supply essential nutrients and confer protection against environmental fluctuations. A key objective during maturation is the synthesis of extracellular polymeric substances (EPS), which are critical for biofilm structural completion and resistance (Young et al. 2012). Cellulose contributes to the structural firmness and organization of biofilm (Serra et al. 2013).

Once matured, the biofilm enters a maintenance phase to preserve its stability and physiological functionality. The biofilm matrix which is mainly composed of cellulose, polysaccharides, and other components serves as a defense mechanism against antimicrobial agents, drugs, and biotic/abiotic stressors (Agustín et al. 2023). This matrix integrity ensures the biofilm's resilience and operational success (Ziemba et al. 2016). Continuous microbial recruitment sustains the dynamic activity of biofilms (Wang and Newby 2014).

Dispersion is the final stage and is pivotal for biofilm spread and translocation to new sites. This phase is often triggered by environmental stimuli such as oxygen/temperature imbalances, nutrient scarcity, moisture fluctuations, or external disturbances (Dumitrache et al. 2013). Cellulases, enzymes that degrade cellulose, play a central role in dispersing the biofilm by enabling bacterial transition from a communal lifestyle to a planktonic state. These dispersed cells, already resistant, can seed new, robust biofilms (Mcmanus and Schorey 2023).

7.3 Life Cycle of a Bacterial Biofilm on Fabric

The Extended Derjaguin, Landau, Verwey, and Overbeek (XDLVO) theory provides the primary mechanism governing planktonic cell attachment to textile surfaces (Wang and Newby 2014). This framework posits that bacterial species adhere more effectively to hydrophobic textiles, such as cotton, due to interfacial energy interactions. Bacterial biofilm formation on textiles is not a linear process but occurs cyclically as fabrics are worn, dried, washed, and reworn repeatedly.

The lifecycle begins when axillary bacteria are introduced into the fabric during activities such as heavy sweating, exercise, hyperthermia, or seasonal changes (Møllebjerg et al. 2021). Sweat absorption into the textile facilitates bacterial suspension and subsequent adherence which is the critical first step in biofilm development. Once adhered, bacterial growth progresses, though nutrient availability often limits biofilm expansion. Apocrine sweat serves as a primary nutrient source for biofilms (Shelley et al. 1953), while eccrine sweat contributes water, electrolytes, and carbohydrates (Robinson and Robinson 1954). Sebum is a lipid-rich secretion which further supplies triglycerides, fatty acids, and cholesterol esters (Stefaniak and Harvey 2006).

When sweating ceases, sweat absorption diminishes, and evaporation reduces bacterial load. However, textiles that retain moisture and resist complete drying enable irreversible bacterial adherence, fostering biofilm formation. This not only leads to textile discoloration but also poses significant challenges for laundering industries (Mohana et al. 2013). Established biofilms act as substrates for secondary colonization by other axillary bacterial species, amplifying microbial complexity (Fig. 7.2).

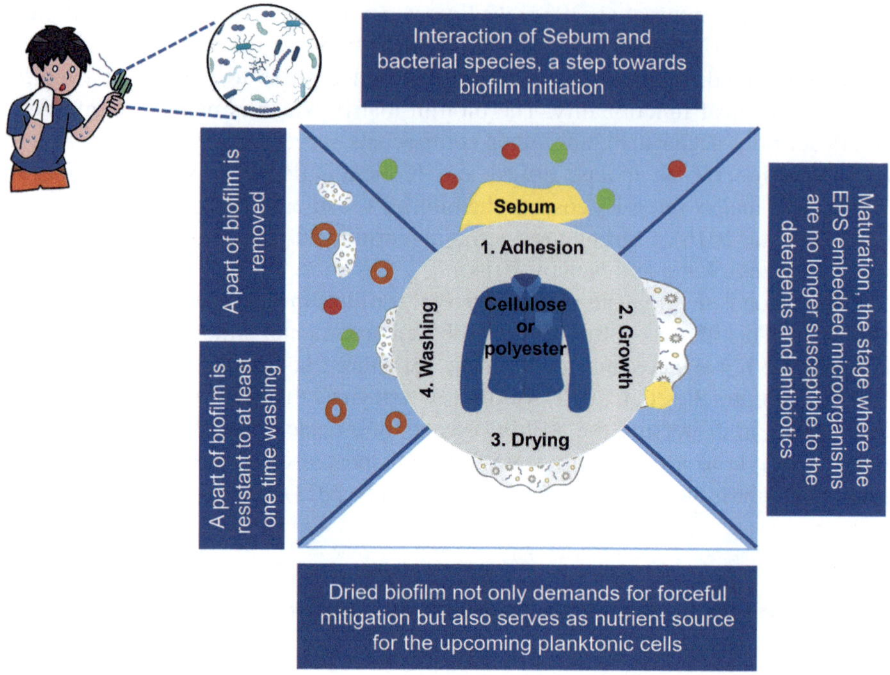

Fig. 7.2 Biofilm dynamics on textile surfaces

In healthcare settings, hospital textiles such as gowns, bed linens, and curtains can serve as potent vectors for infectious disease transmission. Studies reveal a strong correlation between skin microbiota and bacteria colonizing textile fibers (Dixit et al. 2023). Conventional laundering often fails to eliminate biofilm-associated pathogens, as some bacteria migrate between fibers and persist post-wash. This underscores the growing concern about textiles as reservoirs of contamination in clinical environments.

Addressing these challenges requires rigorous hygiene protocols and infection control measures. Establishing effective standards for healthcare textiles demands a holistic evaluation of contributing factors, including fabric properties, microbial ecology, and laundering efficacy (Dixit et al. 2023). Figure 7.3 summarizes this cyclical biofilm lifecycle.

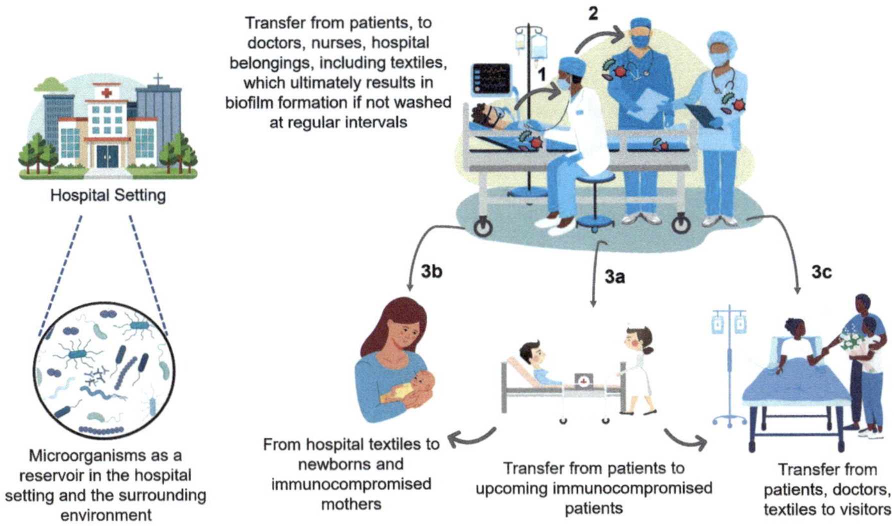

Fig. 7.3 Transmission pathways of nosocomial pathogens via contaminated textiles

7.4 Factors Affecting Biofilm Formation

Biofilm formation on textile surfaces is influenced by a diverse array of factors, primarily categorized into three domains: textile material characteristics, environmental conditions, and microbial traits. Among these, textile-based factors are pivotal due to their direct impact on fabric-microbe interactions. The composition of textile fibers plays a crucial role, as natural fibers like cotton which are hydrophilic by nature can absorb moisture, creating a microenvironment conducive to microbial proliferation. In contrast, synthetic fibers such as polyester and nylon, with their hydrophobic properties, often reduce microbial adhesion. Surface roughness further amplifies biofilm development, as textured fabrics provide micro-niches and increased surface area for microbial attachment (Varshney et al. 2021).

Environmental conditions, including pH and oxygen availability, also significantly shape biofilm dynamics. For instance, *Staphylococcus aureus* demonstrates enhanced growth in hypoxic or carbon dioxide-rich environments, underscoring the role of species-specific adaptations (Liu et al. 2020). Microbial traits, such as the production of extracellular polymeric substances (EPS), further enhance biofilm resilience by stabilizing bacterial adhesion and conferring resistance to cleaning processes (Toyofuku et al. 2016). A nuanced understanding of these interconnected factors is essential not only for designing strategies to mitigate biofilm formation in healthcare and industrial textiles but also for innovating materials with tailored anti-adhesive or antimicrobial properties.

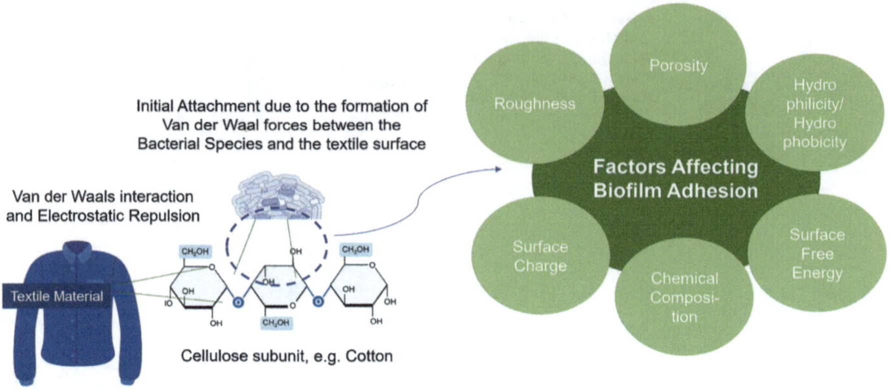

Fig. 7.4 Factors governing bacterial adhesion to textile fibers

7.4.1 Textile-Based Factors Influencing Biofilm Formation

The initial attachment of microorganisms to textile surfaces that is a critical step in biofilm development is greatly influenced by the presence of conditioning films (Goel et al. 2021). Textiles encompass a broad range of materials, from natural fibers like cotton, linen, hemp, jute, wool, and silk to synthetic fibers such as polyester, polyamide, polypropylene, and polyacrylonitrile, each with distinct physicochemical properties. Conditioning films modify surface characteristics (e.g., hydrophobicity, charge), enhancing microbial adhesion (Varshney et al. 2021). Once adhered, microorganisms secrete extracellular polymeric substances (EPS) to stabilize their attachment and establish a scaffold for biofilm maturation (Toyofuku et al. 2016). Understanding these early-stage interactions is pivotal for devising strategies to disrupt biofilm resilience (Fig. 7.4). The following sections detail key textile-based factors governing this process.

7.4.1.1 Hydrophilicity/Hydrophobicity

The hydrophilicity or hydrophobicity of textiles which is a core physicochemical property and plays a contested yet central role in bacterial adhesion. While some studies assert that hydrophilic materials favor microbial attachment (Hemmatian et al. 2021), others argue that hydrophobic surfaces promote stronger bacterial adhesion (Koziarz and Yamazaki 1998). For instance, hydrophobic synthetic fibers (e.g., polyester, polypropylene) exhibit higher bacterial retention compared to hydrophilic natural fibers like cotton (Koziarz and Yamazaki 1998). This phenomenon is partly explained by the Extended Derjaguin, Landau, Verwey, and Overbeek (XDLVO) theory, which posits that hydrophobic surfaces lower the energy barrier for bacterial attachment. Hydrophobic textiles also absorb sebum; a lipid-rich secretion containing triglycerides and fatty acids which serves as a nutrient source for colonizing bacteria such as *Escherichia coli* and *Staphylococcus hominis*, accelerating microcolony formation (Møllebjerg et al. 2021).

Conversely, modifying cotton to enhance hydrophilicity (e.g., via thermo-responsive microgels) reduces bacterial adhesion by forming a hydration layer that physically blocks microbial attachment (Gu et al. 2019). These findings underscore a dual dynamic: hydrophobic surfaces promote adhesion through nutrient retention and reduced energy barriers, while hydrophilic surfaces resist adhesion via hydration-driven repellency.

7.4.1.2 Surface Free Energy

Surface free energy (SFE) refers to the excess energy present at the outermost layer of a material compared to its bulk, arising from unbalanced intermolecular forces between surface atoms or molecules. This energy plays a pivotal role in bacterial adhesion to textile fibers by governing interfacial interactions such as Lifshitz-van der Waals forces and electron donor-acceptor exchanges. Hydrophilic fibers (e.g., cotton) typically exhibit higher SFE, while hydrophobic fibers (e.g., polyester) have lower SFE. However, the relationship between SFE and microbial adhesion is not linear. For example, modifying cellulose fibers with polydimethylsiloxane (PDMS) reduced their SFE from 52 mJ/m^2 to 29 mJ/m^2, significantly inhibiting *Escherichia coli* adhesion due to "passive antimicrobial activity." Further reduction to 14.5 mJ/m^2 using PDMS combined with perfluorooctyltriethoxysilane amplified this effect (Vilcnik et al. 2009). Conversely, *Pseudomonas putida* adhered rapidly to hydrophilic cellulose fibers with SFE >50 mJ/m^2, demonstrating that hydrophilicity alone does not universally deter bacterial attachment (Rochex et al. 2004). These contrasting findings underscore the complexity of SFE's role, which depends on microbial species, environmental conditions, and synergistic interactions with other physicochemical properties like hydrophobicity.

7.4.1.3 Surface Charge

The zeta potential is a measure of the electrical potential at the fiber-liquid interface and directly influences microbial adhesion by dictating electrostatic interactions between textile surfaces and bacterial cells. Surface charge is determined by the molecular and supramolecular structure of fibers, adsorbate concentration, and medium composition. For instance, *Pseudomonas aeruginosa* exhibited stronger adhesion to polyamide fibers (moderate hydrophobicity) than to highly hydrophobic polyester fibers in a negatively charged medium. However, increasing ionic strength reversed the zeta potential of polyamide from negative to positive, enhancing adhesion through electrostatic attraction to negatively charged bacterial cells. Polyester, retaining its negative charge under the same conditions, repelled bacteria (Roy et al. 2019). This highlights how surface charge modulation (e.g., via ionic strength) can override hydrophobicity's effects, emphasizing the need to consider electrostatic forces in biofilm prevention strategies.

7.4.1.4 Chemical Composition

The chemical composition of textile fibres integrates hydrophobicity, surface charge, and SFE into a cohesive framework. Natural fibres like cellulose (cotton) are hydrophilic due to hydroxyl groups that interact with water, while synthetic fibers like polyester and polypropylene are hydrophobic owing to aromatic benzene rings, high crystallinity, and chemical inertness (Bawn 1987). According to Derjaguin, Landau, Verwey, and Overbeek (DLVO) theory, hydrophilic cellulose fibers should theoretically resist biofilm formation due to strong hydration layers. However, their high SFE and porosity often counteract this advantage, enabling microbial colonization. This paradox illustrates how chemical composition alone cannot predict biofilm susceptibility; instead, it interacts dynamically with structural and environmental factors.

7.4.1.5 Roughness

Nanoscale surface roughness amplifies bacterial adhesion by increasing the effective surface area and creating micro-niches for microbial anchoring. For example, cotton fibers with inherently rough surfaces showed higher *E. coli* adhesion compared to smoother polyester and polyester-cotton blends (Bajpai et al. 2011). Similarly, viscose fibers, characterized by high nano-roughness, exhibited maximal bacterial binding, while silk with the lowest roughness demonstrated minimal adhesion (Varshney et al. 2021). Silk's resistance is attributed to its smooth surface and inherent antimicrobial peptides (Krasowska and Sigler 2014), underscoring how roughness interacts with material-specific properties to modulate biofilm formation.

7.4.1.6 Porosity

Porosity is defined by the size, distribution, and volume of voids within textiles. Porosity profoundly impacts biofilm dynamics. Porous materials, such as loosely woven fabrics, mimic medical implants where high porosity correlates with increased infection risk (Katsikogianni and Missirlis 2004). Bacteria preferentially adhere to pores matching their size (1–5 μm), as these cavities enhance contact points and mechanical anchoring. Conversely, larger pores (>10 μm) behave like flat surfaces, offering fewer adhesion sites (Edwards and Rutenberg 2001). Manufacturing techniques (e.g., knitting density, fiber blending) can modulate porosity, providing a tunable parameter for designing biofilm-resistant textiles (Hemmatian et al. 2021).

7.4.2 Environmental Factors Influencing Biofilm Formation

Alongside conditioning film factors, various environmental variables enhance or diminish bacterial growth. Key factors include temperature, pH, moisture, humidity, extracellular polymeric substances (EPS), oxygen availability, and nutrient accessibility. For instance, *Pseudomonas aeruginosa*, a prominent biofilm-forming bacterium, exhibits distinct structural adaptations under varying oxygen conditions. Under normal oxygen levels, it forms mushroom-shaped biofilms composed of rod-shaped cells interconnected by nutrient channels. Conversely, under anoxic conditions, it develops a three-dimensional mesh-like structure of elongated filamentous cells, with gaps separating colonies (Hamada et al. 2014).

The role of moisture in biofilm development has been previously discussed. In a study investigating *Salmonella enterica* serotype Kentucky, biofilm formation was assessed across temperatures of 4, 10, 25, 37, and 42 °C. Results revealed robust biofilm growth between 25–42 °C, with slower progression at lower temperatures. The same study highlighted pH sensitivity; optimal biofilm maturation for *S. enterica* occurred within a pH range of 6.0–8.0 (Roy et al. 2021).

EPS comprises diverse molecules, including carbohydrates, lipids, proteins, and nucleic acids. For example, in *Staphylococcus aureus* biofilms, controlled cell lysis mediated by *cidA* facilitates the release of genomic DNA, which serves as a critical structural component (Flemming et al. 2007). Similarly, environmental DNA (e-DNA) is integral to *P. aeruginosa* biofilms, functioning as a scaffold that connects planktonic cells and stabilizes the matrix. The release of e-DNA in *P. aeruginosa* is governed by quorum-sensing systems.

7.4.3 Microbial Factors Influencing Biofilm Formation

Microbial factors such as bacterial appendages, cell-to-cell communication, EPS production, and inoculum size significantly influence biofilm dynamics. Surface attachment, the initial step in biofilm formation, has been extensively studied in *Pseudomonas aeruginosa*, a model organism for such processes. Bacterial cells reversibly attach to surfaces via polar appendages, a transient phase termed reversible attachment (O'toole and Kolter 1998). Subsequent interactions mediated by surface proteins transition this into irreversible attachment (Toyofuku et al. 2016).

Cell-to-cell communication, particularly quorum sensing (QS), regulates EPS production and virulence expression in mature biofilms. QS enables bacterial populations to coordinate structural and functional adaptations, often resulting in complex biofilm architectures. For example, *P. aeruginosa* employs QS systems to regulate swarming motility and extracellular DNA release, both of which directly influence biofilm morphology (Toyofuku et al. 2016). These signaling mechanisms underscore the interplay between microbial behavior and biofilm structural integrity.

7.5 Challenges of Biofilm Formation on Textile Surfaces in Healthcare, Daily Use and Industrial Sector

Biofilm formation on textile surfaces presents complex challenges across healthcare, daily use, and industrial sectors. In healthcare settings, biofilms on hospital textiles—such as gowns, bed linens, surgical drapes, and fabrics in contact with human skin—facilitate pathogen transmission, contributing to hospital-acquired infections (HAIs) and reducing the efficacy of disinfecting agents (Hu et al. 2015). In daily use, biofilms on clothing and household fabrics cause malodors, stains from pigmented microbial compounds, and reduced textile lifespan (Mohana et al. 2013). Similarly, in industrial contexts, biofilm proliferation on textiles used in manufacturing, processing, and transportation compromises material performance, cleanliness, and durability (Kim et al. 2023). Addressing these challenges requires sector-specific strategies, including advanced antimicrobial treatments, optimized hygiene protocols, and innovative material engineering (Ng et al. 2021).

7.5.1 Challenges in the Healthcare Sector

Biofilm formation on healthcare textiles poses significant risks due to the interplay between microbial communities and material properties. Persistent contamination from biofilms elevates the likelihood of healthcare-associated infections (HAIs), driven by the inherent characteristics of textiles. Natural fibers like cotton and wool, despite their hydrophobicity, exhibit hygroscopic properties that retain moisture, promoting bacterial adhesion and biofilm development. Surface texture, moisture retention capacity, and bacterial specificity further influence this affinity (Ivankovic et al. 2022). For example, *Pseudomonas aeruginosa* adheres robustly to textiles, degrading fibers and posing health risks (Pietrzak et al. 2016).

Environmental conditions in healthcare facilities, such as elevated humidity and temperature, exacerbate biofilm proliferation. Moisture-rich environments—common in bedding, uniforms, and patient care textiles—not only sustain bacterial growth but also enhance biofilm resistance to cleaning agents. This resilience stems from extracellular polymeric substances (EPS), which shield biofilms from antimicrobial treatments (Funari and Shen 2022; Agustín et al. 2023).

Biofilms on healthcare textiles often harbor multidrug-resistant pathogens. Coagulase-negative staphylococci (CoNS), prevalent in hospital environments, form persistent biofilms linked to HAIs (Kalantar-Neyestanaki et al. 2023). Comparative studies reveal that cotton and polyester fabrics support *Candida albicans* biofilm growth more effectively than hard surfaces like glass or metal, complicating infection control (Traoré et al. 2002). Traditional disinfection methods frequently fail to penetrate biofilms, necessitating novel antimicrobial coatings and surface modifications (Ledwoch et al. 2021; Maillard and Centeleghe 2023). Current research emphasizes materials engineered to resist microbial adhesion and disrupt biofilm viability (Ng et al. 2021).

7.5.2 Challenges During Daily Usage of Fabrics

During perspiration, skin-associated bacteria migrate into textiles alongside sweat, leading to malodor and discoloration. Bacterial colonies embedded in biofilms resist routine laundering, diminishing textile quality and user comfort. Sweat-induced sebum secretion releases compounds that, upon microbial breakdown, generate volatile organic compounds (VOCs) responsible for persistent odors (Møllebjerg et al. 2021). Irreversibly adhered bacteria retain malodorous compounds and pigments, causing irreversible staining (Mohana et al. 2013). Over time, biofilm-associated metabolic activity degrades fibers, reducing tensile strength and fabric durability (Sun 2016; Dixit et al. 2023).

Winter wear garments, which are infrequently laundered and rarely exposed to sunlight, face heightened risks. Prolonged storage under humid conditions promotes microbial proliferation, accelerating fabric degradation (Gulati et al. 2022). These challenges are particularly acute in high-performance textiles used in sports and healthcare, where durability and hygiene are critical.

7.5.3 Challenges in the Industrial Sector

Industrial biofilms clog textile filters, reduce permeability, and cause biofouling, increasing operational costs and impairing processes like water filtration and food processing (Kim et al. 2023). Contaminated food-contact textiles facilitate bacterial transfer to products, posing global public health risks. While biodegradable polymers like polybutylene succinate (PBS), polylactide (PLA), and polypropylene (PP) are used in food packaging, biofilm formation on these materials can compromise functionality. For instance, zinc nanoparticles (ZnNPs) and silver nanoparticles (AgNPs) have demonstrated efficacy in inhibiting biofilm formation by Gram-positive and Gram-negative bacteria on polyester (Fontecha-Umaña et al. 2020). Contaminated packaging materials degrade food quality, taste, and shelf life, while also posing infection risks (Pleva et al. 2021).

7.6 Applications Focusing on Biofilm Removal from Textile Materials

Textiles are a fundamental necessity across human activities, whether in daily use, healthcare (e.g., surgical gowns, patient attire), or industrial applications (e.g., filtration, packaging). Their ubiquity underscores the critical need to address biofilm formation, as textiles often serve as conditioning films that facilitate microbial colonization, posing risks of disease transmission and material degradation (Kalantar-Neyestanaki et al. 2023; Kim et al. 2023). Figure 7.5 outlines the strategies discussed

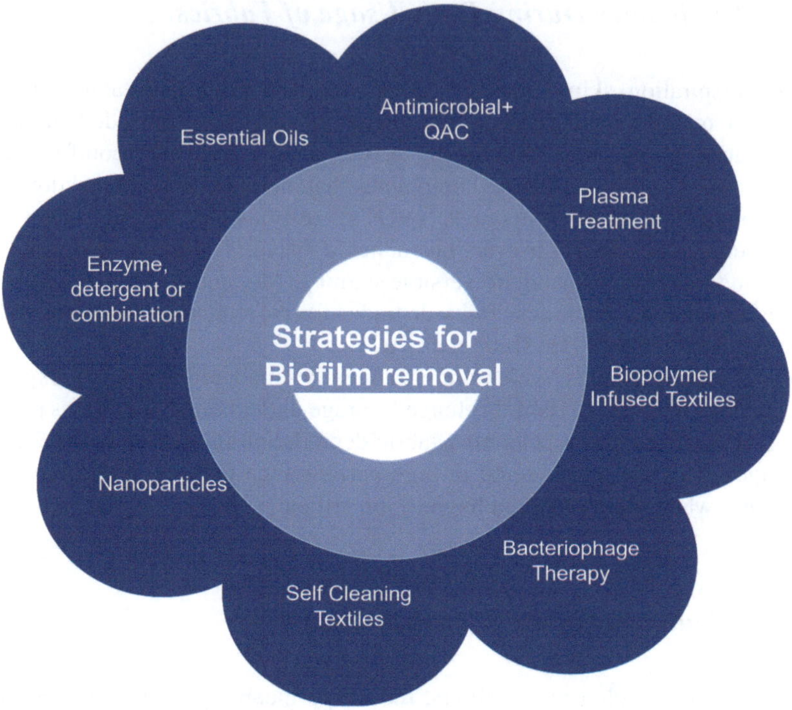

Fig. 7.5 Interdisciplinary strategies for biofilm mitigation in textiles

below, which broadly fall into two categories: preventive measures and treatment approaches. Treatments can further be classified as biostatic (inhibiting microbial growth) or biocidal (killing microbes). Conventional methods for biofilm removal include chemical agents, detergents, and enzymes (Agustín et al. 2023). However, many of these approaches have limitations, such as compromising textile durability, elasticity, or structural integrity (Kim et al. 2023).

Antimicrobial agents like nanoparticles (e.g., silver, zinc, copper), quaternary ammonium compounds (QACs), polybiguanides, N-halamines, chitosan, triclosan, and bioactive plant-derived products (BOA's) are widely used for textile treatment (Agustín et al. 2023). Basic biofilm removal techniques often involve detergents or enzymes. However, studies reveal that mature biofilms resist elimination through single washing cycles, highlighting the need for advanced solutions (Møllebjerg et al. 2021).

Enzymes such as proteases and cellulases destabilize biofilms by degrading extracellular polymeric substances (EPS) and microbial adhesins. For example, proteases and cellulases demonstrated significant antibiofilm activity against

Pseudomonas fluorescens biofilms, except for protease Polarzyme (Molobela et al. 2010). Enhanced efficacy is observed when enzymes are combined with detergents. A study targeting *P. aeruginosa* and *P. fluorescens* biofilms reported synergistic effects, achieving near-complete biofilm removal through combined enzymatic and detergent treatments (Agustín et al. 2023).

Healthcare textiles require specialized strategies to prevent biofilm-associated infections. Plasma-treated textile materials, for instance, effectively removed *Staphylococcus aureus* biofilms by disrupting microbial adhesion (Fu et al. 2020). Silver-based interventions are particularly promising: silver-impregnated dressings eradicated *S. aureus* biofilms within 48 h (Toy and Macera 2011), while silver-coated gelling fibers eliminated *P. aeruginosa* biofilms in 3 days (Davis et al. 2019). Similarly, silver-coated polyethylene terephthalate (PET) fabrics reduced *Staphylococcus epidermidis* biofilm formation by 31% compared to untreated controls (Klueh et al. 2000).

Durability is critical for clinical applications. A recent study demonstrated that cotton fabrics pretreated with silver nanoparticles (AgNPs) retained antibacterial efficacy against six pathogens including *Escherichia coli*, *Klebsiella pneumoniae*, *Acinetobacter baumannii*, *P. aeruginosa*, methicillin-resistant *S. aureus* (MRSA), and *Enterococcus faecalis* even after 20 washing cycles, confirming long-term functionality (Tahir et al. 2024).

Combination therapies amplify antibiofilm efficacy. For example, sulfonamide-functionalized quaternary ammonium compounds (sulfonamide-QACs) applied to fabrics (e.g., sportswear, medical curtains) inhibited both planktonic and biofilm forms of *E. coli*, *S. aureus*, and *Arthrobacter* species (Caschera et al. 2019). Similarly, cotton fabrics coated with copper nanoparticles grown on graphene sheets exhibited 99% microbial eradication against *E. coli*, *Micrococcus luteus*, and *Corynebacterium xerosis*, while remaining non-cytotoxic, making them suitable for healthcare and sportswear applications (Kim et al. 2023).

Essential oils like thyme, oregano, and carvacrol have demonstrated potent antimicrobial and antibiofilm activity. A study on polystyrene surfaces treated with these oils suppressed *Listeria monocytogenes* biofilm formation for 24 h, suggesting their potential for textile applications (Desai et al. 2012). These natural compounds offer an eco-friendly alternative to synthetic agents, though their long-term stability on textiles requires further exploration. Table 7.1 summarizes these innovative approaches, emphasizing their mechanisms, applications, and limitations.

Table 7.1 Summary of interventions targeting biofilm formation and removal on textiles

Treatment	Target/ function	Targeted bacterial species	References
Detergent & enzymes	Antibiofilm	*Pseudomonas fluorescens*	Møllebjerg et al. (2021)
Protease & cellulase enzymes	Antibiofilm	*Pseudomonas fluorescens*	Molobela et al. (2010)
Combined enzyme and detergent	Antibiofilm	*Pseudomonas aeruginosa, Pseudomonas fluorescens*	Agustín et al. (2023)
Plasma-treated textile pile debridement material	Antibiofilm	*Staphylococcus aureus*	Fu et al. (2020)
Silver-treated dressing	Antibiofilm	*Staphylococcus aureus*	Toy and Macera (2011)
Silver-treated gelling fibre dressing	Antibiofilm	*Pseudomonas aeruginosa*	Davis et al. (2019)
Silver-coated PET fabric	Antibiofilm	*Staphylococcus epidermidis*	Klueh et al. (2000)
Silver nanoparticles (AgNPs) on surgical instruments and cotton fabric	Antibacterial	*Escherichia coli, Klebsiella pneumoniae, Acinetobacter baumannii, Pseudomonas aeruginosa, MRSA, Enterococcus faecalis*	Tahir et al. (2024)
Sulfonamide QACs (quaternary ammonium compounds)	Both	*Escherichia coli, Staphylococcus aureus, Arthrobacter sp.*	Caschera et al. (2019)
Cu nanoparticles on graphene sheets	Both	*Escherichia coli, Staphylococcus aureus, Pseudomonas aeruginosa, Micrococcus luteus, Corynebacterium xerosis*	Kim et al. (2023)
Plant essential oils (thyme, oregano, carvacrol)	Both	*Listeria monocytogenes*	Desai et al. (2012)

7.7 Future Prospects

The future of biofilm management on textiles lies in leveraging advanced insights into biofilm development to design innovative removal strategies. The primary focus is the creation of antimicrobial coatings using materials such as silver, copper, or zinc oxide nanoparticles, which inhibit microbial adhesion through sustained antimicrobial activity (Kim et al. 2023). Smart coatings represent another frontier; these responsive materials release antimicrobial agents when triggered by environmental stimuli (e.g., moisture, pH changes, or temperature fluctuations), enabling real-time biofilm prevention (Agustín et al. 2023).

Enzyme-based cleaners offer a biodegradable and fabric-friendly alternative. Proteases, lipases, and amylases degrade biofilm components like proteins, lipids, and polysaccharides without compromising textile integrity (Møllebjerg et al.

2021). Photodynamic therapy (PDT) employs light-activated photosensitizers (e.g., porphyrins or chlorophyll derivatives) to generate reactive oxygen species (ROS), disrupting biofilm matrices on textiles (Qi et al. 2019).

Quorum sensing inhibitors (QSIs) are under investigation for their ability to block microbial communication, a critical step in biofilm formation. By interfering with signaling molecules like acyl-homoserine lactones (AHLs) or autoinducer peptides, QSIs prevent biofilm maturation and virulence expression (Toyofuku et al. 2016). Cold plasma treatment, already used in medical implant sterilization, generates reactive oxygen and nitrogen species (RONS) that eradicate biofilms. Adapting this technology for textiles could enable non-thermal, chemical-free sterilization (Matthes et al. 2022).

Bacteriophage therapy introduces viruses that specifically lyse biofilm-forming bacteria. Phages embedded in textile coatings could provide targeted, species-specific antibiofilm action without disrupting beneficial microbiota (Pinto et al. 2020). Self-cleaning textiles with superhydrophobic or lotus-leaf-inspired surfaces repel moisture, preventing bacterial adhesion at the initial stage. These textiles, functionalized with nanomaterials like silica or fluoropolymers, minimize biofilm risk in high-moisture environments (Kong et al. 2023).

Biopolymer-infused textiles, such as chitosan-coated bandages, combine biocompatibility with antimicrobial activity. Chitosan's cationic properties disrupt bacterial membranes and inhibit EPS production, reducing biofilm viability (Onbasli 2020). Biodegradable antimicrobial fabrics are emerging as sustainable alternatives, particularly in healthcare settings where disposable textiles contribute to nosocomial infections. Materials like polylactic acid (PLA) blended with natural antimicrobial agents (e.g., essential oils) balance functionality and environmental impact (Ivankovic et al. 2022).

The integration of nanotechnology, biotechnology, and materials science is pivotal for advancing these solutions. For instance, graphene-copper nanocomposites enhance antimicrobial efficacy while maintaining textile flexibility (Kim et al. 2023), and CRISPR-based gene editing could engineer phages or microbes to degrade biofilms selectively. Collaborative innovation across disciplines will drive the development of eco-friendly, durable, and scalable strategies to mitigate biofilm challenges in textiles.

References

Agustín MDR, Stengel P, Kellermeier M, Tücking K-S, Müller M (2023) Monitoring growth and removal of Pseudomonas biofilms on cellulose-based fabrics. Microorganisms 11:892

Bajpai V, Dey A, Ghosh S, Bajpai S, Jha MK (2011) Quantification of bacterial adherence on different textile fabrics. Int Biodeterior Biodegradation 65:1169–1174

Bawn C (1987) Encyclopedia of polymer science and engineering. Polymer 28:1234

Caschera A, Mistry KB, Bedard J, Ronan E, Syed MA, Khan AU et al (2019) Surface-attached sulfonamide containing quaternary ammonium antimicrobials for textiles and plastics. RSC Adv 9:3140–3150

Catovic C, Abbes I, Barreau M, Sauvage C, Follet J, Duclairoir-Poc C et al (2022) Cotton and flax textiles leachables impact differently cutaneous Staphylococcus aureus and Staphylococcus epidermidis biofilm formation and cytotoxicity. Life 12:535

Davis SC, Li J, Gil J, Head C, Valdes J, Glinos GD et al (2019) Preclinical evaluation of a novel silver gelling fiber dressing on Pseudomonas aeruginosa in a porcine wound infection model. Wound Repair Regen 27:360–365

Desai MA, Soni KA, Nannapaneni R, Schilling MW, Silva JL (2012) Reduction of Listeria monocytogenes biofilms on stainless steel and polystyrene surfaces by essential oils. J Food Prot 75:1332–1337

Dixit S, Varshney S, Gupta D, Sharma S (2023) Textiles as fomites in the healthcare system. Appl Microbiol Biotechnol 107:3887–3897

Donlan RM (2001) Biofilm formation: a clinically relevant microbiological process. Clin Infect Dis 33:1387–1392

Dumitrache A, Wolfaardt G, Allen G, Liss SN, Lynd LR (2013) Form and function of Clostridium thermocellum biofilms. Appl Environ Microbiol 79:231–239

Edwards KJ, Rutenberg AD (2001) Microbial response to surface microtopography: the role of metabolism in localized mineral dissolution. Chem Geol 180:19–32

Flemming HC, Neu TR, Wozniak DJ (2007) The EPS matrix: the “house of biofilm cells”. J Bacteriol 189:7945–7947

Fontecha-Umaña F, Ríos-Castillo AG, Ripolles-Avila C, Rodríguez-Jerez JJ (2020) Antimicrobial activity and prevention of bacterial biofilm formation of silver and zinc oxide nanoparticle-containing polyester surfaces at various concentrations for use. Foods 9:442

Fu Y, An Q, Cheng Y, Yang Y, Wang L, Zhang H et al (2020) A textile pile debridement material consisting of polyester fibers for in vitro removal of biofilm. Polymers 12:1360

Funari R, Shen AQ (2022) Detection and characterization of bacterial biofilms and biofilm-based sensors. ACS Sens 7:347–357

Goel N, Fatima SW, Kumar S, Sinha R, Khare SK (2021) Antimicrobial resistance in biofilms: exploring marine actinobacteria as a potential source of antibiotics and biofilm inhibitors. Biotechnol Rep 30:e00613

Gu P, Fan N, Wang Y, Wang J, Müller-Buschbaum P, Zhong Q (2019) Linear control of moisture permeability and anti-adhesion of bacteria in a broad temperature region realized by cross-linking thermoresponsive microgels onto cotton fabrics. ACS Appl Mater Interfaces 11:30269–30277

Gulati R, Sharma S, Sharma RK (2022) Antimicrobial textile: recent developments and functional perspective. Polym Bull 79:5747–5771

Hamada M, Toyofuku M, Miyano T, Nomura N (2014) cbb3-type cytochrome c oxidases, aerobic respiratory enzymes, impact the anaerobic life of Pseudomonas aeruginosa PAO1. J Bacteriol 196:3881–3889

Hemmatian T, Lee H, Kim J (2021) Bacteria adhesion of textiles influenced by wettability and pore characteristics of fibrous substrates. Polymers 13:223

Hu H, Johani K, Gosbell IB, Jacombs AS, Almatroudi A, Whiteley GS et al (2015) Intensive care unit environmental surfaces are contaminated by multidrug-resistant bacteria in biofilms: combined results of conventional culture, pyrosequencing, scanning electron microscopy, and confocal laser microscopy. J Hosp Infect 91:35–44

Ivankovic T, Rajic A, Ercegovic Razic S, Rolland Du Roscoat S, Skenderi Z (2022) Antibacterial properties of non-modified wool, determined and discussed in relation to ISO 20645:2004 standard. Molecules 27:1876

Kalantar-Neyestanaki D, Mansouri S, Tadjrobehkar O, Isaei E (2023) The frequency of adherence, biofilm-associated, Arginine Catabolic Mobile element genes, and biofilm formation in clinical and healthcare worker coagulase-negative staphylococci isolates. BMC Microbiol 23:222

Katsikogianni M, Missirlis Y (2004) Concise review of mechanisms of bacterial adhesion to biomaterials and of techniques used in estimating bacteria-material interactions. Eur Cell Mater 8:37–57

Kim J, Kang SH, Choi Y, Lee W, Kim N, Tanaka M et al (2023) Antibacterial and biofilm-inhibiting cotton fabrics decorated with copper nanoparticles grown on graphene nanosheets. Sci Rep 13:11947

Klueh U, Wagner V, Kelly S, Johnson A, Bryers J (2000) Efficacy of silver-coated fabric to prevent bacterial colonization and subsequent device-based biofilm formation. J Biomed Mater Res 53:621–631

Kong Q, Ren X, Li Z (2023) Three-dimensional porous structure on cotton fabric through the breath figure method with functions of self-cleaning and oil/water separation. Cellulose 30:3915–3930

Kostakioti M, Hadjifrangiskou M, Hultgren SJ (2013) Bacterial biofilms: development, dispersal, and therapeutic strategies in the dawn of the postantibiotic era. Cold Spring Harb Perspect Med 3:a010306

Koziarz J, Yamazaki H (1998) Immobilization of bacteria on to polyester cloth. Biotechnol Tech 12:407–410

Krasowska A, Sigler K (2014) How microorganisms use hydrophobicity and what does this mean for human needs? Front Cell Infect Microbiol 4:112

Ledwoch K, Magoga M, Williams D, Fabbri S, Walsh J, Maillard JY (2021) Is a reduction in viability enough to determine biofilm susceptibility to a biocide? Infect Control Hosp Epidemiol 42:1486–1492

Liu Y, Zhang J, Ji Y (2020) Environmental factors modulate biofilm formation by Staphylococcus aureus. Sci Prog 103:36850419898659

Maillard JY, Centeleghe I (2023) How biofilm changes our understanding of cleaning and disinfection. Antimicrob Resist Infect Control 12:95

Matthes R, Jablonowski L, Pitchika V, Holtfreter B, Eberhard C, Seifert L et al (2022) Efficiency of biofilm removal by combination of water jet and cold plasma: an in-vitro study. BMC Oral Health 22:157

Mcmanus WR, Schorey JS (2023) Comparison of ultrastructure, extracellular matrix, and drug susceptibility in *M. Avium* subs. *Hominissuis biofilms*. Pathogens 12:1427

Mitchell A, Spencer M, Edmiston C Jr (2015) Role of healthcare apparel and other healthcare textiles in the transmission of pathogens: a review of the literature. J Hosp Infect 90:285–292

Mohana DC, Thippeswamy S, Abhishek RU (2013) Antioxidant, antibacterial, and ultraviolet-protective properties of carotenoids isolated from Micrococcus spp. Radiat Prot Environ 36:168–174

Møllebjerg A, Palmén LG, Gori K, Meyer RL (2021) The bacterial life cycle in textiles is governed by fiber hydrophobicity. Microbiol Spectr 9:e0118521

Molobela IP, Cloete TE, Beukes M (2010) Protease and amylase enzymes for biofilm removal and degradation of extracellular polymeric substances (EPS) produced by Pseudomonas fluorescens bacteria. Afr J Microbiol Res 4:1515–1524

Ng G, Judzewitsch P, Li M, Pester CW, Jung K, Boyer C (2021) Synthesis of polymer brushes Via SI-PET-RAFT for photodynamic inactivation of bacteria. Macromol Rapid Commun 42:e2100106

O'toole GA, Kolter R (1998) Initiation of biofilm formation in Pseudomonas fluorescens WCS365 proceeds via multiple, convergent signalling pathways: a genetic analysis. Mol Microbiol 28:449–461

Onbasli D (2020) Chitosan biopolymer and its anti-biofilm activity. Int J Pharm Sci Rev Res 64:50–54

Pietrzak K, Otlewska A, Puchalski M, Gutarowska B, Guiamet P (2016) Antimicrobial properties of silver nanoparticles against biofilm formation by Pseudomonas aeruginosa on archaeological textiles. Appl Environ Biotechnol 2:1–9

Pinto AM, Cerqueira MA, Bañobre-Lópes M, Pastrana LM, Sillankorva S (2020) Bacteriophages for chronic wound treatment: from traditional to novel delivery systems. Viruses 12:235

Pleva P, Bartošová L, Máčalová D, Zálešáková L, Sedlaříková J, Janalíková M (2021) Biofilm formation reduction by eugenol and thymol on biodegradable food packaging material. Foods 11:2

Qi M, Chi M, Sun X, Xie X, Weir MD, Oates TW et al (2019) Novel nanomaterial-based antibacterial photodynamic therapies to combat oral bacterial biofilms and infectious diseases. Int J Nanomedicine 14:6937–6956

Robinson S, Robinson AH (1954) Chemical composition of sweat. Physiol Rev 34:202–220

Rochex A, Lecouturier D, Pezron I, Lebeault JM (2004) Adhesion of a Pseudomonas putida strain isolated from a paper machine to cellulose fibres. Appl Microbiol Biotechnol 65:727–733

Roy S, Ghosh S, Bhowmick N, Roy Choudhury PK (2019) Role of physicochemical factors on bacterial attachment in textile fibrous media. Water Environ J 33:21–30

Roy PK, Ha AJ, Mizan MFR, Hossain MI, Ashrafudoulla M, Toushik SH et al (2021) Effects of environmental conditions (temperature, pH, and glucose) on biofilm formation of Salmonella enterica serotype Kentucky and virulence gene expression. Poult Sci 100:101209

Sauer K, Stoodley P, Goeres DM, Hall-Stoodley L, Burmølle M, Stewart PS et al (2022) The biofilm life cycle: expanding the conceptual model of biofilm formation. Nat Rev Microbiol 20:608–620

Serra DO, Richter AM, Hengge R (2013) Cellulose as an architectural element in spatially structured Escherichia coli biofilms. J Bacteriol 195:5540–5554

Shelley WB, Hurley HJ Jr, Nichols AC (1953) Axillary odor; experimental study of the role of bacteria, apocrine sweat, and deodorants. AMA Arch Derm Syphilol 68:430–446

Stefaniak AB, Harvey CJ (2006) Dissolution of materials in artificial skin surface film liquids. Toxicol In Vitro 20:1265–1283

Sun G (2016) Antimicrobial textiles. Elsevier, Amsterdam, pp 319–336

Tahir I, Amina SJ, Ahmed NM, Janjua HA (2024) Antimicrobial coating of biologically synthesized silver nanoparticles on surgical fabric and surgical blade to prevent nosocomial infections. Heliyon 10:e35968

Toy LW, Macera L (2011) Evidence-based review of silver dressing use on chronic wounds. J Am Acad Nurse Pract 23:183–192

Toyofuku M, Inaba T, Kiyokawa T, Obana N, Yawata Y, Nomura N (2016) Environmental factors that shape biofilm formation. Biosci Biotechnol Biochem 80:7–12

Traoré O, Springthorpe VS, Sattar SA (2002) A quantitative study of the survival of two species of Candida on porous and non-porous environmental surfaces and hands. J Appl Microbiol 92:549–555

Varshney S, Sain A, Gupta D, Sharma S (2021) Factors affecting bacterial adhesion on selected textile fibres. Indian J Microbiol 61:31–37

Vilcnik A, Jerman I, Surca Vuk A, Kozelj M, Orcl B, Tomsic B et al (2009) Structural properties and antibacterial effects of hydrophobic and oleophobic sol-gel coatings for cotton fabrics. Langmuir 25:5869–5880

Wang H, Newby BM (2014) Applicability of the extended derjaguin-landau-verwey-overbeek theory on the adsorption of bovine serum albumin on solid surfaces. Biointerphases 9:041006

Yin W, Wang Y, Liu L, He J (2019) Biofilms: The microbial "Protective Clothing" in extreme environments. Int J Mol Sci 20:3423

Young JM, Leschine SB, Reguera G (2012) Reversible control of biofilm formation by Cellulomonas spp. in response to nitrogen availability. Environ Microbiol 14:594–604

Ziemba C, Shabtai Y, Piatkovsky M, Herzberg M (2016) Cellulose effects on morphology and elasticity of Vibrio fischeri biofilms. NPJ Biofilms Microbi 2:1

Chapter 8
Antimicrobial Textiles: Latest Developments and Functional Dimensions

Fatima Mujahid, Azalfah Ibrar, Hasan Ejaz, Zeeshan Nawaz, Ahsen Taqveem, Zia Ashraf, Khadeeja Nasir, and Mohsin Khurshid

Abstract During the past few years, extensive research has been conducted on textiles with antimicrobial properties, owing to their indispensable role in diverse applications such as healthcare, medicine, athletic wear, personal protective equipment (PPE), footwear, and medical dressings. A wide range of antimicrobial agents has been utilized for fabricating these textiles, including natural and naturally derived compounds, organic and synthetic reagents (e.g., triclosan, quaternary ammonium compounds [QACs], and nitrogen-halamines), as well as metals and metal oxide nanoparticles. The efficacy and applicability of these agents have been rigorously evaluated across various contexts. Modern processing techniques, such as surface coating, chemical bonding, nanotechnology, and microencapsulation, are employed to integrate these agents into textiles during manufacturing. While synthetic antimicrobial agents demonstrate strong pathogen-neutralizing capabilities and robust adherence to textile substrates, their use raises potential environmental, health, and safety concerns. These challenges have spurred the advancement of next-generation antimicrobial coatings, such as plant extract-derived silver and zinc

F. Mujahid · A. Ibrar · Z. Nawaz · M. Khurshid (✉)
Institute of Microbiology, Government College University Faisalabad, Faisalabad, Pakistan
e-mail: zeeshannawaz@gcuf.edu.pk; mohsinkhurshid@gcuf.edu.pk

H. Ejaz
Department of Clinical Laboratory Sciences, College of Applied Medical Sciences, Jouf University, Sakaka, Saudi Arabia
e-mail: hetariq@ju.edu.sa

A. Taqveem
School of Allied Health Sciences, Faculty of Sciences, Multan University of Science and Technology, Multan, Pakistan

Z. Ashraf
College of Allied Health Professionals, Government College University Faisalabad, Faisalabad, Pakistan
e-mail: ziaashraf@gcuf.edu.pk

K. Nasir
Faculty of Allied Health Sciences, Superior University, Lahore, Pakistan

M. Khurshid et al. (eds.), *Microbial Biotechnology in Textile Innovation*, SDGs and Textiles, https://doi.org/10.1007/978-981-95-0347-6_8

nanoparticles, chitosan-based formulations, and biopolymer-integrated systems, which leverage green synthesis methodologies. These innovations aim to minimize environmental impact while preserving antimicrobial performance and improving the durability of treated textiles.

Keywords Polymers · Surface coating · Chemical bonding · Nanotechnology · Microencapsulation · Environmental impact

8.1 Introduction

Microorganisms are ubiquitous in the environment and can proliferate rapidly under suitable conditions such as moisture, nutrient availability, and optimal temperatures. Their presence on textiles often leads to undesirable effects, including discoloration, odor formation, staining, heightened contamination risks, and reduced mechanical strength (Gao and Cranston 2008). Such microbial contamination poses threats to both living organisms and inanimate materials. A comprehensive understanding of these microorganisms, including their sources, growth mechanisms, and interactions with specific materials is essential for effective management and mitigation of their adverse impacts on textiles, as well as for preserving textile integrity during use and storage. Furthermore, the prevalence of nosocomial infections in healthcare settings and surface contamination by microorganisms, particularly in textiles that directly contact the human body, underscores the critical importance of antimicrobial finishing (Szostak-Kotowa 2004).

Historically, clothing has evolved from a basic survival tool to a cornerstone of human society. The interplay between microbes and textiles holds profound significance across diverse fields, including palaeo-history, medicine, fashion, and public health, with wide-ranging implications (Sanders et al. 2021). Ancient civilizations pioneered the use of natural antimicrobial agents on textiles. For example, Egyptians employed spices and herbal coatings on mummy wraps, while the Chinese integrated bamboo fibers for their inherent antimicrobial compound "bamboo kun" into housing structures. These fibers exhibit natural antibacterial and antifungal properties, attributed to bioactive compounds such as 2,6-dimethoxy-p-benzoquinone. The mid-twentieth century saw the adaptation of antibiotics, initially developed during World War II, for textile preservation. Military fabrics like tarpaulins, tents, and truck covers required protection against microbial degradation in harsh environments to prevent fiber damage and infections. Treatments involving antimony salts, chlorinated waxes, and copper compounds were employed, yielding stiffened fabrics with distinct functional properties (Ahmad et al. 2020).

In contemporary applications, antimicrobial agents are widely used to counteract textile degradation, including staining, odor generation, color fading, and associated health hazards. Durable antimicrobial finishes or the integration of biocides into synthetic fibers during manufacturing processes have proven effective in mitigating these issues (Afraz et al. 2019; Abdul-Reda Hussein et al. 2023). Consequently, managing microbial risks has become a global priority, driven by the escalating prevalence and diversity of microbial threats—from localized outbreaks to pandemics such as the plague, SARS, COVID-19, cholera, HIV/AIDS, Ebola, and others. While frequent laundering remains the most reliable method to reduce microbial load on textiles, this approach is often impractical in hospitals due to continuous use. Antimicrobial textiles thus offer significant advantages for high-risk environments, including healthcare, sanitary services, and sewage management.

Advanced surface modification techniques, such as plasma treatment, microencapsulation, polymerization, sol-gel methods, nanotechnology, and electrospinning, have enabled the development of textiles with multifunctional properties, including water repellency, flame resistance, and antibacterial efficacy (Nadi et al. 2018).

The demand for antimicrobial textiles has surged exponentially in recent years, with the market valued at USD 10.7 billion in 2021 and projected to grow by 50% by 2026, reflecting a compound annual growth rate (CAGR) of 6.5%. These textiles are categorized based on criteria such as efficacy, mechanism of action, durability, toxicity, and fiber composition. Key categories include wearable and wash-resistant textiles, textiles composed of natural fibers such as cotton, silk, wool, and linen, as well as synthetic fibers like polyethylene (PE), polyester (PES), and polypropylene (PP). Blended materials, including cotton-PES, cotton-elastane, and wool-acrylic combinations, are also widely utilized. Further classifications encompass biocidal, biostatic, and antibiofilm textiles, textiles engineered to release bioactive compounds, and those featuring either leaching or bound antimicrobial finishes (Gulati et al. 2022).

The primary objective of this chapter is to provide an extensive summary of synthetic, chemical, and natural antimicrobial agents currently available in the market for textile applications, along with the advanced treatment methodologies employed to develop antimicrobial textiles. This chapter emphasizes the functional dimensions of various antimicrobial textile types, addressing their unique properties and applications. Additionally, it explores the latest advancements in the field, including innovations in material science, sustainable production techniques, and the integration of multifunctional finishes. By analyzing these developments, the chapter aims to shed light on emerging trends that enhance antimicrobial efficacy, durability, and environmental compatibility, ultimately contributing to safer and more sustainable textile solutions in healthcare, consumer goods, and industrial settings.

8.2 Bacterial Growth Mechanism on Textile Fabrics

Understanding the growth mechanisms of bacteria on textile fabrics is critical for effectively preventing their proliferation. Textiles, garments, and apparel are typically composed of synthetic or natural fibers, both of which consist of polymers. These polymers which are synthesized from either chemical or natural monomers depending on the fiber type provide a surface for airborne bacteria to adhere to and act as a nutrient source that facilitates bacterial colonization and growth (Puvvada et al. 2019; Roy et al. 2019). Notably, bacterial species such as *Flavobacterium*, *Micrococcus*, and *Pseudomonas* (particularly *Pseudomonas aeruginosa*) are among the most prevalent microorganisms colonizing textiles (Gutarowska et al. 2017).

The absorption of sweat and atmospheric moisture by textile fibers further accelerates bacterial proliferation by creating ideal humid conditions. Once established, bacterial colonies metabolize polymers as nutrients, slowing the conversion of polymers to monomers within the fabric and releasing bacterial metabolites. These metabolites, produced during bacterial activity, play a critical role in degrading the fiber's polymeric structure (Gupta et al. 2017; Varshney et al. 2020). Consequently, the synergistic interaction between bacteria and their metabolites contributes to polymer degradation, fabric deterioration, and the unchecked expansion of bacterial colonies (Lamba et al. 2017). When pathogenic bacteria are involved, this process can lead to skin irritation or infections in humans (Mondal and Saha 2019).

Natural fibers are more vulnerable to bacterial invasion than synthetic fibers, as their polymers provide abundant carbohydrates that support microbial growth. In contrast, synthetic polymers derived from chemical processes are less hospitable to bacterial cells (Andra et al. 2021). Therefore, preventing bacterial adhesion to polymeric fabrics is essential to curbing their proliferation. Integrating antibacterial agents into these polymers offers an effective strategy to inhibit bacterial growth and preserve textile integrity.

8.3 Challenges Posed by Antimicrobial Agents

Antimicrobial agents exhibit diverse modes of action. Most function through leaching or migration from the application surface. While effective, leaching technologies can compromise durability and product lifespan and introduce risks when incorporated into garments, such as adverse effects on skin microbiota, potential skin rashes, irritations, and breaches of the skin barrier (Di Martino 2022; Amrhar et al. 2025). Other antibacterial agents operate by inducing protein coagulation or physically extracting microbes, directly targeting and eradicating pathogens. However, their efficacy is often short-lived (Mousavi Khaneghah et al. 2018; Ke et al. 2021).

When antimicrobial agents are attached to textile surfaces or integrated into fibers, their effectiveness diminishes significantly due to restricted availability. Biocides may also degrade gradually during routine textile use and laundering, further reducing their longevity (Gulati et al. 2022; Tanasa et al. 2023). Although chemical disinfectants effectively inhibit bacterial growth, their well-documented environmental and human health risks have driven research toward advanced antibacterial solutions that suppress microbial proliferation without fostering resistance or endangering humans and ecosystems. Consequently, high concentrations of biocides must be applied to textiles to balance efficacy and durability.

Common antimicrobial agents used in textiles include metals, halogenated phenols, quaternary ammonium compounds (QACs), metal-organic complexes, and

N-halamine compounds (Emam 2019; Tanasa et al. 2023). QACs, copper, and silver are potent active agents but often leach from fabrics. Conversely, agents like Chitosan exhibit lower efficacy and are prone to removal after two to three wash cycles. Metals and metallic salts integrated into fabrics via ionization processes provide longer-lasting antimicrobial effects than chemical treatments, though this method increases production costs by 20–30% (Bibi et al. 2024).

Environmental concerns further complicate the use of certain agents. For example, triclosan, when washed into aquatic systems, resists integration and accumulates in terrestrial and aquatic environments, obstructing oxygen absorption and disrupting ecosystems. Such risks have prompted stringent regulatory measures to govern the application of these compounds (Dar et al. 2022).

8.4 Synthetic/Chemical Antimicrobial Agents

Antimicrobials encompass a diverse array of chemical compounds and physical agents that interact with microorganisms, including bacteria, fungi, viruses, and protozoa. Their primary function is to eradicate these microorganisms or inhibit their proliferation. However, many synthetic/chemical antimicrobials exhibit significant drawbacks such as the rapid development of microbial resistance shortly after their introduction and various undesirable side effects that can limit or entirely preclude their use. Furthermore, chemical biocides may pose risks to environmental integrity and human health if improperly managed or utilized (Tanasa et al. 2023).

8.4.1 Triclosan

Triclosan (2,4,4′-trichloro-2′-hydroxydiphenyl ether), a chlorinated phenolic compound, is widely employed as an antimicrobial and preservative agent in products such as disinfectants, toothpaste, personal care items, soaps, deodorants, shampoos, and textiles (Yueh and Tukey 2016). It exhibits broad-spectrum antiseptic and disinfectant properties, effectively targeting both gram-negative and gram-positive bacteria. Triclosan demonstrates biocidal activity against eukaryotic and prokaryotic cells in broth and has been extensively used for personal hygiene and disinfection for decades (Kim et al. 2015).

The compound inhibits bacterial growth by disrupting fatty acid synthesis, thereby preventing the production of essential lipids such as lipoproteins, phospholipids, and lipopolysaccharides. Additionally, it blocks the bacterial enzyme enoyl-acyl carrier protein reductase (ENR/ECR) and damages eukaryotic cell membranes, contributing to its dual toxicity and antibacterial effects (Orhan 2020).

Despite its lower toxicity compared to other pollutants, triclosan's persistence in wastewater, biosolids, and environmental matrices remains a concern. Its physicochemical properties, including hydrophobicity and resistance to degradation, hinder its removal from ecosystems. Studies have linked triclosan exposure to adverse health outcomes, including antibiotic resistance development, muscle function impairment, endocrine disruption, and skin irritation (Dhillon et al. 2015).

8.4.2 *Quaternary Ammonium*

Quaternary ammonium compounds (QACs) function as surfactants, accumulating at the interface between bacterial cell membranes (composed of lipids) and the surrounding aqueous environment. They interact with microbes through two primary mechanisms: (1) polar interactions mediated by the cationic nitrogen group and (2) non-polar interactions driven by the hydrophobic alkyl chain. The cationic ammonium group binds to negatively charged bacterial membranes, forming a surfactant-microbe complex. This complex disrupts membrane-associated proteins critical for structural integrity and may interfere with bacterial DNA. Additionally, the hydrophobic chain penetrates microbial cells, impairing essential cellular processes. Notably, extending the alkyl chain length enhances the antibacterial efficacy of QACs (Liu et al. 2015).

While QACs exhibit broad-spectrum activity against bacteria, enveloped viruses, pathogenic fungi, and mycobacteria, they are ineffective against non-enveloped viruses. Low-alcohol formulations are particularly effective, achieving rapid disinfection within 3–5 min of contact. However, incorporating alcohol or solvents into QAC-based disinfectants accelerates surface drying, potentially compromising disinfection efficacy. Cetrimide (alkyl trimethylammonium bromide) and benzalkonium chloride demonstrate antibacterial, antifungal, and antiviral properties specifically targeting enveloped viruses and are safe for topical application on skin or mucous membranes to mitigate infection risks. A key limitation of QACs is their reduced effectiveness in the presence of hard water, anionic detergents, or organic matter. Furthermore, *Pseudomonas* species can metabolize cetrimide as a carbon, nitrogen, and energy source, undermining its antimicrobial utility (Tanasa et al. 2023).

8.4.3 *Zinc Pyrithione*

Zinc pyrithione (ZnPT), a versatile metallo-organic compound, is renowned for its broad-spectrum antimicrobial properties, effectively inhibiting the growth of bacteria, fungi, mildew, and algae. Historically used as fungicide in anti-dandruff shampoos, ZnPT also serves as an antifouling agent in marine coatings, replacing toxic tin compounds. Its applications extend to textiles, paints, adhesives, floor coverings, and wire/cable insulation (Windler et al. 2013).

8.4.4 N-Halamines

N-Halamines are organic compounds characterized by one or more covalent nitrogen-halogen bonds (e.g., chlorine, bromine, or iodine), formed through the halogenation of amines, amides, or imides. These compounds have been extensively employed as biocidal agents for decades. N-Halamine biocides are utilized in a wide range of applications, including water filtration systems, swimming pool sanitizers, medical equipment, and textiles (Sun et al. 2012; Hui and Debiemme-Chouvy 2013).

Textiles treated with N-Halamines demonstrate exceptional antimicrobial efficacy. The treatment process often involves a bleaching step, enabling the fabric to rapidly eliminate a broad spectrum of microorganisms, such as Gram-negative and Gram-positive bacteria, within short timeframes. When the oxidative halogen (e.g., chlorine) is depleted, the antimicrobial activity of N-Halamine-treated fabrics can be fully regenerated by re-exposing the material to diluted household bleach, restoring its biocidal properties (Dong et al. 2017). Notably, microorganisms do not develop resistance to N-Halamines, making them a sustainable antimicrobial option. However, the antibacterial efficacy of N-Halamines is highly dependent on the surface properties of the treated material, which must be carefully optimized for maximum performance (Mahira et al. 2019).

Despite their advantages, N-Halamines face practical limitations. For example, certain N-Halamine derivatives require organic solvents for dissolution, increasing the overall cost of treatment. Additionally, studies have identified potential toxicity risks to humans, which restricts their widespread adoption in the production of antibacterial cotton textiles (Ma et al. 2019). In recent years, advancements in synthetic polymer design have focused on mimicking the physicochemical properties of host defense peptides. These synthetic (co)polymers exhibit broad-spectrum antimicrobial activity, rapid bactericidal action, and a reduced likelihood of resistance development, offering promising alternatives to traditional biocides (Ergene et al. 2018).

8.5 Metals and Metal Oxide Nanoparticles

Throughout history, metals and metallic compounds have played a pivotal role in medical applications. Many heavy metals exhibit toxicity toward diverse microbial groups, even at minimal concentrations, whether in elemental form or as compounds. Their mechanism of action involves binding to intracellular proteins, thereby inactivating them (Fig. 8.1). While various metals have been employed for antimicrobial purposes, silver remains the most extensively used in the textile industry. Silver diffuses into fiber surfaces during use, releasing silver ions (Ag^+) upon exposure to moisture. The rate of silver ion release depends on the fiber's chemical composition, physical properties, and silver content, with slower release

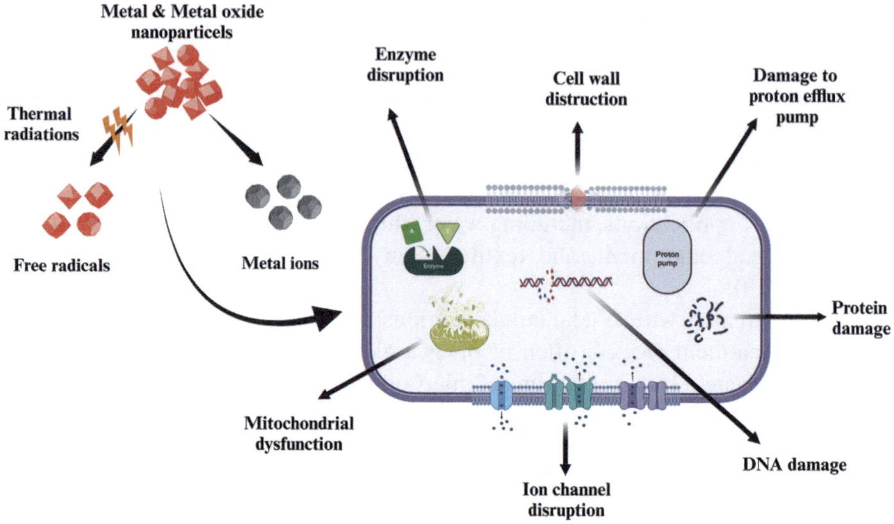

Fig. 8.1 Mechanism of Action (MOA) of metal nanoparticles in antimicrobial textiles

rates correlating to prolonged biocidal efficacy (Ghoranneviss and Shahidi 2013; Evans and Kavanagh 2021).

Beyond pure metals, metal oxide nanoparticles such as titanium dioxide (TiO_2), tin oxide (SnO_2), zinc oxide (ZnO), gold (Au), and copper oxide (CuO) have been applied to natural and synthetic textiles. Silver nanoparticles (AgNPs), typically around 10 nm in size, demonstrate potent toxicity against a broad spectrum of microorganisms while exhibiting relatively low toxicity to human cells. These nanoparticles are prized for their durability, enhanced dye uptake, and antiviral activity against pathogens like SARS-CoV-2 (responsible for COVID-19). Copper oxide nanoparticles (CuONPs) also show significant antimicrobial potential, with efficacy against Gram-positive and Gram-negative bacteria attributed to three mechanisms: (1) direct nanoparticle-bacterial contact, (2) release of copper ions (Cu^{2+}), and (3) generation of reactive oxygen species (ROS) (Butola and Mohammad 2016; Jeremiah et al. 2020; Román et al. 2020).

Market data from European manufacturers indicate that silver salts (e.g., nano silver chloride) dominate textile applications, accounting for 79% of silver usage. Silver ion exchangers and metallic silver represent 8% and 13%, respectively (Windler et al. 2013). Zinc oxide nanoparticles (ZnO NPs), a class of inorganic compounds, are valued not only for their antibacterial properties but also for their ultraviolet (UV) blocking capabilities. Their effectiveness stems from a direct band gap of 3.37 eV and a high exciton binding energy of 60 meV, enabling broad-spectrum UV absorption (Pandimurugan and Thambidurai 2017).

Despite their advantages, metal ion-based textile treatments face technical and environmental challenges that hinder commercial scalability. While recognized for their efficacy and unique antimicrobial benefits, concerns persist regarding toxicity, environmental leaching, and disposal impacts. Leaching not only restricts the scope and volume of applications but paradoxically enhances biocidal activity by increasing metal ion availability (Owen and Laird 2021; Ramzan et al. 2022).

Recent advances focus on the biosynthesis and application of nanostructured inorganic materials. For example, cotton fabrics treated with nanostructured coatings have demonstrated activity against *Bacillus subtilis*. Eco-friendly nanohybrid materials such as oligochitosan (derived from crab shells) combined with nanosilica (sourced from rice husks) exhibit resistance to fungal infections like *Phytophthora infestans*, highlighting their potential in sustainable agriculture (Kim et al. 2019; Sadalage et al. 2020).

A persistent challenge in nanoparticle research lies in preventing agglomeration, achieving the desired morphology, and ensuring uniformity in nanoparticle size. These factors are critical to optimize their functional properties. To mitigate these challenges, the use of ligands as capping agents has become a widely adopted strategy. Numerous natural capping agents have been explored and successfully integrated into textile applications, offering sustainable and biocompatible alternatives to synthetic counterparts. For instance, seaweed-capped zinc oxide nanoparticles (SW-ZnO NPs) were applied to cotton fabric using the pad-dry-cure technique, demonstrating robust antibacterial activity against Gram-positive bacteria such as *Staphylococcus aureus* and *Streptococcus pyogenes*, as well as Gram-negative bacteria including *Escherichia coli* and *Klebsiella aerogenes*. Similarly, date seed extract; a renewable and eco-friendly resource was employed as a natural capping agent to synthesize capped ZnO nanoparticles. These nanoparticles not only exhibited antimicrobial efficacy but also imparted ultraviolet (UV) protective qualities when incorporated into cotton fabrics, showcasing their multifunctional utility. Furthermore, amino-capped titanium dioxide (TiO_2) nanoparticles were fabricated on functional cotton fabric using a two-step sol-gel and hydrothermal synthesis approach. This hybrid material demonstrated notable antibacterial activity against *S. aureus*, underscoring the potential of tailored capping agents to enhance nanoparticle performance in textile applications (Pandimurugan and Thambidurai 2017; El-Naggar et al. 2018; Zhang et al. 2019). These innovations highlight the growing emphasis on green chemistry and sustainable material design in addressing both technical challenges and environmental concerns.

8.6 Natural Agents

A diverse array of natural bioactive compounds has been isolated and utilized in the development of antimicrobial textiles. Plant extracts, animal-derived substances (e.g., honey), and essential oils have been incorporated into fabrics to combat wound infections and microbial colonization. Additionally, natural dyes, mordants, and pigments have been extensively studied for their inherent antimicrobial properties, offering sustainable alternatives to synthetic counterparts (Gulati et al. 2022).

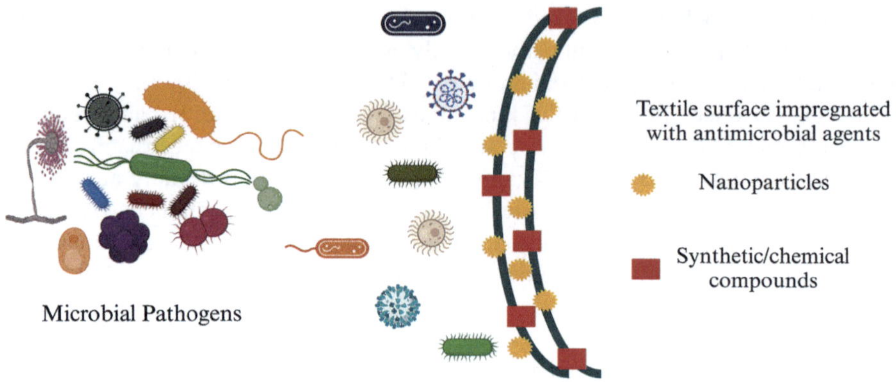

Fig. 8.2 Natural compounds as sustainable antimicrobial agents for textiles

Natural compounds are particularly advantageous for meeting the biocidal requirements of textile-based materials, as they provide critical attributes such as non-toxicity, environmental sustainability, and renewability. These bioactive agents inhibit microorganisms including bacteria and fungi that naturally colonize textiles (Fig. 8.2). For instance, cellulose-based fibrous scaffolds fabricated via electrospinning have demonstrated the ability to effectively encapsulate essential oils derived from cinnamon, lemongrass, and peppermint. These scaffolds exhibit potent biocidal resistance against *Escherichia coli* (a Gram-negative bacterium) even at low concentrations, highlighting their potential for topical therapeutic applications such as wound dressings or antimicrobial bandages (Liakos et al. 2015).

8.6.1 Cyclodextrins

Cyclodextrins, a class of natural cyclic oligosaccharides, are increasingly utilized in antimicrobial textile finishing due to their unique structural and functional properties. Characterized by a hydrophilic exterior and a hydrophobic inner cavity, cyclodextrins can form inclusion complexes with diverse molecules, making them versatile agents in textile applications. Derived from the enzymatic conversion of starch via cyclodextrin glycosyltransferase, these compounds consist of glucopyranoside subunits linked by α-1,4 glycosidic bonds. The three primary types—α-cyclodextrin (α-CD, six subunits), β-cyclodextrin (β-CD, seven subunits), and γ-cyclodextrin (γ-CD, eight subunits)—vary in cavity size, with β-CD being the most extensively studied (Sahu and Singh 2019).

Cyclodextrins offer notable advantages in textile functionalization, including eco-friendliness, cost-efficiency, and compatibility with sustainable practices. Their ability to encapsulate active agents such as antimicrobials, fragrances, and insecticides within their hydrophobic cavity enables controlled release mechanisms, enhancing durability and long-term efficacy. Additionally, cyclodextrins improve dye uptake, chelate metal ions, and facilitate drug delivery, broadening their utility in technical textiles. In antimicrobial finishing, cyclodextrins enhance fabric properties by stabilizing bioactive compounds; for example, β-cyclodextrin complexes with essential oils or biocides impart durable antimicrobial activity while minimizing environmental impact (Shahid and Mohammad 2013; Bezerra et al. 2020). Beyond antimicrobial applications, cyclodextrins serve as carriers for UV-blocking agents and insecticides, underscoring their multifunctional role in advancing textile innovation.

8.6.2 *Sericin*

Sericin, a natural protein derived from the silkworm *Bombyx mori*, exhibits a range of valuable properties, including biocompatibility, biodegradability, UV radiation resistance, oxidation resistance, moisture retention efficacy, antibacterial activity, gel-forming tendencies, and adhesive characteristics. When applied as a coating on cotton fabric, sericin significantly enhances antimicrobial performance against bacterial strains such as *Escherichia coli* and *Staphylococcus aureus*, as demonstrated in controlled studies (Rajendran et al. 2012; Shahid and Mohammad 2013). These properties make sericin a promising candidate for functional textile finishes that prioritize sustainability and safety.

8.6.3 *Chitosan*

Chitosan, a cationic polysaccharide obtained through the alkaline deacetylation of chitin, is widely employed in antimicrobial textile processing due to its inherent antibacterial and anti-odor properties. When applied as a hydrogel to cellulosic fabrics, chitosan demonstrates potent antibacterial activity against pathogens such as *Escherichia coli*, *Staphylococcus aureus*, and *Listeria monocytogenes* (Benltoufa et al. 2020). Its antimicrobial mechanism involves ionic interactions between positively charged amine groups in the chitosan polymer backbone and negatively charged bacterial cell walls. This interaction disrupts membrane integrity, causing intracellular electrolyte leakage, cell death, and inhibition of bacterial colony formation. The cationic charge of chitosan is generated either through protonation of amine groups in acidic conditions or via structural modifications, both of which are critical for its biocidal efficacy. However, studies indicate that the antibacterial performance of chitosan is highly concentration-dependent (Goy et al. 2016).

In the textile industry, chitosan is utilized in diverse applications, including fiber production for enhancing cotton, wool, silk, and polypropylene fabrics, functional finishing processes such as durable press, anti-static, and deodorizing treatments, and medical textiles where its antibacterial properties are critical for wound dressings and surgical fabrics. Additionally, chitosan improves dye affinity and print durability in textile printing applications, offering enhanced color retention and design precision (Elamri et al. 2023). However, achieving effective microbial resistance often requires high chitosan concentrations, which can compromise fabric performance by reducing air permeability, increasing stiffness, and diminishing durability over repeated use (Hameed et al. 2022).

8.6.4 Natural Dyes

Natural dyes and pigments have long been used in fabric dyeing, offering dual benefits of coloration and antimicrobial functionality. Curcumin, a primary bioactive compound in turmeric, serves as a safe, non-toxic natural dye. Studies employing pad dyeing techniques demonstrate that curcumin application on wool and cotton not only imparts vibrant hues but also confers robust antimicrobial properties. These dyed fabrics exhibit high durability against standard laundering processes compared to untreated materials, maintaining both colorfastness and microbial resistance. Furthermore, natural dyes derived from plant sources such as *Terminalia catappa*, *Morinda citrifolia*, *Tectona grandis*, and *Artocarpus heterophyllus* have been shown to deliver significant antimicrobial activity against diverse microorganisms, including bacteria and fungi (Prusty et al. 2010; Reddy et al. 2013). Similarly, myrobalan which is a natural mordant has been combined with a dye extracted from the fruit of *Opuntia ficus-indica* to enhance the antimicrobial characteristics of silk fabric, illustrating the synergistic potential of plant-derived compounds in textile functionalization (Ganesan and Karthik 2017).

8.6.5 Microbial Pigments

In addition to plant-based dyes, microbial pigments have gained traction for coloring cotton, glace cotton, silk, and rayon. A red pigment derived from *Rhodonellum psychrophilum* contains a complex mixture of bioactive compounds, including prodigiosin, 2-methyl-3-butyl-prodigiosin, 2-methyl-3-hexyl-prodigiosin, anhydrorhodovibrin, 3,4-didehydrorhodopsin, tetradecanoyl-hexadecanoyl, and alloxanthin. This pigment demonstrates broad-spectrum antimicrobial activity, exhibiting antibacterial effects against *Escherichia coli* and *Staphylococcus aureus*, as well as antifungal properties against *Candida albicans* and *Saccharomyces cerevisiae* (Bisht et al. 2020).

8.6.6 Plant Extracts

Recent years have seen intensified research into natural antimicrobial agents such as eucalyptus, aloe vera, capsaicin, and neem, driven by their eco-friendly profiles and renewable sourcing. These plant extracts show efficacy against diverse pathogens, including bacteria, viruses, and fungi. However, challenges such as limited durability and variable potency under real-world conditions persist. To address these limitations, researchers are exploring advanced application techniques including nanoencapsulation, chemical grafting, and cross-linking to enhance the stability and longevity of these natural biocides in textile finishes (Tawiah et al. 2016).

8.6.7 Essential Oils

Essential oils, aromatic compounds extracted from plants, serve dual roles as antimicrobial agents and fragrance enhancers in textiles. Historically valued for their therapeutic and psychological benefits, these oils are now increasingly applied to fabrics using technologies like microencapsulation and electrospinning. Such methods improve retention and controlled release, ensuring sustained antimicrobial activity against microbes while maintaining the pleasant sensory properties of oils. Studies confirm their effectiveness when optimized application strategies are employed, positioning them as versatile solutions for functional textile design (Tawiah et al. 2016).

8.7 Types of Antimicrobial Textiles

While some fabrics possess inherent antimicrobial properties, their efficacy is often limited to specific microbial strains, resulting in a narrow antimicrobial spectrum. To address this, the incorporation of external antimicrobial agents such as metallic nanoparticles, quaternary ammonium compounds, or natural extracts into textiles enhances their functionality, enabling broad-spectrum activity against diverse pathogens. This approach allows customization for various fabric types, including natural fibers (e.g., cotton, wool), synthetic fibers (e.g., polyester, polypropylene), and blended materials, ensuring versatility across applications like healthcare, sportswear, and protective gear (Andra et al. 2021) (Fig. 8.3).

8.7.1 Antibacterial Textiles

The textile industry employs a wide range of antibacterial agents, including antibiotics such as ciprofloxacin, disinfectants like triclosan, polysaccharides such as chitosan, and metal-based coatings incorporating silver, zinc, or copper, to imbue fabrics with microbial resistance (Mahltig et al. 2013; Windler et al. 2013). These agents function through two primary mechanisms: active action, where they directly kill bacteria or inhibit their growth, and passive action, where they prevent bacterial colonization by obstructing adhesion to textile fibers (Höfer 2006). To assess efficacy, broth microdilution remains the conventional and reliable method for

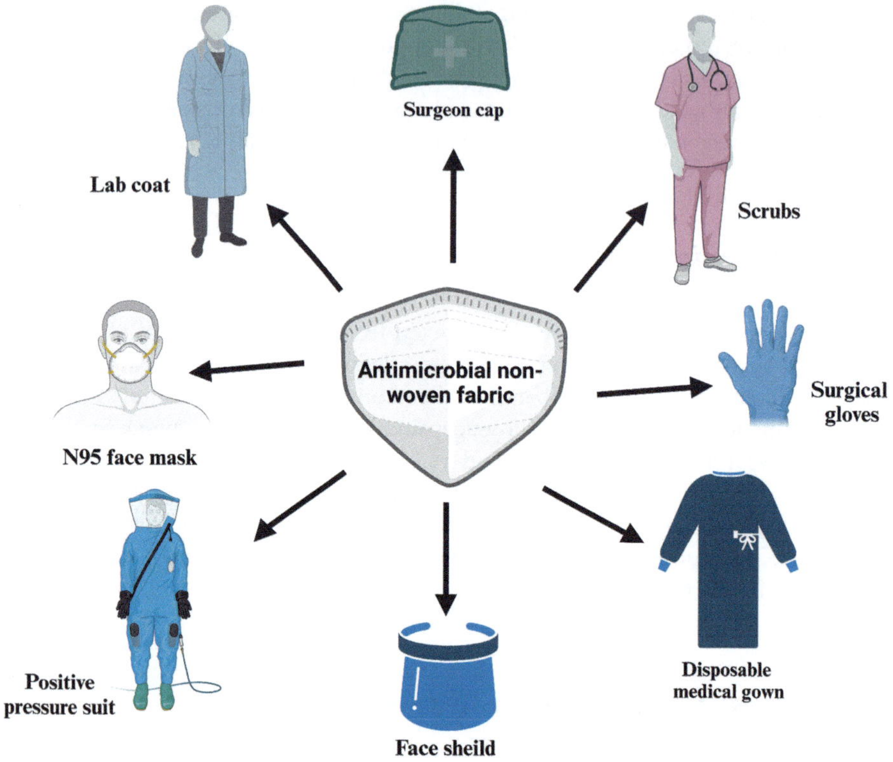

Fig. 8.3 Applications of antimicrobial textiles in medical settings

determining the minimum inhibitory concentration (MIC) and minimum bactericidal concentration (MBC). MIC represents the lowest concentration of an agent that visibly inhibits bacterial growth, while MBC is identified by subculturing treated bacterial cells in fresh growth medium and observing the concentration at which no growth occurs (Bancalari et al. 2020).

Most studies focus on direct application to maximize adherence, often using leaching-type fabrics, though nanoparticle impregnation based on MIC values has also been explored. For instance, antibacterial bandages incorporating silver (Ag), zinc oxide (ZnO), and Ag/ZnO nanoparticles demonstrated superior efficacy against *Acinetobacter baumannii* and *Pseudomonas aeruginosa*, with Ag nanoparticles showing the highest potency (Khatami et al. 2018). Another key evaluation method, the Fabric Disc Diffusion Assay, involves placing a textile sample disc on an agar plate inoculated with bacteria; a visible zone of inhibition post-incubation indicates antimicrobial activity. This method validated the antibacterial effects of *Helichrysum arenarium* methanolic extract on *Staphylococcus aureus* and *Escherichia coli*, with dyed wool fabrics exhibiting significant microbial resistance (Cuce et al. 2022). Certain agents exhibit selective antibacterial properties, as seen in nano selenium brooms tested against *S. aureus*, *Bacillus subtilis*, *P. aeruginosa*, *E. coli*, and *Proteus vulgaris*, which showed activity exclusively against *B. subtilis*. Such specificity highlights their potential use in surgical settings to reduce *Bacillus*-related bacteremia, as evidenced by MIC/MBC analyses (Sadalage et al. 2020). These methodologies and agents collectively underscore the multifaceted strategies employed to develop textiles with tailored antimicrobial functionalities.

8.7.2 Antiviral Textiles

The evaluation of antiviral compounds typically involves administering test compounds to viruses and incubating the mixture with susceptible host cells for plaque assay analysis. For textile samples, studies have demonstrated that antiviral properties can be assessed by filtering viral suspensions through the treated fabric and incubating the filtrate with virus-sensitive cells. For instance, cotton fabrics treated with 3% sodium pentaborate pentahydrate, 0.03% triclosan, and 7% glucapon exhibited broad-spectrum antimicrobial activity against bacteria, yeast, and fungi, alongside significant antiviral efficacy against adenovirus type 5 and poliovirus type 1 (Iyigundogdu et al. 2017). Similarly, nonwoven filters coated with 1-chloro-2,2,5,5-tetramethyl-4-imidazolidinone were evaluated using the sandwich test method, where surviving avian influenza viruses in embryonated chicken eggs confirmed the antiviral activity of material (Ren et al. 2018).

Moreover, antiviral textiles have been developed using nanoparticle-coated fabrics. For example, silver nanoparticles synthesized via radiochemical methods were immobilized onto cotton fabric surfaces, resulting in modified textiles with demonstrated antiviral activity against Influenza A and Feline Calicivirus, a surrogate for human norovirus (Seino et al. 2016). These advancements highlight the potential of functionalized textiles to mitigate viral transmission in medical, public health, and consumer applications.

8.7.3 Antifungal Textiles

The fungicidal and fungistatic properties of textiles can be evaluated using methods such as disk diffusion and tube dilution techniques. Filamentous fungi, recognized as primary agents of fabric deterioration due to their ability to thrive on natural substrates, include species such as *Cladosporium*, *Alternaria*, and *Aspergillus*. Recent advancements in antimicrobial textiles have demonstrated promising antifungal efficacy. For example, fabrics functionalized with modified silica incorporating silver and carbon additives showed significant suppression of fungal growth, highlighting their potential for durable applications (Igal and Vázquez 2020). Similarly, zinc oxide nanoparticles biosynthesized by the fungus *Phanerochaete chrysosporium* were integrated into cellulose fibers, yielding materials that effectively inhibited the growth of *Aspergillus niger* (an opportunistic pathogen),

Geotrichum candidum (a yeast), and even the cellulolytic fungus *Phanerochaete chrysosporium* itself, which is typically associated with wood degradation (Sharma et al. 2021).

Further innovations include cotton fabrics treated with guanazole complexes containing silver (Ag^{+}) and zinc (Zn^{2+}) ions, which exhibited broad-spectrum antimicrobial activity against bacterial pathogens and fungal species such as *A. niger*, *Fusarium chlamydosporum*, and *Penicillium* spp. Additionally, these treated fabrics demonstrated flame-retardant properties, underscoring their multifunctionality for protective textiles in high-risk environments (Nabipour et al. 2020). These examples illustrate the synergy of material science and biotechnology in developing textiles that address both microbial degradation and safety hazards.

8.8 Synthesis of Antimicrobial Textiles and Latest Developments

The development of antimicrobial textiles with functional capabilities is a critical area of research, where processing techniques play a pivotal role in ensuring durability and efficacy. Key methods include spraying, padding, surface coating, grafting, and crosslinking (Tania et al. 2021). Modern approaches such as plasma treatment, colloidal solutions, microencapsulation, sol-gel processing, and the *in-situ* synthesis of antimicrobial agents on textiles further enhance performance and stability (Gulati et al. 2022) (Fig. 8.4).

8.8.1 Surface Coating

Among the diverse techniques for synthesizing antimicrobial textiles (AMT), antimicrobial coatings have proven instrumental in inhibiting harmful microbial growth and biofilm formation on textile surfaces, particularly in medical applications (Leyland et al. 2016). These coatings are often synthesized using nanoparticles due to their potent antimicrobial properties. Common strategies involve integrating bactericidal agents such as silver ions, antibiotics, or natural compounds into hybrid or composite materials. For example, silver-ion-embedded coatings on medical fabrics have demonstrated sustained antimicrobial activity against multidrug-resistant pathogens (Ahmadi et al. 2020).

8.8.2 Sol-Gel Method

The sol-gel method is a widely adopted technique for depositing antimicrobial coatings on textiles, particularly through the integration of nanoparticles. This process enables the creation of functionalized surfaces with tailored antimicrobial

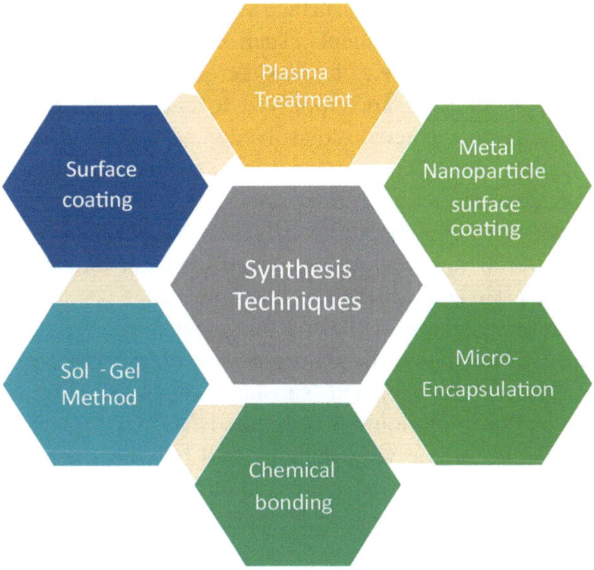

Fig. 8.4 Synthesis techniques for antimicrobial textiles

properties. Precursors such as zinc acetate (for zinc oxide-based coatings) or metal alkoxides like tetraethoxysilane (TEOS) are dissolved in a solvent blend commonly water, ethanol, and acids (e.g., hydrochloric acid) to facilitate hydrolysis and condensation.

In one study, Ag:ZnO nanoparticles were synthesized via hydrolysis of zinc acetate. A hybrid Ag:ZnO/chitosan (CS) coating was formulated using 3-glycidyloxypropyltrimethoxysilane, chitosan, and TEOS via sol-gel processing. This nanocomposite was applied to cotton/polyester (50/50) and 100% cotton fabrics, resulting in coatings with robust antimicrobial activity against both Gram-positive (e.g., *Staphylococcus aureus*) and Gram-negative (e.g., *Escherichia coli*) bacteria. The study highlights the potential of sol-gel-derived nanocomposites for advanced medical textiles (Buşilă et al. 2015).

8.8.3 Chemical Bonding

Chemical bonding is a pivotal approach for enhancing the antimicrobial properties of textiles, ensuring durability and efficacy in resisting microbial contamination. The production of antimicrobial textiles (AMT) through chemical bonding often involves integrating synthetic organic antimicrobial agents with polymers that inherently possess antimicrobial properties. The chemical structure of the antimicrobial agent plays a critical role in determining the synthesis method and final performance (Akca 2020).

The process typically begins with surface activation to increase textile reactivity. This involves chemical pre-treatments, such as sodium hydroxide application, which amplifies reactive groups (e.g., hydroxyl groups) on fabric surfaces. Following activation, grafting techniques are employed to chemically attach antimicrobial agents to the textile. Here, the agent is dissolved in a compatible solvent and applied under controlled pH and temperature conditions, facilitating covalent bond formation between the agent's functional groups and textile fibers (Leyland et al. 2016). For instance, zwitterionic materials grafted via covalent bonds create durable, biocompatible antibacterial surfaces, making them ideal for bioengineering and biomedical applications (He et al. 2016).

Recent advancements, such as plasma treatment, have expanded the scope of chemical bonding by enabling the incorporation of diverse antimicrobial compounds. These include halogenated substances (triclosan, chlorine, fluorine, N-halamines), quaternary ammonium compounds (QACs), guanidine-based polymers (e.g., polyhexamethylene vinyguanide), and polymers incorporating sulpho- and phospho-derivatives. These innovations enhance the versatility and functionality of AMT, addressing evolving industrial and medical demands (Karypidis et al. 2023).

8.8.4 *Microencapsulation*

Microencapsulation involves encasing antimicrobial agents within a protective coating to enable controlled release and minimize potential toxicity, representing a significant advancement in material science by merging functionality with enhanced safety. This technique ensures sustained antimicrobial efficacy over prolonged periods through the gradual release of active agents. Commonly encapsulated agents include cannabidiol (noted for its emerging antimicrobial potential), methyl salicylate (valued for its dual antimicrobial and aromatic properties), and chitosan (a natural biopolymer with inherent antimicrobial activity), all of which enhance the performance of cellulosic fibers. Key application methods include spraying, padding, pad-drying, and coating. Among these, pad-drying (or mechanical-thermal fixation) is particularly effective for micro- and nanoparticulate coatings with low affinity for textile substrates. This process involves brief thermal treatment (1–5 min at 100–150 °C) to induce cross-linking, termed *thermal fixation*, which ensures durable adhesion of the coating. The efficiency and simplicity of pad-drying, coupled with its ability to stabilize coatings on diverse fabrics, make it a preferred method in industrial settings (Saha and Mondal 2021).

8.8.5 *Coating*

Coating is a widely utilized procedure applicable to diverse materials, including yarns and fabrics composed of natural or synthetic fibers, as well as woven, nonwoven, and knitted textiles (Boh Podgornik et al. 2021). Direct coating techniques such as calendaring, roller, or knife methods require the finishing agent to be sufficiently viscous for proper application (Boh Podgornik et al. 2021). In contrast, spray-coating employs an airbrush with less viscous solutions, enabling the deposition of nanoparticles and facilitating uniform coverage on complex textile structures (Sarathi and Thilagavathi 2009).

The coating method has been pivotal in developing medical textiles with tailored antimicrobial properties. For instance, covalent bonding of reactive organo-selenium compounds to cotton fabrics achieved a 99% reduction in *Staphylococcus aureus* and moderate efficacy against *Escherichia coli*, alongside fungicidal activity against *Candida albicans* (Hoque et al. 2023). Similarly, thiazole-derived agents applied via coating demonstrated notable antibacterial effects against both Gram-positive and Gram-negative bacteria (Mohamed et al. 2018).

Recent innovations focus on multifunctional coatings that integrate multiple active agents to confer antibacterial, antiviral, and UV-protective properties simultaneously (Liu et al. 2022). These advancements are critical in medical settings requiring high-level protection. Beyond antimicrobial performance, coatings now enable functionalities such as antifouling, photochromism, thermochromism, easy-clean surfaces, and non-toxicity (Szadkowski et al. 2024).

8.9 Durability Challenges and Advances in Antimicrobial Textiles

The durability of antimicrobial textiles remains a significant challenge, particularly due to the risk of leaching, which compromises long-term antimicrobial efficacy and protective benefits (Vats et al. 2023). Textiles synthesized via nanotechnology, while exhibiting potent antimicrobial properties, often face leaching issues as nanoparticles detach from fibers, undermining durability and functionality. However, advancements in nanomaterial engineering have mitigated these limitations. For instance, zinc-infused textiles retain effectiveness even after extensive washing, demonstrating enhanced durability for medical applications (Broadhead et al. 2021).

Natural antimicrobial compounds, such as plant-derived agents, face additional challenges due to their inherent volatility, sensitivity to oxygen/light, and weak bonding with textile substrates (Soroh et al. 2021). To address this, microencapsulation has emerged as an innovative method, shielding core materials (e.g., essential oils, natural extracts) from environmental degradation. This technique is particularly valuable for medical textiles, enabling sustained release of active agents while maintaining fabric integrity (Vats et al. 2023).

Safety and practicality are equally critical. Antimicrobial textiles must be non-irritating and hypoallergenic for direct skin contact. Recent evaluations confirm that advanced treatments, such as covalently bonded coatings or biocompatible microcapsules, meet these safety criteria without provoking adverse reactions. In healthcare, these innovations represent a transformative leap in infection control. By integrating durability, functionality, and safety, antimicrobial textiles enhance patient safety and operational efficiency. Ongoing research continues to expand their applications, offering creative solutions to persistent challenges in medical settings, from surgical gowns to wound dressings (Vojnits et al. 2024). Future developments are expected to prioritize multifunctional textiles combining antimicrobial, antiviral, and self-cleaning properties, further solidifying their role in modern healthcare.

8.10 Future Prospects

The synthesis of antimicrobial textiles has advanced significantly, leveraging both natural and synthetic fibers. Antimicrobial properties are imparted through diverse agents, including chemicals, plant- and animal-derived compounds, metal-based nanoparticles, dyes, and mordants (Gulati et al. 2022). However, recent advancements underscore the growing emphasis on minimizing environmental impacts. Innovations in controlled-release technologies now enable the fabrication of antimicrobial textiles (AMT) with enhanced biocidal efficacy at lower active-agent concentrations, reducing ecological footprints while maintaining performance. Notably, modern coatings exhibit improved leaching resistance, retaining antimicrobial activity even after multiple washing cycles, which addresses historical durability challenges.

In light of the documented environmental and health risks associated with conventional agents like triclosan and quaternary ammonium compounds (QACs), there is an urgent need to prioritize research into safer, non-toxic alternatives (Boh Podgornik et al. 2021). Future efforts should focus on developing novel antimicrobial agents derived from renewable sources, such as microorganisms (e.g., bacteriocins, microbial pigments) or plant-based compounds (e.g., polyphenols, essential oils) (Szadkowski et al. 2024). While these natural agents currently exhibit lower efficacy compared to synthetic counterparts, their potential for optimization through nanoformulation, combinatorial synergies, or structural modification remains vast (Yılmaz 2023).

Sustainability is now a cornerstone of textile innovation. Next-generation antimicrobial textiles must balance environmental and human safety with performance, durability, ease of application, and cost-effectiveness. Emerging strategies include green synthesis of nanoparticles, biodegradable coatings, and circular design principles to minimize waste. While the ideal formulation remains elusive, breakthroughs in biofunctionalized fibers, stimuli-responsive coatings, and AI-driven material design signal a promising trajectory (Karypidis et al. 2023). By integrating multidisciplinary approaches ranging from synthetic biology to advanced manufacturing, the field is poised to deliver textiles that not only combat pathogens but also align with global sustainability goals, revolutionizing healthcare, sportswear, and protective gear.

8.11 Conclusions

In recent years, escalating consumer demand for comfort, hygiene, and well-being particularly in protection against pathogens has driven rapid growth in the global antimicrobial textile market. These textiles are engineered using both synthetic and natural fibers, with antimicrobial properties imparted through diverse agents such as chemicals, metal-based nanoparticles, plant- and animal-derived compounds, dyes, and mordants. Depending on the treatment method, fabrics may exhibit narrow or broad-spectrum activity against bacteria, fungi, and viruses.

While antimicrobial textiles offer significant benefits, their long-term advantages must be carefully weighed against potential challenges, including environmental persistence, microbial resistance, and human health risks. Synthetic agents like triclosan and quaternary ammonium compounds, despite their efficacy, raise sustainability concerns due to toxicity and non-biodegradability. Conversely, natural alternatives such as chitosan, essential oils, and lignin derivatives demonstrate promise as eco-friendly solutions but require optimization to match the performance of synthetic counterparts.

Future innovation must prioritize the development of "green" antimicrobial agents and sustainable manufacturing processes. Advances in microencapsulation, biofunctionalized coatings, and nanotechnology can enhance durability while minimizing leaching and environmental impact. Additionally, interdisciplinary research into biobased materials, circular design principles, and AI-driven formulation will be critical to balancing efficacy, safety, and scalability. By aligning antimicrobial performance with planetary health, the next generation of textiles will not only meet consumer needs but also contribute to a safer, more sustainable future for healthcare, fashion, and industrial applications.

References

Abdul-Reda Hussein U, Mahmoud Z, Alaziz KA, Alid M, Yasin Y, Ali F et al (2023) Antimicrobial finishing of textiles using nanomaterials. Braz J Biol 84:e264947

Afraz N, Uddin F, Syed U, Mahmood A (2019) Antimicrobial finishes for textiles. Curr Trends Fash Technol Text Eng 4:87–94

Ahmad S, Ullah T, Ziauddin. (2020) Fibers for technical textiles. Springer, Cham

Ahmadi Y, Siddiqui MT, Haq QMR, Ahmad S (2020) Synthesis and characterization of surface-active antimicrobial hyperbranched polyurethane coatings based on oleo-ethers of boric acid. Arab J Chem 13:2689–2701

Akca C (2020) Antimicrobial finishing. In: Waste in textile and leather sectors. IntechOpen, London, p 21

Amrhar R, Singh J, Eesaee M, Carrière P, Saidi A, Nguyen-Tri P (2025) Polymeric nanocomposites-based advanced coatings for antimicrobial and antiviral applications: a comprehensive overview. Results Surf Interfaces:100497. https://doi.org/10.1016/j.rsurfi.2025.100497

Andra S, Balu SK, Jeevanandam J, Muthalagu M (2021) Emerging nanomaterials for antibacterial textile fabrication. Naunyn Schmiedeberg's Arch Pharmacol 394:1355–1382

Bancalari E, Martelli F, Bernini V, Neviani E, Gatti M (2020) Bacteriostatic or bactericidal? Impedometric measurements to test the antimicrobial activity of Arthrospira platensis extract. Food Control 118:107380

Benltoufa S, Miled W, Trad M, Slama RB, Fayala F (2020) Chitosan hydrogel-coated cellulosic fabric for medical end-use: antibacterial properties, basic mechanical and comfort properties. Carbohydr Polym 227:115352

Bezerra FM, Lis MJ, Firmino HB, Dias Da Silva JG, Curto Valle RDCS, Borges Valle JA et al (2020) The role of cyclodextrin in the textile industry—review. Molecules 25:3624

Bibi A, Afza G, Afzal Z, Farid M, Sumrra SH, Hanif MA et al (2024) Synthetic vs. natural antimicrobial agents for safer textiles: a comparative review. RSC Adv 14:30688–30706

Bisht G, Srivastava S, Kulshreshtha R, Sourirajan A, Baumler DJ, Dev K (2020) Applications of red pigments from psychrophilic Rhodonellum psychrophilum GL8 in health, food and antimicrobial finishes on textiles. Process Biochem 94:15–29

Boh Podgornik B, Šandrić S, Kert M (2021) Microencapsulation for functional textile coatings with emphasis on biodegradability—a systematic review. Coatings 11:1371

Broadhead R, Craeye L, Callewaert C (2021) The future of functional clothing for an improved skin and textile microbiome relationship. Microorganisms 9:1192

Buşilă M, Muşat V, Textor T, Mahltig B (2015) Synthesis and characterization of antimicrobial textile finishing based on Ag: ZnO nanoparticles/chitosan biocomposites. RSC Adv 5:21562–21571

Butola B, Mohammad F (2016) Silver nanomaterials as future colorants and potential antimicrobial agents for natural and synthetic textile materials. RSC Adv 6:44232–44247

Cuce M, Kilinc M, Kilinc N, Tekin A, Kut D (2022) Color, fastness, and antimicrobial properties of wool fabrics dyed with Helichrysum arenarium subsp. aucheri extract. J Nat Fibers 19:797–809

Dar OI, Aslam R, Pan D, Sharma S, Andotra M, Kaur A et al (2022) Source, bioaccumulation, degradability and toxicity of triclosan in aquatic environments: a review. Environ Technol Innov 25:102122

Dhillon GS, Kaur S, Pulicharla R, Brar SK, Cledón M, Verma M et al (2015) Triclosan: current status, occurrence, environmental risks and bioaccumulation potential. Int J Environ Res Public Health 12:5657–5684

Di Martino P (2022) Antimicrobial agents and microbial ecology. AIMS Microbiol 8:1–4

Dong A, Wang Y-J, Gao Y, Gao T, Gao G (2017) Chemical insights into antibacterial N-halamines. Chem Rev 117:4806–4862

Elamri A, Zdiri K, Hamdaoui M, Harzallah O (2023) Chitosan: a biopolymer for textile processes and products. Text Res J 93:1456–1484

El-Naggar ME, Shaarawy S, Hebeish A (2018) Multifunctional properties of cotton fabrics coated with in situ synthesis of zinc oxide nanoparticles capped with date seed extract. Carbohydr Polym 181:307–316

Emam HE (2019) Antimicrobial cellulosic textiles based on organic compounds. 3 Biotech 9:29

Ergene C, Yasuhara K, Palermo EF (2018) Biomimetic antimicrobial polymers: recent advances in molecular design. Polym Chem 9:2407–2427

Evans A, Kavanagh KA (2021) Evaluation of metal-based antimicrobial compounds for the treatment of bacterial pathogens. J Med Microbiol 70:001363

Ganesan P, Karthik T (2017) Analysis of colour strength, colour fastness and antimicrobial properties of silk fabric dyed with natural dye from red prickly pear fruit. J Text Inst 108:1173–1179

Gao Y, Cranston R (2008) Recent advances in antimicrobial treatments of textiles. Text Res J 78:60–72

Ghoranneviss M, Shahidi S (2013) Effect of various metallic salts on antibacterial activity and physical properties of cotton fabrics. J Ind Text 42:193–203

Goy RC, Morais ST, Assis OB (2016) Evaluation of the antimicrobial activity of chitosan and its quaternized derivative on E. coli and S. aureus growth. Rev Bras 26:122–127

Gulati R, Sharma S, Sharma RK (2022) Antimicrobial textile: recent developments and functional perspective. Polym Bull 79:5747–5771

Gupta P, Bairagi N, Priyadarshini R, Singh A, Chauhan D, Gupta D (2017) Bacterial contamination of nurses' white coats after first and second shift. Am J Infect Control 45:86–88

Gutarowska B, Pietrzak K, Machnowski W, Milczarek JM (2017) Historical textiles–a review of microbial deterioration analysis and disinfection methods. Text Res J 87:2388–2406

Hameed AZ, Raj SA, Kandasamy J, Baghdadi MA, Shahzad MA (2022) Chitosan: a sustainable material for multifarious applications. Polymers 14:2335

He L, Li S, Chung CT, Gao C, Xin JH (2016) Constructing safe and durable antibacterial textile surfaces using a robust graft-to strategy via covalent bond formation. Sci Rep 6:36327

Höfer D (2006) Antimicrobial textiles–evaluation of their effectiveness and safety. Curr Probl Dermatol 33:42–50

Hoque E, Tran P, Jacobo U, Bergfeld N, Acharya S, Shamshina JL et al (2023) Antimicrobial coatings for medical textiles via reactive organo-selenium compounds. Molecules 28:6381

Hui F, Debiemme-Chouvy C (2013) Antimicrobial N-halamine polymers and coatings: a review of their synthesis, characterization, and applications. Biomacromolecules 14:585–601

Igal K, Vázquez P (2020) Waste in textile and leather sectors. IntechOpen, London, pp 1–20

Iyigundogdu ZU, Demir O, Asutay AB, Sahin F (2017) Developing novel antimicrobial and antiviral textile products. Appl Biochem Biotechnol 181:1155–1166

Jeremiah SS, Miyakawa K, Morita T, Yamaoka Y, Ryo A (2020) Potent antiviral effect of silver nanoparticles on SARS-CoV-2. Biochem Biophys Res Commun 533:195–200

Karypidis M, Karanikas E, Papadaki A, Andriotis EG (2023) A mini-review of synthetic organic and nanoparticle antimicrobial agents for coatings in textile applications. Coatings 13:693

Ke CL, Deng FS, Chuang CY, Lin CH (2021) Antimicrobial actions and applications of chitosan. Polymers 13:904

Khatami M, Sharifi I, Nobre MA, Zafarnia N, Aflatoonian MR (2018) Waste-grass-mediated green synthesis of silver nanoparticles and evaluation of their anticancer, antifungal and antibacterial activity. Green Chem Lett Rev 11:125–134

Kim SA, Moon H, Lee K, Rhee MS (2015) Bactericidal effects of triclosan in soap both in vitro and in vivo. J Antimicrob Chemother 70:3345–3352

Kim B-M, Minh NV, Choi H-Y, Kim W-G (2019) Coralmycin derivatives with potent anti-gram negative activity produced by the myxobacteria Corallococcus coralloides M23. Molecules 24:1390

Lamba NM, Herson DS, Jindani R, King MW (2017) Evaluation of antimicrobial-treated fabric properties. AATCC J Res 4:14–21

Leyland NS, Podporska-Carroll J, Browne J, Hinder SJ, Quilty B, Pillai SC (2016) Highly efficient F, Cu doped TiO2 anti-bacterial visible light active photocatalytic coatings to combat hospital-acquired infections. Sci Rep 6:24770

Liakos I, Rizzello L, Hajiali H, Brunetti V, Carzino R, Pompa P et al (2015) Fibrous wound dressings encapsulating essential oils as natural antimicrobial agents. J Mater Chem B 3:1583–1589

Liu Y, Ren X, Liang J (2015) Antimicrobial modification review. Bioresources 10:1964–1985

Liu J, Qiu S, Qi P, Sun J, Li H, Gu X et al (2022) Constructing a fully biobased coating to improve the flame retardancy, antibacterial properties, and UV resistance of polyamide 6 fabrics. ACS Appl Eng Mater 1:268–277

Ma W, Li L, Lin X, Wang Y, Ren X, Huang TS (2019) Novel ZnO/N-halamine-mediated multifunctional dressings as quick antibacterial agent for biomedical applications. ACS Appl Mater Interfaces 11:31411–31420

Mahira S, Jain A, Khan W, Domb AJ (2019) In: Domb AJ, Kunduru KR, Farah S (eds) Antimicrobial materials for biomedical applications. The Royal Society of Chemistry, London, p 0

Mahltig B, Soltmann U, Haase H (2013) Modification of algae with zinc, copper and silver ions for usage as natural composite for antibacterial applications. Mater Sci Eng C 33:979–983

Mohamed FA, Abd El-Megied SA, Bashandy MS, Ibrahim HM (2018) Synthesis, application and antibacterial activity of new reactive dyes based on thiazole moiety. Pigm Resin Technol 47:246–254

Mondal MIH, Saha J (2019) Antimicrobial, UV resistant and thermal comfort properties of chitosan-and Aloe vera-modified cotton woven fabric. J Polym Environ 27:405–420

Mousavi Khaneghah A, Hashemi SMB, Eş I, Fracassetti D, Limbo S (2018) Efficacy of antimicrobial agents for food contact applications: biological activity, incorporation into packaging, and assessment methods: a review. J Food Prot 81:1142–1156

Nabipour H, Wang X, Rahman MZ, Song L, Hu Y (2020) An environmentally friendly approach to fabricating flame retardant, antibacterial and antifungal cotton fabrics via self-assembly of guanazole-metal complex. J Clean Prod 273:122832

Nadi A, Boukhriss A, Bentis A, Jabrane E, Gmouh S (2018) Evolution in the surface modification of textiles: a review. Text Prog 50:67–108

Orhan M (2020) Triclosan applications for biocidal functionalization of polyester and cotton surfaces. J Eng Fiber Fabr 15:1558925020940104

Owen L, Laird K (2021) Development of a silver-based dual-function antimicrobial laundry additive and textile coating for the decontamination of healthcare laundry. J Appl Microbiol 130:1012–1022

Pandimurugan R, Thambidurai S (2017) UV protection and antibacterial properties of seaweed capped ZnO nanoparticles coated cotton fabrics. Int J Biol Macromol 105:788–795

Prusty A, Das T, Nayak A, Das N (2010) Colourimetric analysis and antimicrobial study of natural dyes and dyed silk. J Clean Prod 18:1750–1756

Puvvada RU, Wooding JP, Bellavia MC, Mcguinness EK, Sulchek TA, Losego MD (2019) Bacterial growth and death on cotton fabrics conformally coated with ZnO thin films of varying thicknesses via atomic layer deposition (ALD). JOM 71:178–184

Rajendran R, Balakumar C, Sivakumar R, Amruta T, Devaki N (2012) Extraction and application of natural silk protein sericin from Bombyx mori as antimicrobial finish for cotton fabrics. J Text Inst 103:458–462

Ramzan U, Majeed W, Hussain AA, Qurashi F, Qamar SUR, Naeem M et al (2022) New insights for exploring the risks of bioaccumulation, molecular mechanisms, and cellular toxicities of AgNPs in aquatic ecosystem. Water 14:2192

Reddy N, Han S, Zhao Y, Yang Y (2013) Antimicrobial activity of cotton fabrics treated with curcumin. J Appl Polym Sci 127:2698–2702

Ren T, Dormitorio TV, Qiao M, Huang T-S, Weese J (2018) N-halamine incorporated antimicrobial nonwoven fabrics for use against avian influenza virus. Vet Microbiol 218:78–83

Román LE, Gomez ED, Solís JL, Gómez MM (2020) Antibacterial cotton fabric functionalized with copper oxide nanoparticles. Molecules 25:5802

Roy S, Shankar S, Rhim J-W (2019) Melanin-mediated synthesis of silver nanoparticle and its use for the preparation of carrageenan-based antibacterial films. Food Hydrocoll 88:237–246

Sadalage PS, Nimbalkar MS, Sharma KKK, Patil PS, Pawar KD (2020) Sustainable approach to almond skin mediated synthesis of tunable selenium microstructures for coating cotton fabric to impart specific antibacterial activity. J Colloid Interface Sci 569:346–357

Saha J, Mondal MIH (2021) Antimicrobial textiles from natural resources. Elsevier, Amsterdam, pp 1–43

Sahu O, Singh N (2019) The impact and prospects of green chemistry for textile technology. Elsevier, Amsterdam, pp 367–416

Sanders D, Grunden A, Dunn RR (2021) A review of clothing microbiology: the history of clothing and the role of microbes in textiles. Biol Lett 17:20200700

Sarathi P, Thilagavathi G (2009) Synthesis and characterization of titanium dioxide nano-particles and their applications to textiles for microbe resistance. J Text Appar Technol Manag 6

Seino S, Imoto Y, Kosaka T, Nishida T, Nakagawa T, Yamamoto TA (2016) Antiviral activity of silver nanoparticles immobilized onto textile fabrics synthesized by radiochemical process. MRS Adv 1:705–710

Shahid M, Mohammad F (2013) Green chemistry approaches to develop antimicrobial textiles based on sustainable biopolymers a review. Ind Eng Chem Res 52:5245–5260

Sharma JL, Dhayal V, Sharma RK (2021) White-rot fungus mediated green synthesis of zinc oxide nanoparticles and their impregnation on cellulose to develop environmental friendly antimicrobial fibers. 3 Biotech 11:269

Soroh A, Owen L, Rahim N, Masania J, Abioye A, Qutachi O et al (2021) Microemulsification of essential oils for the development of antimicrobial and mosquito repellent functional coatings for textiles. J Appl Microbiol 131:2808–2820
Sun X, Cao Z, Porteous N, Sun Y (2012) An N-halamine-based rechargeable antimicrobial and biofilm controlling polyurethane. Acta Biomater 8:1498–1506
Szadkowski B, Śliwka-Kaszyńska M, Marzec A (2024) Bioactive and biodegradable cotton fabrics produced via synergic effect of plant extracts and essential oils in chitosan coating system. Sci Rep 14:8530
Szostak-Kotowa J (2004) Biodeterioration of textiles. Int Biodeterior Biodegrad 53:165–170
Tanasa F, Teaca C-A, Nechifor M, Ignat M, Duceac IA, Ignat L (2023) Highly specialized textiles with antimicrobial functionality—advances and challenges. Text 3:219–245
Tania IS, Ali M, Arafat MT (2021) Antimicrobial textiles from natural resources. Elsevier, Amsterdam, pp 189–215
Tawiah B, Badoe W, Fu S (2016) Advances in the development of antimicrobial agents for textiles: the quest for natural products. Review. Fibres Text East Eur 24:136–149
Varshney S, Pandey P, Gupta D, Sharma S (2020) Role of fabric properties, moisture and friction in transfer of bacteria from fabric to fabric. Text Res J 90:478–485
Vats M, Swami M, Sheokand B, Rawat V, Negi DS, Kumar A (2023) Antiviral and antimicrobial coatings based on functionalized nanomaterials. Elsevier, Amsterdam, pp 389–420
Vojnits K, Mohseni M, Parvinzadeh Gashti M, Nadaraja AV, Karimianghadim R, Crowther B et al (2024) Advancing antimicrobial textiles: a comprehensive study on combating ESKAPE pathogens and ensuring user safety. Materials 17:383
Windler L, Height M, Nowack B (2013) Comparative evaluation of antimicrobials for textile applications. Environ Int 53:62–73
Yılmaz F (2023) Application of Achillea millefolium as a natural antibacterial agent in finishing of textile. Fibers Polym 24:3175–3182
Yueh MF, Tukey RH (2016) Triclosan: a widespread environmental toxicant with many biological effects. Annu Rev Pharmacol Toxicol 56:251–272
Zhang G, Wang D, Yan J, Xiao Y, Gu W, Zang C (2019) Study on the photocatalytic and antibacterial properties of TiO2 nanoparticles-coated cotton fabrics. Materials 12:2010

Chapter 9
Biodegradable Textiles: Innovations and Applications

Maryam Zain, Farah Deeba, Maryam Ayyaz, Memona Yasmin, Saher Mahmood, Ali Haider, and Muhammad Shafiq

Abstract Biodegradable textiles are emerging as a critical innovation in the fashion and textile industries, designed to naturally decompose at the end of their lifecycle. These materials are primarily derived from renewable sources such as plants, algae, and microorganisms, which significantly reduce landfill waste and help mitigate microplastic pollution. As environmental concerns intensify, the demand for sustainable alternatives to conventional textiles has grown, fostering various innovations in this area. This chapter explores the pivotal role of biodegradable textiles in sustainable development and their potential to address the environmental challenges associated with traditional textile production. It discusses the innovative contributions of microbial biotechnology in enhancing the biodegradability of textiles and the development of smart textiles with environmentally friendly properties.

M. Zain (✉) · F. Deeba
Department of Biochemistry and Biotechnology, The Women University Multan, Multan, Pakistan
e-mail: maryam.9002@wum.edu.pk; farah.9003@wum.edu.pk

M. Ayyaz
Atta-ur-Rahman School of Applied Biosciences, National University of Sciences and Technology, Islamabad, Pakistan

M. Yasmin
Knowledge Research and Innovation Centre (KRIC), Lahore, Punjab, Pakistan

S. Mahmood
Department of Microbiology and Molecular Genetics, The Women University Multan, Multan, Pakistan
e-mail: saher.9006@wum.edu.pk

A. Haider
University Institute of Medical Laboratory Technology (UIMLT), The University of Lahore, Lahore, Pakistan

M. Shafiq
Institute of Clinical Pharmacy, Department of Pharmacology, Shantou University Medical College, Shantou, Guangdong, China
e-mail: drshafiq@stu.edu.cn

M. Khurshid et al. (eds.), *Microbial Biotechnology in Textile Innovation*, SDGs and Textiles, https://doi.org/10.1007/978-981-95-0347-6_9

Additionally, the chapter highlights the diverse applications of biodegradable textiles across sectors such as fashion, healthcare, agriculture, and broader sustainability efforts. Specifically, it addresses the growing trend of biodegradable textiles in fashion to reduce waste, their use in healthcare for disposable medical products, and their role in agriculture to improve soil quality. The chapter offers an in-depth exploration of the advancements in biodegradable textile technology and its promising potential for a more sustainable future.

Keywords Biodegradable textiles · Sustainable fashion · Smart textiles · Sustainable development · Waste reduction · Agricultural textiles · Biodegradable fabrics

9.1 Introduction

The textile industry is a significant contributor to global pollution due to its high consumption of resources such as energy, water, and chemicals. Textile processing involves various chemical-intensive procedures, including scouring, bleaching, dyeing, desizing, and finishing. These processes generate substantial waste and release hazardous chemicals into the environment, adversely affecting ecosystems and human health. To address these environmental challenges, industry professionals and researchers have been exploring biotechnological solutions, particularly the use of microorganisms, to enhance or replace conventional textile processing methods. Microorganisms such as algae, bacteria, and fungi produce enzymes that can catalyze reactions in a more environmentally friendly and efficient manner. Enzymatic degradation processes reduce reliance on toxic chemicals and improve the biodegradability of textile components, ultimately enhancing the industry's sustainability (Leal Filho et al. 2022; Uddin et al. 2023).

Biodegradable textiles are defined as textiles that break down with the help of microorganisms, primarily fungi and certain bacteria, converting them into biomass, carbon dioxide, and water. This process of biodegradation plays a crucial role in reducing textile waste, minimizing microplastic pollution, and significantly lowering the environmental impact of textile manufacturing. Biodegradable textiles are primarily composed of natural fabrics such as cotton and silk, as well as innovative biodegradable biopolymers. However, these natural alternatives often lack the stability and performance required for various textile applications (Alaswad et al. 2022; Brunšek et al. 2023). Therefore, researchers are continuously exploring ways to enhance these textiles and develop eco-friendly methods for their synthesis.

Several factors influence the synthesis and biodegradability of textiles. Fiber composition plays a crucial role, as natural fibers such as cotton, wool, and silk are inherently biodegradable and eco-friendly due to their cellulose- and protein-based structure. In contrast, synthetic fibers like nylon and polyester require specific environmental conditions to break down and often contribute to microplastic pollution. Additionally, textile processing methods, including dyeing and finishing treatments, significantly impact their biodegradability (Bao et al. 2024; Varnaitė-Žuravliova and Baltušnikaitė-Guzaitienė 2025). Enzymatic treatments, for instance, enhance the breakdown process by decomposing complex compounds and eliminating non-biodegradable components. Environmental conditions also play a vital role in textile degradation. Factors such as adequate oxygen levels, humidity, and bacterial composting activity facilitate decomposition. In contrast, landfill environments often lack these conditions, which limits the degradation process and prolongs the persistence of textile waste in the environment (Mehta 2023; Varnaitė-Žuravliova and Baltušnikaitė-Guzaitienė 2025).

Biodegradable textiles are materials that can be broken down into harmless, non-toxic components through microbial activity. Common types of biodegradable textiles include natural fibers, bio-based polymers, and protein-based fibers. Natural fibers, such as cotton, hemp, and bamboo, decay relatively quickly, especially in their unprocessed forms. Bio-based polymers, including polyhydroxyalkanoates (PHA), polylactic acid (PLA), and cellulose-based fibers, are utilized for their eco-friendly characteristics. Protein-based fibers, derived from natural sources like wool, casein, and silk, not only offer biodegradability but also enhance fabric properties (Lee 2016; Kumar et al. 2019).

Biodegradable textiles have diverse applications across various fields, including fashion, medicine, and environmental engineering. In the fashion and apparel industry, many brands are incorporating biodegradable materials to develop more environmentally friendly clothing lines, thereby reducing their carbon footprint (Mehta 2023; Rahaman et al. 2024). In medical applications, biodegradable textiles are widely used in surgical products such as implants, wound dressings, and scaffolds. For instance, biodegradable scaffolds made from hyaluronic acid are being explored for their compatibility with human tissues (Sharma et al. 2023; Zhang et al. 2024). This chapter aims to explore the challenges associated with producing biodegradable textiles and examine innovative approaches to enhance their sustainability. Additionally, the chapter evaluates the impact of dyes and finishing treatments on biodegradability. By the end of this discussion, readers will gain insights into natural methods for manufacturing biodegradable textiles, along with an understanding of the key challenges and future prospects that can improve their practical applications. Furthermore, the chapter highlights additional objectives and emerging applications of biodegradable textiles across various industries.

9.2 Biodegradable Textiles in Sustainable Development

Biodegradable fabrics have the potential to transform the textile industry by reducing its environmental footprint and aligning with sustainability goals. These fabrics offer promising solutions to the challenges posed by traditional textile waste. As ongoing research addresses current limitations, these materials may become even more crucial for ecological development (Rahaman et al. 2024). Biodegradable textiles play a vital role in minimizing the environmental impact of the textile industry.

Unlike synthetic fibers, which take decades to degrade and release harmful chemicals, biodegradable textiles decompose naturally, reducing landfill waste and pollution. Furthermore, producing textiles from recycled resources contributes to a more sustainable economy. Recent innovations, such as biofiber spinning machines, enable the creation of sustainable textiles from renewable animal and plant sources (Daria et al. 2020; Egan and Salmon 2021).

These textiles also drive advancements in material science by fostering the development of new resources, such as biopolymers, including polyhydroxyalkanoates (PHA) and polylactic acid (PLA). These polymers are increasingly used in textile manufacturing for their eco-friendly properties, particularly in sustainable packaging. Additionally, biodegradable textiles support the principles of a circular economy by allowing products to re-enter the ecosystem as non-toxic components at the end of their lifecycle. This minimizes waste and promotes environmental sustainability. With growing consumer awareness of sustainability, biodegradable textiles offer companies an opportunity to innovate and establish a unique market presence. Researchers are also exploring the use of biodegradable materials in smart textiles to develop eco-friendly wearable technology. Moreover, as environmental regulations become more stringent, adopting biodegradable textiles enables companies to meet green mandates while enhancing their brand image. This aligns with consumer preferences for eco-friendly products and supports regulatory compliance (Samir et al. 2022; Mehta 2023).

Another significant advantage of biodegradable textiles is their health and safety benefits. Many eco-friendly fabrics are non-toxic, making them safer for both workers and consumers. Additionally, novel antimicrobial fabrics synthesized from biopolymers reduce reliance on toxic chemicals while maintaining safety and functionality. By addressing these aspects, biodegradable textiles significantly contribute to sustainable development in the textile industry. This research not only reduces environmental impact but also fosters innovation and aligns with consumer preferences, ultimately supporting the transition to a more circular economy (Mehta 2023).

Sustainability in textile innovation is crucial for addressing the environmental and social challenges posed by the traditional textile industry. The creation of textiles often involves excessive water usage, chemical contamination, and substantial carbon emissions, contributing to ecological degradation (Harsanto et al. 2023). By focusing on sustainability, textile innovators can decrease the environmental footprint of resources, improve resource efficiency, and promote ethical practices within the industry. Sustainable innovations, such as the development of biodegradable fibers, closed-loop reprocessing systems, and eco-friendly dyeing methods, not only help reduce environmental impact but also meet the increasing consumer demand for eco-friendly goods. Furthermore, sustainable practices in industry promote greater transparency and social responsibility, benefiting workers and communities by encouraging fair labor practices and reducing health risks. Ultimately, sustainability in textile innovation is crucial for creating a more resilient and accountable industry that aligns with global ecological goals and the well-being of future generations (Blackburn 2009).

9.3 Environmental Challenges in the Textile Industry

Industrial advancements have significantly disrupted the global ecosystem, leading to widespread pollution. The textile industry is a major contributor to water contamination, releasing harmful pollutants such as chemicals, organic waste, and heavy metals into aquatic ecosystems. Toxic substances like mercury, lead, and arsenic accumulate in water bodies, posing serious risks to both human health and marine life (Sharma et al. 2021; Mitra et al. 2022). Textile dyeing processes discharge large volumes of synthetic dyes and hazardous chemicals, altering water composition, reducing photosynthesis, and endangering aquatic organisms.

Moreover, industrial runoff rich in phosphorus and nitrogen exacerbates eutrophication depleting oxygen levels and creating "dead zones" where marine life cannot survive. While environmental regulations such as the Clean Water Act in the United States aim to control industrial pollution, enforcement remains a challenge, particularly in developing nations where wastewater treatment infrastructure is inadequate. (Berradi et al. 2019; Oladimeji et al. 2024).

Beyond water pollution, textile manufacturing also contributes to air contamination through emissions of sulfur dioxide and nitrogen oxides, which lead to acid rain and respiratory health issues (Sharma et al. 2021; Mitra et al. 2022; Singh et al. 2023). As industrialization continues to expand, adopting sustainable practices and enforcing stricter environmental regulations are crucial to mitigating the industry's ecological impact.

9.4 Microbial Biotechnology in Enhancing Biodegradability

9.4.1 Microbial Enzymes and Textile Degradation

Biodegradable polymers have recently attracted significant attention as a potential solution to the growing problem of post-consumer textile waste (PCTW) accumulation, thereby promoting sustainable textile usage. Textile products primarily consist of fibers derived from either natural or synthetic polymers (Mehta 2023). Each fiber type presents distinct biodegradation challenges. Natural fibers are inherently biodegradable and renewable, ensuring that their PCTW does not persist in the environment for extended periods. However, the addition of synthetic dyes, surface treatments, and other chemical modifications during production to enhance functionality and performance may often impact their biodegradability. Moreover, the degradation of these fibers in natural soil or compost, under both abiotic and biotic conditions, is influenced by various factors such as structural composition, morphology, exposure duration, pH, moisture levels, and temperature (Young et al. 2022).

Conversely, petrochemical-derived synthetic fibers dominate the global textile fiber market, with an annual production of approximately 113 million tons. These fibers play a crucial role in the textile industry due to their superior durability and functionality, which many natural fibers lack. Biodegradation presents an opportunity to divert textile waste from landfills by utilizing alternative methods such as composting, anaerobic digestion, or bio-catalytically enhanced recycling processes. For instance, research has demonstrated that cotton fibers degrade more efficiently when treated with the cellulase-producing bacterium *Bacillus subtilis*, as opposed to traditional chemical treatments (Fouda et al. 2024). Similarly, under controlled conditions, *Bacillus licheniformis*, a protease-producing bacterium, has been shown to significantly degrade wool fibers. These findings underscore the growing potential of microbial enzymes in mitigating textile waste through environmentally friendly means (Singh et al. 2017).

Cellulose-degrading enzymes, including cellobiohydrolases I and II, endo-1,4-β-glucanases, and β-glucosidase, collectively break down cellulose into glucose. Additionally, newly discovered enzymes such as CBM33 and GH61 have been found to enhance cellulose degradation by interacting with fiber surfaces. The application of cellulases in the textile industry is well-documented, offering advantages such as reduced energy and water consumption while improving the sustainability of textile production. These enzymes are widely used in processes like biopolishing, biostonewashing, and bioscouring (de Souza Moreira et al. 2016).

Advancements in genetic engineering have further enhanced microbial efficiency in degrading textile polymers. For example, researchers have successfully engineered *Escherichia coli* to produce higher levels of PETase, an enzyme capable of breaking down polyethylene terephthalate (PET), a common polyester component (Peng et al. 2018). This genetic modification has shown significant potential in accelerating the decomposition of synthetic fibers, which are typically resistant to natural degradation (Tournier et al. 2023). By enabling the selective depolymerization of synthetic fibers, biotechnology facilitates the recovery of monomers and high-value products, thereby contributing to waste reduction and sustainability efforts (Adeoye et al. 2024). Additionally, the development of bio-based fibers designed to decompose naturally further supports a circular economy and minimizes environmental impact (Deckers et al. 2023).

The microbial breakdown of textile waste occurs through a sequence of steps. Initially, extracellular enzymes secreted by bacteria and fungi cleave large polymer chains in textile fibers into smaller molecules, such as monomers. These monomers are then absorbed by microorganisms and utilized as an energy source. Through metabolic pathways such as the citric acid cycle and glycolysis, these molecules are converted into biomass and byproducts, including carbon dioxide and water (Sun et al. 2018).

9.4.2 Innovations in Biodegradable Textile Production

Innovations in environmentally friendly fiber processing, particularly in sustainable dyeing and finishing techniques, have garnered significant interest from the textile industry. These technologies aim to enhance the quality and functionality of textiles while simultaneously reducing their environmental impact. Natural dyes, such as mahogany sawdust extract, have emerged as a viable alternative to synthetic dyes. The use of metallic mordants can further improve the colorfastness and durability of these natural dyes (Islam et al. 2024). Recent advancements have led to the development of eco-friendly one-step dyeing methods incorporating TiO_2 nanoparticles, which not only impart color but also provide antibacterial and UV-protective properties to textiles (Islam et al. 2024).

The adoption of organic fibers, such as cotton, hemp, and bamboo, reduces the reliance on synthetic chemicals, thereby minimizing air, water, and soil pollution (Anu et al. 2024). Moreover, integrating biodegradable textiles into the production process is essential for lowering industrial waste and reducing the textile sector's carbon footprint (Jadhav et al. 2024). While these innovations offer promising solutions for promoting sustainability in the textile industry, challenges related to scalability, cost-effectiveness, and consumer acceptance must be addressed to facilitate widespread adoption of eco-friendly practices.

Conventional textile dyeing is a major environmental concern due to its high consumption of chemicals and water. However, emerging dyeing techniques are being developed to mitigate these impacts. Enzyme-based dyeing methods, such as those utilizing laccase enzymes, present a more sustainable alternative by significantly reducing energy and water consumption. Additionally, the use of microbes, plants, and algae for the production of natural dyes offers a non-toxic and renewable substitute for synthetic dyes (Chaudhary et al. 2023).

9.4.3 Enzyme Applications in Fiber Processing

Cellulases are commonly used for denim finishing and biopolishing, while amylases assist in desizing cotton, enhancing fabric quality without the need for harsh chemicals (Chakraborty and Dunford 2024). Enzymatic treatments align with green chemistry principles and eco-friendly textile processing by significantly reducing energy and water consumption (Shah et al. 2024). However, despite their advantages, the widespread adoption of enzyme-based treatments is hindered by high costs, prolonged processing times, and the need for specialized equipment (Azzouz 2021). Integrating enzymatic methods into existing textile manufacturing processes remains a challenge, necessitating further research and technological advancements (Azzouz 2021). While enzyme-based treatments offer promising benefits for fiber degradation and durability, overcoming practical and economic barriers is essential for their broader implementation in the industry.

9.4.4 Microbial-Assisted Biopolymer Synthesis

In the textile industry, the microbially assisted synthesis of biopolymers, such as polyhydroxyalkanoates (PHAs), has gained significant attention (Dey et al. 2022). PHAs are biodegradable, biocompatible, and can be engineered with desirable properties for textile applications. These fibers offer a sustainable alternative to conventional plastics, as they can be spun into fabrics with strength, flexibility, and biodegradability (Kopf et al. 2023). Additionally, microbes can be utilized to produce other bioplastics, such as poly (ethylene furanoate) (PEF), presenting a promising approach to replacing petroleum-based plastics in textiles. Advances in microbial biotechnology are addressing key challenges, including high production costs and complex recovery processes, making bioplastics increasingly viable for large-scale use.

9.5 Smart Textiles with Biodegradability

One of the latest advancements in smart textiles is the development of self-healing materials capable of repairing minor damage. Research on textiles embedded with microcapsules containing enzymes or healing agents has demonstrated that fibers can regenerate when damaged. When integrated with conductive and antimicrobial properties, these functional fabrics offer exciting possibilities for wearable electronics while maintaining biodegradability (He et al. 2017). The rapid progress in self-healing and functional textiles is driven by advancements in smart materials and polymer technology. These textiles not only exhibit enhanced durability but also possess unique properties such as bioactivity and environmental responsiveness.

The creation of biodegradable fibers has been facilitated by innovations in spinning technologies, including melt, electro, and wet spinning. Electrospinning, in particular, has garnered attention due to its ability to produce nanofibers with high surface area, promoting faster degradation rates. These fibers are widely used in clothing and filtration materials (Ramkumar et al. 2008). Growing environmental concerns and the demand for sustainable materials have driven the development of biodegradable fibers through various techniques, including electrospinning, melt spinning, and solution-blow spinning, utilizing polymers such as poly(lactic acid) (PLA) and poly(butylene succinate) (PBS) (Fig. 9.1). These technologies not only enable the production of fibers with specific properties but also contribute to reducing plastic pollution.

New antibacterial knitted fabrics are being designed with a three-layer structure that enhances moisture transfer and maintains long-term effectiveness without relying on harmful chemicals (Tian 2023). While these innovations offer promising advancements, challenges remain in their commercialization and widespread adoption, necessitating further research and development to overcome these barriers (Celep et al. 2022).

Biodegradability refers to the ability of a polymer to break down through biological processes into fundamental substances such as water, carbon dioxide, methane, basic elements, biomass, and humic matter (Mudhoo et al. 2011). Microbial activity under aerobic and anaerobic conditions degrades polymer chains, leading to a reduction in molecular weight and a loss of physical, mechanical, and chemical properties (Silva et al. 2023). The rate of biodegradation in different environments is influenced by factors such as temperature, oxygen availability, moisture content,

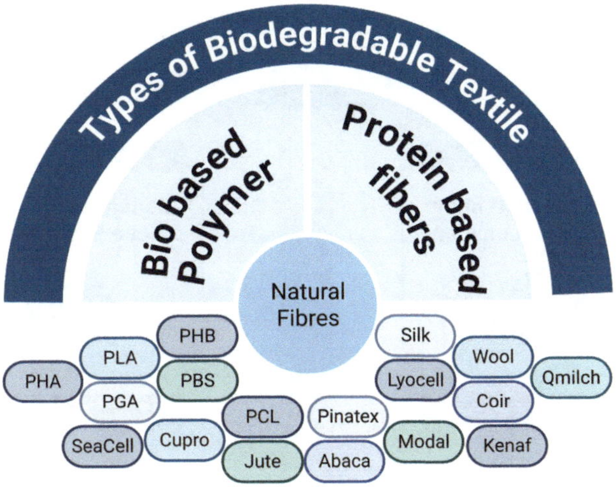

Fig. 9.1 Types of biodegradable textiles. (Created in BioRender. Shafiq, M. (2025) https://BioRender.com/soejvqe)

microbial composition, enzyme activity, and salt concentration (Grima et al. 2000). The key distinction between anaerobic and aerobic biodegradation lies in the presence or absence of oxygen (Fig. 9.2) (Reineke 2001).

Biodegradable materials are now being used to manufacture nonwoven fabrics, which are widely employed in hygiene and medical products. Techniques such as spun bonding and melt blowing have been adapted for processing polymers like PHAs and PLA, enabling the production of eco-friendly nonwoven materials (Ajmeri and Ajmeri 2016). Enzyme-assisted processing plays a crucial role in improving the mechanical properties and biodegradability of fibers. For example, the application of cellulases and lipases during fiber production enhances tensile strength and dye absorption in cotton and polyester fibers. These enzymatic treatments also introduce weak points in the polymer chains, making them more susceptible to microbial degradation (Kothawale et al. 2023).

Microbial treatments during textile production, such as biofilms and microbial inoculants, further enhance fiber biodegradability. These approaches are particularly effective in breaking down resistant fibers like polyester and nylon, which are otherwise difficult to degrade through conventional methods (Tokiwa et al. 2009). As microbial and enzymatic technologies continue to advance, they hold great potential for improving the sustainability of the textile industry.

9.6 Biodegradable Textiles: Applications

Biodegradable textiles have gained significant attention across various sectors, including fashion, healthcare, and agriculture, due to their potential to reduce environmental pollution and promote sustainability (Fig. 9.3).

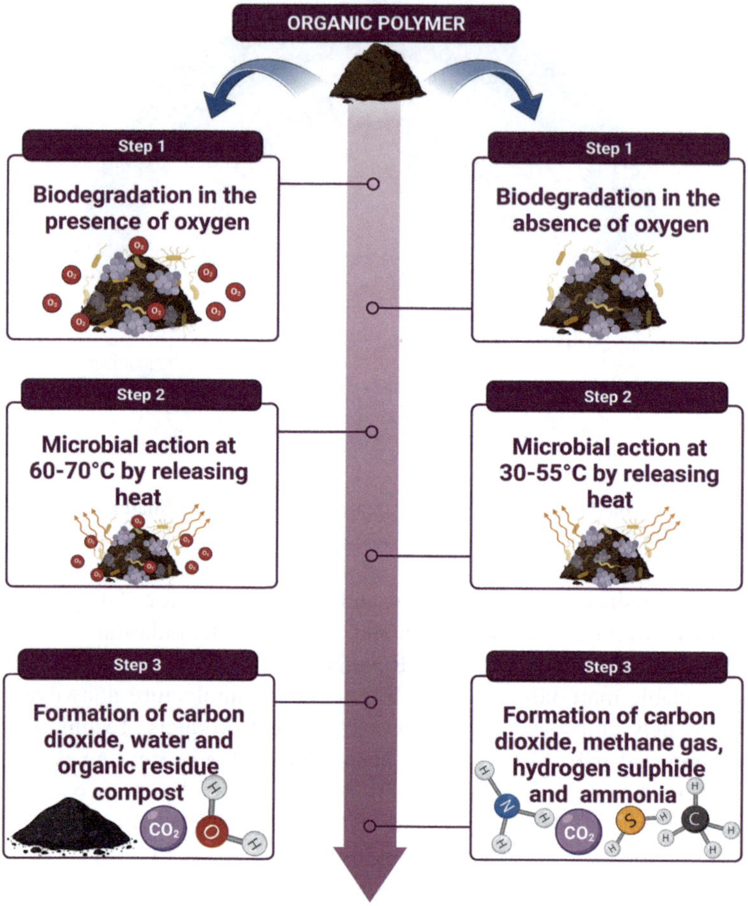

Fig. 9.2 Biodegradation of organic polymers aerobically and anaerobically. (Created in BioRender. Shafiq, M. (2025) https://BioRender.com/pli5kvd)

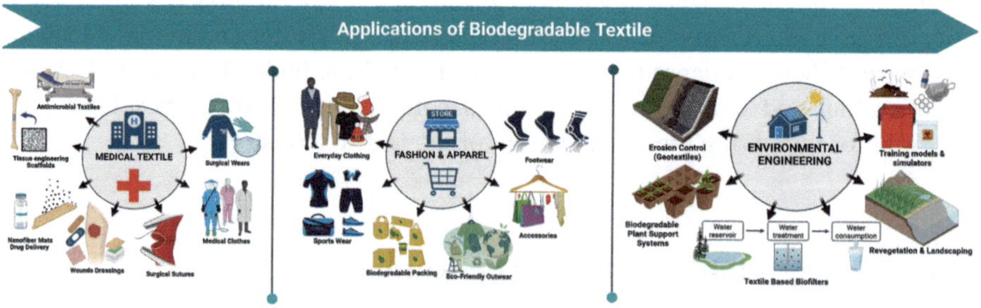

Fig. 9.3 Applications of biodegradable textiles. (Created in BioRender. Shafiq, M. (2025) https://BioRender.com/odi8gem)

9.6.1 Biodegradable Textiles in Fashion

Textiles and apparel are essential to human life; however, the fashion industry is one of the largest contributors to environmental pollution worldwide. Wastewater generated during textile manufacturing, along with fiber waste comprising of natural, inorganic, organic, and synthetic fibers such as nylon, acrylic, and polyester which poses a significant environmental challenge due to its non-degradable nature (Berradi et al. 2019; Azanaw et al. 2022). Furthermore, fast fashion trends have driven the mass production of inexpensive, trendy clothing, encouraging overconsumption and leading to an increase in post-industrial and post-consumer textile waste (Yaseen and Scholz 2019; Lee et al. 2023).

According to the 2018 report by the U.S. Environmental Protection Agency (EPA), textile waste accounted for 5.8% of the total municipal solid waste in landfills. The disposal of textile waste is now a growing global concern, prompting the textile and fashion industries to adopt sustainable strategies that address landfill scarcity and mitigate the environmental impact of textile waste (Yaseen and Scholz 2019).

The use of biodegradable textiles is emerging as a promising alternative to reduce the fashion industry's environmental footprint while simultaneously offering value addition and beneficial applications. Biodegradable textiles, including natural fibers such as silk, wool, cotton, and hemp, as well as innovative bio-based polymers derived from algae, mycelium, and regenerated proteins, present an eco-friendly solution. Additionally, materials like polylactic acid (PLA), produced from renewable sources such as corn starch or sugarcane, align with sustainability principles while providing unique textures, functionality, and style for fashion designers. Many fashion designers and brands have incorporated biodegradable textiles into their collections as part of their commitment to broader sustainability efforts (Rahman et al. 2022).

9.6.2 Biodegradable Textiles in Healthcare

The healthcare industry has undergone a profound transformation by adopting innovative textiles to enhance medical procedures, patient care, and sustainability. Medical textiles have versatile applications throughout the body, providing solutions for various complications, including tissue engineering, orthopedic and cardiovascular surgeries, dentistry, obesity management, and veterinary medicine (Li et al. 2015). To reduce the volume of waste generated in the healthcare sector, manufacturers and healthcare facilities are increasingly adopting sustainable practices by using eco-friendly and biodegradable materials.

Biodegradable polysaccharide fibers are derived from natural sources such as alginate, chitin/chitosan, collagen, catgut, branan ferulate, polylactic acid (PLA), poly(3-hydroxybutyrate-co-3-hydroxyvalerate) (PHBV), polyglycolide (PGA), polycaprolactone (PCL), polydioxanone (PDS), poly-2-hydroxyethyl methacrylate (PHEMA), polylactide (PLA), hyaluronic acid (HA), and polyhydroxy butyrate (PHB), among others (Kennedy and Knill 2006; Li et al. 2015). As these materials gradually degrade within the body, they reduce the stress-shielding effect and eliminate the need for additional procedures for removal. The use of biodegradable materials is gaining particular attention in wound dressings, sutures, and implants as an eco-friendly alternative to conventional petroleum-based materials, ultimately improving patient outcomes (Kennedy and Knill 2006).

Biodegradable materials such as PLA, PGA, PCL, and their copolymers are commonly used to manufacture absorbable sutures. Absorbable sutures, also known as dissolvable stitches, lose their tensile strength within 1–3 weeks and are fully absorbed within approximately 3 months. These sutures provide adequate tensile support during the healing phase, gradually breaking down in the body and preventing complications such as fibrotic processes and stone formation. Additionally, they do not require knots to secure the wound, significantly reducing the potential for scar formation (Kettle et al. 2010; Afewerki et al. 2023).

Another major application of biodegradable materials is in implants, which were first introduced in the 1960s. Today, biodegradable implants are available in various forms, including nails, meshes, pins, arrows, screws, wires, rods, plates, plugs, suture anchors, darts, and drug delivery devices. These implants are primarily composed of polyglycolic acid (PGA) or poly-L-lactic acid (PLLA) and offer financial, clinical, and psychological advantages over traditional metal implants. They provide adequate tensile strength while degrading at a rate similar to new tissue formation, thereby eliminating the need for follow-up visits and surgical removal. Furthermore, biodegradable implants do not produce imaging artifacts, allowing for a clearer assessment of wound healing (Prakasam et al. 2017).

Wound dressings, another crucial medical textile application, are used to cover wounds to prevent secondary injuries and facilitate the natural healing process. Several natural materials, such as polypeptides and polysaccharides, are utilized in wound dressings due to their biocompatibility, biodegradability, and specific molecular structures that promote wound healing. In addition to natural materials, synthetic polymers such as polyvinyl alcohol (PVA), polycaprolactone (PCL), polyurethane (PU), poly(lactic-co-glycolic acid) (PLGA), polylactic acid (PLA), and polyethylene oxide/polyethylene glycol (PEO/PEG) are also widely used. These materials offer functional versatility, as their structure and properties can be tailored to meet various medical needs. They can be fabricated into nanofibers, gels, and sponges that mimic the structure and function of human skin, enhancing their effectiveness in wound care (Kamysz & Kleczkowska, 2025).

The combination of biodegradability and smart technology has led to the development of smart biodegradable textiles. Equipped with sensors, these medical textiles enable real-time monitoring of a patient's physiological condition. They are available in various forms, including wearable sensors, fiber optic sensors, biosensors, nanosensors, and materials for wound care, tissue engineering, and drug delivery systems. Biodegradable sensors can be used to measure various physiological activities, such as glucose levels, body temperature, blood flow, and dissolved oxygen levels (Massaroni et al. 2015).

Synthetic biodegradable polymers, such as polyanhydrides, polyesters, polyurethanes, polyphosphazenes, and poly(glycerol sebacate), are being utilized as scaffolds in tissue engineering. These scaffolds serve as templates for the attachment, proliferation, and differentiation of cells (BaoLin and Ma 2014). Furthermore, these materials have found application in targeted drug delivery systems, addressing the limitations associated with conventional, non-biodegradable drug delivery methods.

In summary, the integration of smart technology with biodegradability has the potential to revolutionize patient care in the fields of patient monitoring, drug delivery, regenerative medicine, and infection prevention. This combination provides safer, more environmentally friendly solutions within the healthcare sector.

9.6.3 Biodegradable Textiles in Agriculture

Biodegradable agro-textiles are increasingly being used in agriculture to enhance productivity and promote environmental sustainability. The remarkable properties of agro-textiles which includes high tensile strength, non-toxicity, resistance to ultraviolet radiation and microorganisms, and biodegradability make them ideal alternatives to conventional petrochemical-based agro-textiles, addressing various agricultural challenges. Their primary functions include filtration, anti-erosion protection, drainage, separation, sealing protection of geotechnical systems, and improving plant growth conditions (Sharma et al. 2022).

These textiles are used in the form of harvest nets to reduce the time needed to collect fruits, shade nets to improve and maintain fruit quality, and mulch mats and crop covers to promote homogeneity, control weeds, and reduce pesticide use. As a result, they help ensure healthier crops and higher yields (Bhavani et al. 2017; Sharma et al. 2022). Another popular application in horticulture is the use of degradable pots for transplantation, which support water, air, and root penetration into the pot's walls. These pots can be directly planted into the soil, thereby minimizing costs, labor, and environmental impact (Sharma et al. 2022).

Agro-textiles are manufactured from natural fibers, such as wool, sisal, coconut, cotton, hemp, flax, bamboo, jute, and palm tree fibers, or from biopolymers like polyhydroxyalkanoates (PHA), polylactide (PLA), poly(ε-caprolactone) (PCL), poly(butylenesuccinate) (PBS), and cellulose acetate (CA) (Daria et al. 2020). PLA is one of the most widely used and cost-effective biopolymers commercially available today (Daria et al. 2020). Furthermore, agro-textiles with specific characteristics and desired biodegradation times can be produced by combining natural fibers with various biopolymers. The optimal selection of agro-textiles depends on their mechanical, physical, hydraulic, and rheological properties (Daria et al. 2020).

Biodegradable agro-textiles not only protect crops but also reduce adverse environmental impacts. Additionally, they provide nutrients and organic matter to plants through gradual biodegradation, thereby improving plant growth. The primary biodegradation mechanism involves the enzymatic activity of soil microorganisms, which can be measured using standardized tests (Song et al. 2009).

The use of biodegradable agro-textiles represents a step towards eco-friendly and sustainable farming practices, enhancing crop productivity, soil conservation, resource optimization, and environmental preservation. These innovations are set to become the cornerstone of sustainable agricultural practices, enabling farmers to adopt regenerative, eco-friendly methods to meet growing agricultural demands.

9.6.4 Biodegradable Textiles for Sustainability

The packaging sector has made significant strides in replacing conventional synthetic polymers with biodegradable polymers, aiming to produce products through sustainable processes that generate minimal waste. These biodegradable packaging materials are now being used across various industries, including food, textiles, biomedicine, cosmetics, and personal care, pharmaceuticals, and e-commerce. A high-quality packaging material should exhibit physical, mechanical, optical, and barrier properties comparable to those of conventional plastic packaging. Several biological materials, such as plant, animal, bacterial, or fungal-derived polysaccharides, polyphenols, proteins, lipids, nucleic acids, and biopolymers like poly(3-hydroxybutyrate) (PHB) or poly(lactic acid) (PLA), are being used to develop biodegradable packaging materials. The functionality of these materials is further enhanced by incorporating bioactive ingredients such as antioxidants, antimicrobials, flavonoids, and vitamins (Shaikh et al. 2021).

Other applications of biodegradable materials include their use in the construction industry as insulation for heating and cooling buildings, helping to address the impacts of climate change. Additionally, due to their high tensile strength, these materials are increasingly used in road construction. Natural fibers are also capable of removing heavy metals, and biodegradable textiles made from these fibers are used for medium- and short-term filtration or drainage purposes (Chen et al. 2024). The automobile industry is incorporating woven and non-woven biodegradable textiles for various functions, including sound and heat insulation, aesthetics, and comfort. Furthermore, biodegradable textiles are used in landscaping for reinforcement or protection, such as providing temporary cover for soil to prevent erosion caused by wind or water (Adak et al. 2025).

Another important application of biodegradable textiles is the development of intelligent geotextiles through the integration of sensors. One example is the incorporation of optical fiber sensors into geotextiles, which, when applied to railways, dams, slopes, and embankments, can provide real-time data on temperature, humidity, and mechanical deformation (Wu et al. 2020).

In summary, biodegradable textiles offer a wide array of applications, demonstrating their versatility in reducing waste, promoting sustainability across various sectors, and supporting ecological health by minimizing pollution. They are poised to play a critical role in developing a healthier planet by encouraging industries to adopt greener practices and fostering a circular economy.

9.7 Environmental Impact of Biodegradable Textiles

The environmental impact and benefits of biodegradable textiles are substantial, especially in tackling the challenges posed by non-biodegradable synthetic materials. Primarily made from biopolymers, biodegradable textiles provide a sustainable alternative that helps mitigate pollution and the accumulation of waste (Nair et al. 2013).

9.7.1 Life Cycle Assessment of Biodegradable Textiles

Life Cycle Assessment (LCA) is an analytical tool used to assess the environmental impact of a product or service throughout its life cycle. It focuses on each phase of the product's journey. LCA helps estimate the environmental consequences at every step, from raw material extraction to disposal (Roy Choudhury 2014). For instance, the extraction of natural fibers, such as cotton and hemp, involves high water consumption and pollution from pesticides and fertilizers. Despite these impacts, natural fibers are considered a more sustainable alternative to synthetic fibers. By applying LCA, brands can preselect materials that are less harmful to the environment during production, thus reducing ecological damage (Chen et al. 2021; Gautam et al. 2024).

The textile processes of spinning, weaving, dyeing, and finishing involve significant energy and water consumption, producing waste and emissions. Traditional dyeing methods, in particular, release toxic chemicals into waterways and require vast amounts of water (Azanaw et al. 2022). Through LCA, textile companies can assess their practices and explore opportunities to adopt cleaner technologies, minimizing waste and energy consumption. The distribution phase, which involves transporting textiles to retailers and consumers, results in substantial carbon emissions. Air transport, for example, has a much larger environmental impact than sea freight. By using LCA, brands can identify these impacts and adopt eco-friendly solutions, such as sustainable packaging and optimized shipping routes, to reduce their carbon footprint (Sivaramakrishnan 2019).

Consumer behavior during the usage phase, such as washing frequency, significantly impacts water and energy consumption. However, sustainable practices, like reducing wash cycles or using cold water, can help reduce these impacts. Educating consumers on appropriate care can increase the sustainability of biodegradable textiles (Gautam et al. 2024).

Finally, the end-of-life management phase addresses the disposal of biodegradable textiles. Ideally, these materials should be composted or recycled rather than sent to landfills, which can slow down the decomposition process. Effective end-of-life strategies are key to maximizing the benefits of biodegradable textiles and minimizing their overall environmental impact (Sivaramakrishnan 2019; Chen et al. 2021).

9.7.2 *Comparison with Traditional Synthetic Textiles*

The issue with traditional synthetic textiles lies in their inability to biodegrade. These textiles are typically made from petroleum-based fibers such as polyester, nylon, and acrylic. For instance, polyethylene terephthalate (PET), a commonly used synthetic fiber, does not biodegrade under standard conditions and poses significant environmental challenges as it accumulates in landfills and oceans (Egan and Salmon 2022). Additionally, synthetic textiles, particularly polyester, are carbon-intensive to produce. Manufacturing a single polyester t-shirt is estimated to emit 5.5 kg CO_2 equivalent (CO_2e), and the textile sector contributes around 10% of global carbon emissions, surpassing even the combined emissions of the aviation and shipping industries. In contrast, biodegradable textiles, such as organic cotton or PLA, typically have a lower Global Warming Potential (GWP). For example, the production of a cotton t-shirt emits about 2.1 kg CO2e, much less than its synthetic counterpart (Leal Filho et al. 2022).

The production of synthetic fibers requires high energy input and predominantly relies on fossil fuels. A significant portion of the carbon footprint of synthetic textiles comes from manufacturing energy alone (Egan and Salmon 2022). While the production of biodegradable textiles can still be energy-intensive, those made from renewable resources like hemp or organic cotton generally have a lower energy requirement (Stasiškienė et al. 2022).

Synthetic textiles are often produced using toxic chemicals that are harmful to ecosystems and human health. The pollutants released from dyeing processes can degrade water quality (Leal Filho et al. 2022). In contrast, biodegradable textiles typically involve fewer and less harmful chemicals, though some concerns remain if chemicals are not properly managed (Stasiškienė et al. 2022). To protect the environment, biodegradable materials must be free of harmful additives. The use of chemicals in conventional synthetic textile production has raised growing concerns about health risks, including skin irritations and respiratory issues for consumers and workers exposed over extended periods. Furthermore, synthetic fabrics release microplastics into the food chain via aquatic organisms, which may pose risks to human health. On the other hand, biodegradable textiles release fewer harmful chemicals during production. Organic cultivation of natural fibers minimizes the use of pesticides and fertilizers, thereby reducing health risks (Collie et al. 2024).

9.7.3 Carbon Footprint and Water Usage of Biodegradable Materials

The carbon footprint (CF) of biodegradable textile production varies significantly depending on the type of materials and processes used. Generally, natural fibers have a lower CF compared to synthetic fibers, and emissions are further reduced when eco-friendly production methods are employed (Leal Filho et al. 2024). Microbial bio-leather and other biodegradable textiles exhibit a lower carbon footprint than conventional leather and cotton, while also showing reduced human toxicity levels (Vinay 2022). Textile products in Pakistan had been found to have a total carbon footprint of 42,624.12 MT CO_2e, which includes the total carbon footprint from the traditional textile production. Biodegradable plastics, for example, produce fewer emissions (21.06–56.86 kg CO_2 equivalent) during production than traditional plastics (52.09–150.36 kg CO2 equivalent), with their emissions being 2.26–6.73 times lower. This makes the lifecycle impacts of biodegradable materials more consistent, reliable, and actionable. Secondary sources indicate that the production phase of traditional plastics accounts for 50.71–50.77% of their total carbon footprint, emphasizing the need for sustainable alternatives to address this issue (Cheng et al. 2023).

However, there is particular concern regarding the water footprint of biodegradable textiles, as their production often requires large amounts of water and can lead to environmental pollution. For instance, it takes approximately 2700 L of water to make a single cotton t-shirt (Compostables 2024). Water pollution from synthetic textiles is primarily caused by the dyeing and finishing processes, contributing to a water scarcity footprint in viscose textile production of about 60.511 m^3 H_2O eq/ton, with most of this impact arising from fiber production (Zhu et al. 2020). While biodegradable textiles also require water for cultivation, such as with organic cotton, their overall water consumption tends to be lower than that of conventional agriculture (Stasiškienė et al. 2022). Hemp, for example, thrives in poorer soils, requires less water than other crops, and is a more sustainable alternative.

On average, around 3400 m^3 of water are needed to produce 1 ton of cotton (Chen et al. 2021). Natural fibers like jute and kenaf, when incorporated into polymer blends, also grow in less fertile soils and have a reduced demand for water, further lowering water usage in textile production. A water footprint assessment suggests that, despite the high water requirement for cotton-based bio-composites, jute and kenaf-based composites have lower water footprints (Korol et al. 2019). This underscores the potential for optimizing material choices based on their environmental impacts.

9.7.4 Contribution to Waste Reduction

One way to address the issue of landfill waste and the harmful effects of microplastics is through the adoption of natural or biopolymer-based biodegradable textiles. By shifting towards biodegradable materials like polylactic acid (PLA) and natural fibers, the textile industry can contribute to a circular economy and help mitigate the growing problem of textile waste, which is expected to reach 4.5 billion tons by 2040 (Razzaq et al. 2024). he biodegradation process not only minimizes pollution but also promotes the health of ecosystems by preventing the accumulation of non-biodegradable waste (Harper and Moody 2023).

A practical way to manage silk waste, for example, is by reusing it to produce biocomposites and bioplastics. This approach not only benefits the environment but also enhances resource management (Qayoom and Manzoor 2024). Biodegradable plastics decompose more rapidly than traditional plastics due to the action of natural bacteria and other microorganisms in soil or water. Materials such as PLA break down into non-toxic components within months, offering a solution to reduce microplastic formation (Pooja et al. 2023).

Microplastic pollution poses one of the biggest environmental threats, impacting both aquatic ecosystems and human health. The persistence of microplastics in the environment is primarily caused by traditional plastics, which take centuries to decompose. A promising solution to curb microplastic pollution is the use of biodegradable materials, such as bioplastics and natural fibers (Mutmainna et al. 2024). Research has shown that certain fungi, such as *Penicillium brevicompactum*, can effectively degrade microplastics, achieving a mass reduction of up to 75% in just 14 days (Ferreira-Filipe et al. 2024). Furthermore, bacteria, algae, and other microorganisms have demonstrated the ability to break down microplastics, suggesting that the biodegradation of microplastics can complement the use of biodegradable materials (Sridhar et al. 2024).

The wastewater treatment process, when enhanced by biodegradable plastics, can play a crucial role in reducing microplastic pollution. These systems can capture and degrade microplastics before they enter aquatic environments, preventing ecological damage in rivers and oceans (Samuel et al. 2024). Additionally, the use of bio-based materials for plastic production results in biodegradable plastics that can reduce greenhouse gas emissions by approximately 45% compared to conventional fossil-based plastics (Zuiderveen et al. 2023).

9.7.5 Biodegradable Textiles in the Context of Circular Economy

Biodegradable textiles, such as PLA and PHA polymers, support the circular economy by replacing traditional textiles with more sustainable alternatives. Through strategies like reduction, reuse, and recycling, biodegradable textiles help address the growing issue of textile waste, which is expected to reach 4.5 billion tons by 2040. These textiles provide a green solution to the environmental challenges of the textile industry by reducing waste, promoting resource efficiency, and facilitating recycling and composting. As they naturally decompose, biodegradable textiles also help to reduce landfill waste. For instance, in 2018, 11 million tons of textile waste were landfilled in the US alone (Razzaq et al. 2024).

By being designed to break down naturally, biodegradable textiles contribute to a reduction in post-consumer textile waste (PCTW), minimizing landfill contributions (Mehta 2023). In general, these textiles require fewer resources and generate less energy during production compared to traditional textiles. This is because biodegradable polymers often benefit from less energy-intensive manufacturing processes (Kalita and Hakkarainen 2023).

Moreover, biodegradable textiles can be recycled and composted within a circular economy. The reuse and recycling of these textiles result in less waste and a lower environmental impact than conventional recycling methods (Abagnato et al. 2024; Ramírez-Escamilla et al. 2024). Composting these textiles offers an eco-friendly end-of-life option, enabling materials to reintegrate into the ecosystem as valuable nutrients that promote soil health. By incorporating compostable textiles into circular systems, it is possible to prevent long-term waste accumulation (Kalita and Hakkarainen 2023).

9.8 Economic and Technological Barriers to Biodegradable Textiles

The transition to biodegradable textiles is crucial for sustainability and reducing plastic pollution. However, this shift faces several challenges, including high production costs, processing complexity, and significantly lower market acceptance compared to synthetic fibers. One of the primary economic barriers to adopting biodegradable textiles is their high production cost. Manufacturers are often hesitant to use biodegradable polymers because they tend to be more expensive to produce than petroleum-based alternatives (Gupta 2024).

Despite the growing demand for biodegradable textiles, many consumers remain unaware of their environmental benefits. This lack of knowledge hinders investments in biodegradable technologies, reducing market demand and making it difficult for manufacturers to justify production (Stasiškienė et al. 2022). Furthermore, current waste management systems are not easily equipped to handle biodegradable materials. Even if recycling and processing facilities were developed, they could be prohibitively expensive for many companies to implement (Mehta 2023).

The development of technologies for producing and processing biodegradable materials is ongoing. Biopolymer processing techniques may differ from traditional textile manufacturing methods, complicating the adoption of biodegradable textiles (Aminoddin Abdellatif and Abdellatif 2021). While biodegradable textiles offer environmental benefits, they may also present performance challenges, such as lower moisture resistance, durability, and lifespan, when compared to synthetic textiles (Sharma et al. 2023). Many consumers prioritize performance characteristics when purchasing textiles, which could limit the acceptance of biodegradable options (Egan and Salmon 2022).

To gain a competitive edge in the market, biodegradable materials need improvements in performance through technological advancements. Additionally, the varying chemical properties of biodegradable plastics complicate recycling using conventional methods. Depending on environmental conditions, biodegradation may or may not occur. For example, PLA fibers do not degrade in soil at temperatures below 23 °C for over 12 weeks, while PHA fibers degrade under composting conditions at 58 °C (Borelbach et al. 2023). Due to the complexity of polymer blends, recycling biodegradable textiles remains challenging, and high costs are associated with producing lower-quality recycled products (Jönsson et al. 2021).

9.8.1 High Cost of Biodegradable Fibers Vs. Conventional Fibers

Due to high production costs and complex manufacturing processes, biodegradable fibers are not yet cost-effective. However, they are essential for transitioning from conventional synthetic fibers to more sustainable alternatives. Companies are unlikely to invest in biodegradable technologies unless consumer awareness of their environmental benefits increases, which remains relatively low (Stasiškienė et al. 2022).

Biodegradable fibers are typically made from natural polymers such as polylactic acid and thermoplastic starch, which are significantly more expensive than petroleum-based fibers (Mudoi and Sinha 2022). For instance, the production cost of bioplastics and other biodegradable materials ranges from €2/kg to €5/kg, whereas traditional petrochemical fibers cost only €1.20/kg (Song et al. 2009). Limited consumer demand for environmentally friendly alternatives further restricts the widespread adoption of bioplastics, particularly in cost-sensitive markets. As a result, manufacturers in such economies struggle to incorporate these materials due to their higher costs. Nonetheless, increasing support for biotechnology-driven advancements in bioplastics could contribute to achieving global sustainable development goals, which may eventually justify their higher production expenses (Gupta 2024).

9.8.2 Challenges in Scaling Up Microbial Technologies to Produce Biodegradable Textiles

Utilizing microbial technologies for the manufacture of biodegradable textiles holds significant potential; however, several challenges must be addressed to scale these technologies from the laboratory to industrial production. The cultivation of microbial cellulose and biopolymers requires highly specific and complex growth conditions, along with specialized substrates. Consequently, production costs remain high, making large-scale cultivation through genetic manipulation of microorganisms more challenging (Nayak et al. 2024).

Research and development (R&D) efforts, along with the industrial scaling of microbial processes for biodegradable textiles, demand substantial capital investment, typically ranging between $100 million and $1 billion. This cost encompasses biological production, bioprocessing, labor, pilot production, demonstration-scale operations, and the construction and commissioning of manufacturing facilities. Scaling up from laboratory processes to industrial production incurs significantly higher expenses (Crater and Lievense 2018).

Despite the potential benefits, these high costs make biodegradable options less viable in price-sensitive and competitive markets. Additionally, the complexities of fermentation and extraction techniques necessitate further optimization, which remains a major barrier to scalability (Mazotto et al. 2021; Shukla et al. 2021). Caling fermentation processes is further complicated by variability in conditions, leading to inconsistencies in product quality and yield. Contamination by other microorganisms poses another significant risk, potentially compromising product integrity and resulting in substantial financial losses. Quality control issues in large-scale production can undermine consumer confidence and slow market adoption. Furthermore, inadequate waste management infrastructure, including the lack of dedicated composting and recycling facilities for biodegradable textiles, raises concerns about proper disposal (Stasiškienė et al. 2022). The environmental benefits of biodegradable textiles can only be fully realized if effective disposal channels are established.

To advance microbial technologies, upfront R&D investment is necessary for developing new microbial strains and optimizing production processes (Ortelli et al. 2019). Additionally, the transition from laboratory-scale to industrial-scale production is often hindered by prolonged testing and validation phases, which can deter potential investors (Kumar et al. 2019).

9.8.3 Durability and Performance Trade-offs in Biodegradable Textiles

The shift to biodegradable textiles is driven by sustainability; however, it comes with trade-offs in durability, strength, and wearability. While biodegradable fibers offer environmental benefits over conventional synthetic fibers, their performance is often limited by a lack of structural stability.

9.8.3.1 Issues with Strength, Durability, and Wearability

Natural textiles, such as cotton, typically exhibit high moisture absorption, which can lead to swelling and degradation, ultimately reducing their durability (Islam et al. 2021). The biodegradability of natural fibers varies; for example, fibers derived from *Typha domingensis* demonstrate different levels of resistance to biodegradation. However, when treated with waterproofing resins, these fibers exhibit enhanced resistance to environmental stress and become significantly more durable than untreated fibers (Holanda et al. 2024).

Biopolymers such as polylactide (PLA) and soy protein contribute to the development of biodegradable textiles that perform satisfactorily while remaining environmentally friendly. However, their physical properties differ from those of traditional synthetic fibers (Yang et al. 2017). Additionally, incorporating recycled materials like biorPET can enhance biodegradability while maintaining textile integrity (Citron 2020). Such biodegradable textiles may retain sufficient mechanical strength, making them suitable for applications such as geocomposites, which absorb water and gradually decompose to support plant growth (Marczak et al. 2020).

9.8.3.2 Balancing Biodegradability with Product Longevity in Textiles

The challenge of creating textiles that are both biodegradable and durable lies in integrating these functionalities without compromising biodegradability or performance. Natural fiber-based biodegradable textiles, such as cellulose, and synthetic polymer textiles exhibit different degradation rates, each with distinct environmental impacts (Mehta 2023). Biodegradation is influenced by chemical processes involved in textile production; for instance, treatments that enhance durability may inhibit biodegradability.

In the short term, biodegradable materials can perform comparably to conventional textiles, but their mechanical properties may degrade over time (Vieira et al. 2013). Recent innovations include biodegradable textiles treated for properties such as fire resistance and water repellency while remaining eco-friendly (Umaquinga-Criollo et al. 2021). A key challenge in developing biodegradable polymers is balancing longevity with environmental safety, highlighting the need for further research into the mechanisms and microorganisms involved in biodegradation (Rana et al. 2014). Advances in enzymatic treatments and modern spinning techniques are enhancing the biodegradability of textiles without compromising their performance (Mukopadhyay 2010).

9.9 Consumer Perception and Adoption of Biodegradable Textiles

Awareness, material characteristics, and socio-economic factors all influence consumer perception and the adoption of biodegradable textiles. Research interest in this area is growing, particularly among younger consumers who are more familiar with environmental issues. Increased awareness of the environmental impact of textiles tends to drive greater interest in biodegradable options (Venkatraman et al.

2021). Individuals with higher education levels are more likely to be informed about eco-friendly textiles, as sustainability-related information significantly shapes consumer attitudes toward biodegradable products (Kang et al. 2013).

Generational differences also play a role in environmental awareness and purchasing intentions. Brands with strong social missions tend to attract millennials, while Generation Z places greater emphasis on the perceived value and risks associated with sustainable products (Lin and Chen 2022). Consumer preferences are influenced by factors such as age, gender, and background, with younger consumers generally prioritizing sustainability more than older generations. To encourage the widespread adoption of biodegradable textiles, effective sustainability marketing and positive consumer perception can build trust and influence purchasing decisions (Diandri and Yeshika 2024).

Additionally, consumer acceptance of biodegradable materials, such as bacterial cellulose, is highly dependent on sensory attributes, particularly texture and appearance (Lee 2016). A major barrier to adoption remains the high cost of biodegradable textiles, which poses a challenge in price-sensitive markets (Thakur et al. 2023). However, the growing interest in sustainable materials suggests significant market potential if production costs decrease.

9.9.1 Regulatory and Certification Challenges in Biodegradability Claims

Although specific requirements vary by region, regulatory frameworks for biodegradability are complex, encompassing national and international standards, testing protocols, and certification processes. These regulations ensure that materials degrade under specified conditions, facilitating safe disposal and market adoption. In the U.S. and Europe, key standards such as ASTM D6400 and ISO 17088 for industrial composting, as well as ASTM D7081 for marine environments, establish criteria for biodegradation, composting, and ecotoxicity (Greene 2014).

The European Economic Community (EEC) employs a tiered approach to assessing chemical biodegradability, using both screening and simulation tests to ensure logical and economically sound evaluations (Nyholm 1991). As a result, biodegradable plastics are subject to stringent regulations; they must undergo rigorous testing and certification to validate their claims and demonstrate compliance with established specifications (Müller 2005).

Eco-labeling helps consumers identify durable and environmentally friendly products. However, existing regulations often favor fossil-based materials, creating an uneven market. To foster fair competition for bio-based alternatives, outdated standards must be revised to align with current scientific advancements and industry practices (Ladu and Vrins 2019).

While regulatory frameworks are essential for ensuring biodegradability, they also present challenges. The lack of harmonization across different regions and industries can slow the adoption of biodegradable materials, highlighting the need for more standardized global policies.

9.10 Future Directions and Opportunities

9.10.1 Innovations in Biodegradable Textiles

9.10.1.1 Emerging Technologies (e.g., 3D Printing with Biodegradable Materials)

The biodegradable textile industry has been revolutionized by emerging technologies that offer sustainable and innovative solutions across various sectors. One of the most significant advancements is 3D printing with biodegradable materials, which has had a profound impact on industries such as medical textiles, sustainable fashion, and interior design (Bănică et al. 2024).

Biodegradable 3D-printed textiles utilize a range of materials, including biopolymers such as polyhydroxyalkanoates (PHA), polylactic acid (PLA), and starch-based polymers, as well as natural fibers like wool, silk, and cellulose-based fibers. Additionally, composite materials such as alginate, lignin, and chitosan are used to enhance functionality (Andrew and Dhakal 2022; Fijoł et al. 2022; Sadeghianmaryan et al. 2024). Key 3D printing techniques employed in biodegradable textile production include stereolithography (SLA), fused deposition modeling (FDM), and direct ink writing (DIW) (Alarifi 2024; Lodhi et al. 2024). These techniques address challenges like the mechanical strength of the fiber as it lacks flexibility and strength and the cost of the biodegradable material as it is quite expensive. However, challenges such as the mechanical strength and flexibility of biodegradable fibers, as well as the high cost of biodegradable materials, remain key limitations (Andrew and Dhakal 2022; Iftekar et al. 2023).

To address these challenges, bacterial cellulose and algal polymers are being explored as renewable alternatives to petrochemical-based synthetic materials, offering high-strength biodegradable fibers (Raj et al. 2023; Sarkar et al. 2023; Younes et al. 2023). The development of smart biodegradable fabrics integrates advanced functionalities such as moisture-wicking, self-healing, and temperature regulation, making them particularly suitable for medical applications and wearable technology (Lv et al. 2023).

Another breakthrough in biodegradability is the incorporation of enzyme-assisted decomposition systems into textiles. These systems facilitate polyethylene terephthalate (PET) degradation, allowing textiles to break down efficiently under controlled conditions, thereby reducing landfill waste (Magalhães et al. 2021).

Nanotechnology further enhances the durability and performance of biodegradable textiles by improving water resistance, antibacterial properties, and mechanical strength through nanoparticles such as cellulose nanocrystals (Dalbanjan et al. 2024). Meanwhile, chemical recycling processes support biodegradability by minimizing reliance on virgin resources, preserving material quality, and breaking polymers down into their monomeric forms (Harmsen et al. 2021).

Advancements in genetic engineering also contribute to sustainability by enhancing the environmental performance, biodegradability, and pest resistance of natural fibers such as wool and cotton (Šantek and Vujasinović 2023). Additionally, the use of naturally modified microorganisms in microbial dyeing has introduced eco-friendly natural dyes, positively impacting water recycling and material conservation (Pranta and Rahaman 2024).

The integration of AI-powered software in textile design has further improved material efficiency, particularly in the production of biodegradable textiles (Singh 2024). Moreover, biodegradable polymers engineered to degrade under specific environmental stimuli such as UV light, moisture, and pH changes offer enhanced biodegradability across diverse conditions without requiring industrial composting.

To achieve a circular economy, industries are increasingly integrating biosynthesis, nanotechnology, chemical recycling, and genetic engineering to develop high-performance sustainable materials that can replace traditional textiles (Bher et al. 2023; Harun-Ur-Rashid and Imran 2024).

9.10.1.2 Research in Bio-inspired and Synthetic Biology-Driven Textiles

Synthetic biology-driven and bio-inspired materials, grounded in biology, material science, and engineering principles, have been employed to create eco-friendly materials with advanced functions such as environmental responsiveness and self-cleaning (Byrne et al. 2018). Bioinspired fabrics mimic superhydrophobicity by leveraging the Wenzel and Cassie–Baxter models, inspired by the lotus effect (Li et al. 2024a, b). Bioprocess fabrication mimics natural biomineralization processes to produce sustainable products (Li et al. 2024a, b).

Advancements in synthetic biology have enabled innovations such as BioLace and living materials, which incorporate programmed cells for functional differentiation (Keune 2018). pecific techniques like inkjet printing and stereolithographic techniques have been employed to replicate biological structures, resulting in fabrics with enhanced capabilities and eco-friendliness (Alarifi 2024; Lodhi et al. 2024).

Recent developments have focused on merging synthetic biology with 3D printing to overcome challenges related to economic feasibility, material properties, and sustainability. This combination holds the potential to revolutionize the textile industry (Liu et al. 2019).

9.10.1.3 Self-Healing Biodegradable Textiles

Self-healing biodegradable textiles combine biodegradability with repair mechanisms, offering transformative potential in both medical and consumer applications. These textiles utilize intrinsic self-healing, relying on the material's innate repair properties, or extrinsic methods, which employ external agents or stimuli (Ganewatta et al. 2021).

In the medical field, biodegradable textiles such as hyaluronan-based fabrics are ideal for wound healing due to their biocompatibility, moisture management, and non-toxicity. Additionally, bioactive agent-releasing textiles, such as those infused with propolis, aid in tissue regeneration and infection prevention (Kubíčková et al. 2021).

Ongoing developments in biodegradability aim to balance the material's mechanical properties while enhancing its performance for broader use in diverse environments. This innovative process holds transformative potential for the creation of durable, high-performance materials (Abdel-Rahman and Abdel-Mohsen 2023).

9.10.1.4 Smart Biodegradable Textiles in Wearable Technology

Smart biodegradable fabrics, used in both medical applications and environmental protection, are integrated with self-adjustable apparel made from biodegradable fibers. These all-natural fabrics, such as alginate, silk, and viscose, offer numerous advantages due to their environmentally friendly nature and compatibility with the human body (Azam et al. 2023).

Through reactive structuring and deliberate conductive enhancements, such as the integration of liquid metal or graphene, these fibers and textiles can exhibit specific characteristics, including electromagnetic interference shielding and multiple sensory properties (Ding et al. 2024). These features pave the way for a new generation of textile materials that can enhance wearable healthcare, environmental sensing, energy harvesting, and other applications (Li et al. 2020).

However, several major challenges remain, including increased fabrication costs, poor durability for extended use, and insufficient material collection, manufacturing, processing methods, and disposal techniques. Addressing these issues could significantly expand the use of smart biodegradable textiles, improving both work efficiency and eco-friendliness (Rahaman et al. 2024).

9.10.2 Collaboration Between Biotechnology and Textile Engineering

Textile engineering has undergone a revolution due to bioengineering, with the adoption of various resource-saving practices replacing conventional technologies (Pedroso-Roussado 2023; Gudi et al. 2024). One such method includes the use of enzymatic treatments in fabric processing, dyeing, and finishing, which reduces the environmental impact by decreasing the consumption of water and energy. These practices have also facilitated the development of new fibers, including BT cotton, colored silk, and spider silk non-wovens. These materials not only enhance durability but also meet original equipment manufacturer (OEM) standards and promote environmental sustainability (Eskandar 2023).

Moreover, these advancements contribute to reducing manufacturing costs, making them an essential part of the global drive toward sustainable development. However, policies opposing the adoption of AI in emerging markets still persist among policymakers and economic operators, highlighting the need for interdisciplinary approaches and partnerships with industries (Singh 2024).

Interdisciplinary research is driving breakthroughs in the development of biodegradable textiles, uniting materials science, biotechnology, and environmental disciplines. Innovations, such as plant and animal waste-based films and bacterial cellulose-based textiles, showcase the power of collaboration in creating biodegradable and aesthetically appealing materials (Gausa and Tucci 2024; Thomas et al. 2024). Biotechnology-driven enzymatic processes further optimize energy-efficient textile production while aligning with the principles of the circular economy.

This holistic approach positions biotechnology as a cornerstone for transforming the textile sector into a more sustainable and efficient industry. However, widespread adoption requires strategic actions and increased awareness (Magalhães et al. 2021; Roadkasamsri et al. 2024; Voukkali et al. 2024).

9.11 Circular Economy and Sustainable Practices

9.11.1 Role of Biodegradable Textiles in a Circular Textile Economy

Biodegradable textiles play a crucial role in establishing the closed-loop system necessary for a circular textile economy to succeed in reducing waste and pollution. Aligning environmental, social, and governance (ESG) factors, the science-based targets initiative (SBTI), and life cycle assessment (LCA) is essential to achieving this goal (Yousaf and Aqsa 2023). This aligns with the principles of sustainability, as biodegradable fabrics, unlike synthetic ones that remain in landfills for many years, break down into their constituent parts once their lifespan is over (Kalita and Hakkarainen 2023; Bao et al. 2024).

For instance, recent developments have shown that bacterial cellulose produced from food industry waste supports the idea that interdisciplinary cooperation can lead to the creation of environmentally friendly fabrics. Both biodegradable polymers and eco-friendly dyes contribute to sustainable production processes by minimizing textile waste after consumption (Bao et al. 2024; Hossen et al. 2024).

However, technological constraints and low consumer awareness are major challenges. Progress in this field will require further research, creativity, and interdisciplinary collaboration, as well as targeted educational campaigns and favorable policy measures to encourage the adoption of this technology across industries (Ramírez-Escamilla et al. 2024).

9.11.2 Integration of Biodegradability with Recycling and Reuse

Biodegradability, when combined with recycling and reusability, represents one of the most innovative solutions for sustainable textiles. Among the various bio-based textiles, bioplastics emerge as natural alternatives to conventional plastics, as they are both recyclable and biodegradable (Rasheed et al. 2024).

Recycling methods for biodegradable materials include mechanical, chemical, and organic recycling, all of which increase the functionality of the material while reducing its environmental impact. For example, mechanical recycling focuses on textile recycling, chemical recycling breaks polymers down into their monomers, and composting helps reduce methane emissions in landfills (Okolie et al. 2023; Kothawade et al. 2024).

However, challenges such as underdeveloped recycling systems and the need for efficient material separation must be addressed to improve sustainability. Multidisciplinary and innovative recycling approaches should be viewed as essential strategies in achieving both biodegradability and reusability within a circular economy, which is crucial for combating climate change (Ferreira et al. 2023; Patel et al. 2023).

9.11.3 Policy and Regulatory Frameworks

Biodegradable textiles can help prevent post-consumer textile waste (PCTW), but this requires the establishment of international regulations. Currently, issues include PCTW being dumped in landfills and the technical complexities involved in reprocessing blended garments, underscoring the need for clear policies (Mehta 2023).

Regulatory measures must be tailored to the specific environments where biodegradation occurs, as most products like PLA (polylactic acid) do not decompose under typical conditions (Pinaeva and Noskov 2024). Additionally, incorporating biodegradable textiles into waste management practices especially for organic recycling and fiber reclamation requires policy guidance to prevent unintended contributions to microplastic pollution (Plakantonaki et al. 2023).

International standards should promote the development of recycling technologies for textile products and encourage the use of biodegradable materials to advance the sustainable textile industry. Establishing robust standards for environmental certifications and biodegradability tests is crucial for aligning sustainability efforts with consumer expectations (Damberg et al. 2024).

Currently, organizations like ISO (International Organization for Standardization) and OECD (Organization for Economic Co-operation and Development) are working to define biodegradability criteria and develop consistent testing standards, such as ISO 14855 and ASTM D5338, which assess how completely and quickly textiles degrade in various environments (Silva et al. 2023; Andréia da Silva et al. 2024).

Certifications like GOTS (Global Organic Textile Standards), OEKO-TEX, and Cradle to Cradle Certified™ help minimize environmental impact, study biodegradation in textiles, and provide consumers with confidence in the sustainability of their purchases (Damberg et al. 2024; Naikwadi 2024; Popescu 2024).

However, biodegradability testing faces several challenges: the lack of standardized procedures and the variability in field conditions. The rate of biodegradation is influenced by factors such as microbial activity, temperature, and moisture (Mehta 2023). Therefore, improvements in testing procedures, along with proper reference to global standards, are essential to ensure accurate assessments of biodegradable textiles (Damberg et al. 2024).

9.12 Conclusions

Advancements in the biodegradable textile industry represent a significant step forward in addressing issues related to traditional fabric materials. These advancements include the use of natural fibers such as cotton, hemp, bamboo, wool, and silk, which are sustainably processed, as well as synthetic biodegradable polymers like polylactic acid (PLA) and polyhydroxyalkanoates (PHA), which are durable and environmentally compatible (Brunšek et al. 2023; Uddin et al. 2024). Additional innovations incorporate both biodegradable and sustainable fibers alongside 3D printing technologies, which enhance the performance of hybrid materials in the most eco-friendly manner (Andrew and Dhakal 2022; Fijoł et al. 2022).

Microbial biotechnology has paved the way for new types of biomass, such as bacterial cellulose and fungal mycelia, produced through fermentation, while direct enzyme applications improve biomass digestibility for effective composting (Saharan et al. 2024). More specifically, these innovations include sustainable dyeing with safe chemicals, enzyme-assisted processes, and smart fabrics with self-healing and sensing functions, which blend cutting-edge technology with green concepts. These advances have applications across industries, including agriculture, fashion, healthcare, and more, promoting the reduction of waste and the creation of a circular economy in which used materials are returned to the environment for reabsorption (Andrew and Dhakal 2022; Khalid et al. 2024).

Microbial biotechnology has become a critical strategy in the development of green textiles, improving biodegradability and reducing waste. Microbial enzymes accelerate the breakdown of textile polymers, allowing them to decompose more rapidly and reducing landfill waste (Pedroso-Roussado 2023). Furthermore, biofibers like bacterial cellulose and fungal mycelium, fermented by microorganisms, are more efficient in resource utilization compared to traditional fibers (Amobonye et al. 2023; Mehrotra et al. 2023; Morrow et al. 2023). Genetic engineering is further enhancing these materials, increasing their durability and strength while maintaining their biodegradability (Gudi et al. 2024).

All these innovations align with the principles of a circular economy, aiming to produce textiles that can cycle back into nature's systems. Environmental challenges such as resource scarcity and pollution are being mitigated through microbial biotechnology, which is positively shaping the textile industry. Recent discoveries suggest that smart textiles with biodegradable components could represent the first step toward creating high-tech textiles that are also environmentally friendly (Kiran et al. 2024; Sabina and Hussain 2024).

The opportunities for using biodegradable textiles are vast, particularly in transforming the fashion industry and enhancing the sustainability of the global environment. These textiles help reduce textile waste and minimize the environmental impact of synthetic fabrics, which can take over 200 years to decompose. Additionally, they help decrease ocean pollution and protect human health by mitigating the negative effects of microplastic pollution. Derived from plant and microbial sources, biodegradable fabrics minimize the use of fossil fuels, reduce carbon emissions, and prevent the degradation of biodiversity. They offer a sustainable path for the fashion industry, encouraging responsible consumption and production practices that integrate well with circular economic frameworks through recycling and composting.

Moreover, these innovations create new economic opportunities by fulfilling the demand for sustainable solutions, supporting industrial growth, and creating jobs, all while meeting both environmental and financial objectives. Biodegradable textiles are therefore a key component in maintaining global sustainability, addressing waste and pollution reduction, and optimizing resource usage. Their implementation will help promote a circular economy based on the urgent environmental needs of the textile sector, while fostering innovation and responsibility. With ongoing technological advancements, these materials are set to revolutionize sustainability, offering increased durability and transforming both the textile sector and global sustainability strategies.

References

Abagnato S, Rigamonti L, Grosso M (2024) Life cycle assessment applications to reuse, recycling and circular practices for textiles: a review. Waste Manag 182:74–90

Abdel-Rahman RM, Abdel-Mohsen A (2023) Marine biomaterials: hyaluronan. Mar Drugs 21:426

Adak B, Chatterjee U, Joshi M (2025) Rubber-based sustainable textiles and potential industrial applications. Textiles, 5(2), 17. https://doi.org/10.3390/textiles5020017

Adeoye AO, Yelwa JM, Imam N, Quadri RO, Lawal OS, Malomo D et al (2024) Dynamics of transportation ecosystem, modeling, and control. Springer, Cham, pp 215–259

Afewerki S, Harb SV, Stocco TD, Ruiz-Esparza GU, Lobo AO (2023) Advanced technologies and polymer materials for surgical sutures. Elsevier, Amsterdam, pp 95–128

Ajmeri J, Ajmeri C (2016) Advances in technical nonwovens. Elsevier, Amsterdam, pp 227–256

Alarifi IM (2024) Revolutionising fabrication advances and applications of 3D printing with composite materials: a review. Virtual Phys Prototyp 19:e2390504

Alaswad SO, Mahmoud AS, Arunachalam P (2022) Recent advances in biodegradable polymers and their biological applications: a brief review. Polymers 14:4924

Aminoddin Abdellatif FHH, Abdellatif MM (2021) Green chemistry for sustainable textiles (Elsevier: Amsterdam) p 453-469

Amobonye A, Lalung J, Awasthi MK, Pillai S (2023) Fungal mycelium as leather alternative: a sustainable biogenic material for the fashion industry. Sustain Mater Technol 38:e00724

AndrélIa Da Silva S, Faccin DBJL, Cardozo NSRM (2024) A kinetic-based criterion for polymer biodegradability applicable to both accelerated and standard long-term composting biodegradation tests. ACS Sustain Chem Eng 12:11856–11865

Andrew JJ, Dhakal H (2022) Sustainable biobased composites for advanced applications: recent trends and future opportunities–a critical review. Compos C Open Access 7:100220

Anu K, Vishnumurthy K, Mahesh A, Natarajan K (2024) Carbon fiber-reinforced, activated carbon-embedded copper oxide nanoparticles/epoxy hybrid composites for EMI shielding in aircraft applications. Polym Bull:1–28. https://doi.org/10.1007/s00289-023-05112-w

Azam F, Ahmad F, Ahmad S, Haji AD (2023) In: Azam F, Ahmad F, Ahmad S, Haji A (eds) Circularity in textiles. Springer, Cham, pp 59–128

Azanaw A, Birlie B, Teshome B, Jemberie M (2022) Textile effluent treatment methods and eco-friendly resolution of textile wastewater. Case Stud Chem Environ Eng 6:100230

Azzouz L (2021) Development and characterisation of novel biocomposites fabricated using natural fibres and rapid prototyping technology PhD thesis, University of Hertfordshire

Bănică C-F, Sover A, Anghel D-C (2024) Printing the future layer by layer: a comprehensive exploration of additive manufacturing in the era of industry 4.0. Appl Sci 14:9919

Bao H, Hong Y, Yan T, Xie X, Zeng X (2024) A systematic review of biodegradable materials in the textile and apparel industry. J Text Inst 115:1173–1192

BaoLin G, Ma PX (2014) Synthetic biodegradable functional polymers for tissue engineering: a brief review. Sci China Chem. 1;57(4):490-500. https://doi.org/10.1007/s11426-014-5086-y. PMID: 25729390; PMCID: PMC4341840

Berradi M, Hsissou R, Khudhair M, Assouag M, Cherkaoui O, El Bachiri A et al (2019) Textile finishing dyes and their impact on aquatic environs. Heliyon 5:e02711

Bhavani K, Mallikarjun N, Sunilkumar NM (2017) Agro textiles-Their applications in agriculture and scope for utilizing natural fibers in agro tech sector. International Journal of Applied Home Science, 4(7), 653–662

Bher A, Cho Y, Auras R (2023) Boosting degradation of biodegradable polymers. Macromol Rapid Commun 44:2200769

Blackburn R (2009) Sustainable textiles: life cycle and environmental impact. Elsevier, Amsterdam

Borelbach P, Kopitzky R, Dahringer J, Gutmann P (2023) Degradation behavior of biodegradable man-made fibers in natural soil and in compost. Polymers 15:2959

Brunšek R, Kopitar D, Schwarz I, Marasović P (2023) Biodegradation properties of cellulose fibers and PLA biopolymer. Polymers 15:3532

Byrne G, Dimitrov D, Monostori L, Teti R, Van Houten F, Wertheim R (2018) Biologicalisation: biological transformation in manufacturing. CIRP J Manuf Sci Technol 21:1–32

Celep G, Tetik GD, Yilmaz F (2022) Next-generation textiles. IntechOpen, London

Chakraborty S, Dunford NT (2024) Algae: nature's renewable resource for fuels and chemicals. Biomass 4:329–348

Chaudhary S, Goyal S, Umar A (2023) Fabrication of biogenic carbon-based materials from coconut husk for the eradication of dye. Chemosphere 340:139823

Chen F, Ji X, Chu J, Xu P, Wang L (2021) A review: life cycle assessment of cotton textiles. Ind Text 72:19–29

Chen, L., Zhang, Y., Chen, Z. et al (2024) Biomaterials technology and policies in the building sector: a review. Environ Chem Lett 22, 715–750. https://doi.org/10.1007/s10311-023-01689-w

Cheng Z, Wang Y, Li N, Mu L, Li J, Tao J et al (2023) Replacing traditional plastics with biodegradable plastics: impact on carbon emissions. Engineering. https://doi.org/10.1016/j.eng.2023.10.002

Citron M (2020) Google Patents

Collie S, Brorens P, Hassan M, Fowler I (2024) Biodegradation behavior of wool and other textile fibers in aerobic composting conditions. Int J Environ Sci Technol:1–13. https://doi.org/10.1007/s13762-024-05802-6

Compostables G (2024) Why Should You Choose Biodegradable Clothing? https://www.greencompostables.com/blog/bioedgradable-clothing
Crater JS, Lievense JC (2018) Scale-up of industrial microbial processes. FEMS Microbiol Lett 365:fny138
Dalbanjan NP, Eelager MP, Korgaonkar K, Gonal BN, Kadapure AJ, Arakera SB et al (2024) Descriptive review on conversion of waste residues into valuable bionanocomposites for a circular bioeconomy. Nano-Struct Nano-Objects 39:101265
Damberg S, Saari UA, Fritz M, Dlugoborskyte V, Božič K (2024) Consumers' purchase behavior of cradle to cradle certified® products—the role of trust and supply chain transparency. Bus Strateg Environ 33:8280–8299
Daria M, Krzysztof L, Jakub M (2020) Characteristics of biodegradable textiles used in environmental engineering: a comprehensive review. J Clean Prod 268:122129
De Souza Moreira LR, Sciuto DL, Ferreira Filho EX (2016) An overview of cellulose-degrading enzymes and their applications in textile industry. In: New and future developments in microbial biotechnology and bioengineering. Elsevier, Amsterdam, pp 165–175
Deckers J, Manshoven S, Fogh L (2023) The role of bio-based textile fibres in a circular and sustainable textiles system. Copenhagen, European Environment Agency
Dey N, Kumar G, Vickram A, Mohan M, Singhania RR, Patel AK et al (2022) Nanotechnology-assisted production of value-added biopotent energy-yielding products from lignocellulosic biomass refinery–a review. Bioresour Technol 344:126171
Diandri PA, Yeshika A (2024) The moderating role of generation differences determinant factors of sustainable apparel behavior intention. Indones J Bus Entrep 10. https://doi.org/10.17358/ijbe.10.1.154
Ding Y, Jiang J, Wu Y, Zhang Y, Zhou J, Zhang Y et al (2024) Porous conductive textiles for wearable electronics. Chem Rev 124:1535–1648
Egan J, Salmon S (2021) Strategies and progress in synthetic textile fiber biodegradability. SN Appl Sci 4:22
Egan J, Salmon S (2022) Strategies and progress in synthetic textile fiber biodegradability. SN Appl Sci 4:1–36
Eskandar K (2023) Revolutionizing biotechnology and bioengineering: unleashing the power of innovation. J Appl Biotechnol Bioeng 10:81–88
Ferreira P, Apolinário A, Forman G (2023) Optimising textile biomaterial selection for sustainable product and circular design: practical guidelines for a greener future. Mater Circ Econ 5:14
Ferreira-Filipe DA, Oliveira L, Paço A, Fernandes AJ, Costa FM, Duarte AC et al (2024) Biodegradation of e-waste microplastics by Penicillium brevicompactum. Sci Total Environ 935:173334
Fijoł N, Aguilar-Sánchez A, Mathew AP (2022) 3D-printable biopolymer-based materials for water treatment: a review. Chem Eng J 430:132964
Fouda A, Alshallash KS, Atta HM, El Gamal MS, Bakry MM, Alawam AS et al (2024) Synthesis, optimization, and characterization of Cellulase enzyme obtained from Thermotolerant Bacillus subtilis F3: an insight into cotton fabric polishing activity. J Microbiol Biotechnol 34:207–223
Ganewatta MS, Wang Z, Tang C (2021) Chemical syntheses of bioinspired and biomimetic polymers toward biobased materials. Nat Rev Chem 5:753–772
Gausa M, Tucci G (2024) In: Gausa M, Tucci G (eds) Knitting food: food and eco-textiles: new perspectives for sustainable urban production systems. Cham, (Springer, pp 103–258
Gautam T, Priya MS, Priya SS (2024) Life cycle assessments of textile products: a study related to evaluating the environmental impact of production, use, and disposal. J Inform Educ Res 4. https://doi.org/10.52783/jier.v4i2.983
Greene J (2014) Starch polymers. Cham, Elsevier, pp 321–356
Grima S, Bellon-Maurel V, Feuilloley P, Silvestre F (2000) Aerobic biodegradation of polymers in solid-state conditions: a review of environmental and physicochemical parameter settings in laboratory simulations. J Polym Environ 8:183–195

Gudi S, Alagappan P, Raigar OP, Halladakeri P, Gowda RS, Kumar P et al (2024) Fashion meets science: how advanced breeding approaches could revolutionize the textile industry. Crit Rev Biotechnol:1–27. https://doi.org/10.1080/07388551.2024.2314309

Gupta H (2024) Biopolymers in the textile industry: opportunities and limitations. Springer, Cham, pp 227–240

Harmsen P, Scheffer M, Bos H (2021) Textiles for circular fashion: the logic behind recycling options. Sustainability 13:9714

Harper R, Moody SC (2023) Filamentous fungi are potential bioremediation agents of semi-synthetic textile waste. J Fungi 9:661

Harsanto B, Primiana I, Sarasi V, Satyakti Y (2023) Sustainability innovation in the textile industry: a systematic review. Sustainability 15:1549

Harun-Ur-Rashid M, Imran AB (2024) Emerging trends in engineering polymers: a paradigm shift in material engineering. Recent Prog Mater 6:1–37

He M, Wang X, Wang Z, Chen L, Lu Y, Zhang X et al (2017) Biocompatible and biodegradable bioplastics constructed from chitin via a "green" pathway for bone repair. ACS Sustain Chem Eng 5:9126–9135

Holanda FSR, Santos LDV, Melo JCR, Boge GM, Sussuchi EM, Nascimento BL et al (2024) Resistance of the fiber-derived geotextile from Typha domingensis submitted to field degradation. Sci Rep 14:8648

Hossen S, Mridha Y, Rahman A, Ouboucetta R, Amin MR (2024) Consumer Perceptions and purchasing trends of eco-friendly textile products in the US market. Int J Bus Econ 1:20–32

Iftekar SF, Aabid A, Amir A, Baig M (2023) Advancements and limitations in 3D printing materials and technologies: a critical review. Polymers 15. https://doi.org/10.3390/polym15112519

Islam JM, Mondal MIH, Das SC (2021) In fundamentals of natural fibres and textiles. Amsterdam, Elsevier, pp 657–690

Islam T, Islam KMR, Hossain S, Jalil MA, Bashar MM (2024) In: Kumar B (ed) In: Dye chemistry - exploring colour from nature to lab. IntechOpen, London

Jadhav HS, Fulke AB, Dasari LN, Dalai A, Haridevi C (2024) Plastic bio-mitigation by Pseudomonas mendocina ABF786 and simultaneous conversion of its CO2 byproduct to microalgal biodiesel. Bioresour Technol 391:129952

Jönsson C, Wei R, Biundo A, Landberg J, Schwarz Bour L, Pezzotti F et al (2021) Biocatalysis in the recycling landscape for synthetic polymers and plastics towards circular textiles. ChemSusChem 14:4028–4040

Kalita NK, Hakkarainen M (2023) Integrating biodegradable polyesters in a circular economy. Curr Opin Green Sustain Chem 40:100751

Kamysz, W., & Kleczkowska, P. (2025). Biological Macromolecule-Based Dressings for Combat Wounds: From Collagen to Growth Factors—A Review. Medical Sciences, 13(3), 106. https://doi.org/10.3390/medsci13030106

Kang J, Liu C, Kim SH (2013) Environmentally sustainable textile and apparel consumption: the role of consumer knowledge, perceived consumer effectiveness and perceived personal relevance. Int J Consum Stud 37:442–452

Kennedy J, Knill C (2006) Medical textiles and biomaterials for healthcare. Amsterdam, Elsevier, pp 3–22

Kettle C, Dowswell T, Ismail KM (2010) Absorbable suture materials for primary repair of episiotomy and second degree tears. Cochrane Database Syst Rev 2010:Cd000006

Keune S (2018) On Textile Farming, University of BORAS

Khalid MY, Arif ZU, Al Rashid A, Bukhari SMZS, Hossain M, Koç M (2024) Shape-memory and self-healing properties of sustainable cellulosic nanofibers-based hybrid materials for novel applications. Giant 19:100299

Kiran S, Rehman FU, Javed S, Abrar S, Mukhtar N, Moied A et al (2024) In: Ahmed S, Shabbir M (eds) Biopolymers in the textile industry: opportunities and limitations. Springer, Cham, pp 241–272

Kopf S, Åkesson D, Hakkarainen M, Skrifvars M (2023) Effect of hydroxyapatite particle morphology on as-spun poly (3-hydroxybutyrate-co-3-hydroxyvalerate)/hydroxyapatite composite fibers. Res Mater 20:100465

Korol J, Hejna A, Burchart-Korol D, Chmielnicki B, Wypiór K (2019) Water footprint assessment of selected polymers, polymer blends, composites, and biocomposites for industrial application. Polymers 11:1791

Kothawade SN, Pande VV, Suryawanshi J, Kumar A, Kumar P, Kumar A (2024) In: Ajay Kumar DK, Rajak PK, Kumar A (eds) Advances in sustainable biomaterials. CRC Press, Boca Raton, FL, pp 92–116

Kothawale SS, Kumar L, Singh SP (2023) Role of organisms and their enzymes in the biodegradation of microplastics and nanoplastics: a review. Environ Res 232:116281

Kubíčková J, Medek T, Husby J, Matonohová J, Vágnerová H, Marholdová L et al (2021) Nonwoven textiles from hyaluronan for wound healing applications. Biomolecules 12:16

Kumar D, Pandit PD, Patel Z, Bhairappanavar SB, Das J (2019) Microbial wastewater treatment. Elsevier, Amsterdam, pp 237–260

Ladu L, Vrins M (2019) Supportive regulations and standards to encourage a level playing field for the bio-based economy. Int J Stand Res 17:58–73

Leal Filho W, Perry P, Heim H, MaP D, Moda H, Ebhuoma E et al (2022) An overview of the contribution of the textiles sector to climate change. Front Environ Sci 10:973102

Leal Filho W, MaP D, Liakh O, Paço A, Dennis K, Shollo F et al (2024) Reducing the carbon footprint of the textile sector: an overview of impacts and solutions. Text Res J. https://doi.org/10.1177/00405175241236971

Lee Y (2016) Case study of renewable bacteria cellulose fiber and biopolymer composites in sustainable design practices. Sustain Fibres Fash Ind 1:141–162

Lee HS, Jung S, Lin K-YA, Kwon EE, Lee J (2023) Upcycling textile waste using pyrolysis process. Sci Total Environ 859:160393

Li G, Li Y, Chen G, He J, Han Y, Wang X et al (2015) Silk-based biomaterials in biomedical textiles and fiber-based implants. Adv Healthc Mater 4:1134–1151

Li Y, Chen W, Lu L (2020) Wearable and biodegradable sensors for human health monitoring. ACS Appl Bio Mater 4:122–139

Li J, Wang Q, Fu Z, Zou Z (2024a) Bioprocessing-inspired fabrication of calcite prismatic layers and growth mechanisms. Cryst Growth Des 24:9502–9510

Li W, Yang L, Huang J, Zheng C, Chen Y, Li Y et al (2024b) Progress on fiber engineering for fabric innovation in ecological hydrophobic design and multifunctional applications. Ind Chem Mater 2:393–423

Lin P-H, Chen W-H (2022) Factors that influence consumers' sustainable apparel purchase intention: the moderating effect of generational cohorts. Sustainability 14:8950

Liu J, Sun L, Xu W, Wang Q, Yu S, Sun J (2019) Current advances and future perspectives of 3D printing natural-derived biopolymers. Carbohydr Polym 207:297–316

Lodhi SK, Gill AY, Hussain I (2024) 3D printing techniques: transforming manufacturing with precision and sustainability. Int J Multidiscip Sci Arts 3:129–138

Lv X, Liu Y, Yu J, Li Z, Ding B (2023) Smart fibers for self-powered electronic skins. Adv Fiber Mater 5:401–428

Magalhães RP, Cunha JM, Sousa SF (2021) Perspectives on the role of enzymatic biocatalysis for the degradation of plastic PET. Int J Mol Sci 22:11257

Marczak D, Lejcuś K, Grzybowska-Pietras J, Biniaś W, Lejcuś I, Misiewicz J (2020) Biodegradation of sustainable nonwovens used in water absorbing geocomposites supporting plants vegetation. Sustain Mater Technol 26:e00235

Massaroni C, Saccomandi P, Schena E (2015) Medical smart textiles based on fiber optic technology: an overview. J Funct Biomater. 13;6(2):204-21. https://doi.org/10.3390/jfb6020204. PMID: 25871010; PMCID: PMC4493508

Mazotto AM, De Ramos SJ, De Brito LAA, Rocha NU, De Souza SA (2021) How can microbiology help to improve sustainability in the fashion industry? Environ Technol Innov 23:101760

Mehrotra R, Sharma S, Shree N, Kaur K (2023) Bacterial cellulose: an ecological alternative as a biotextile. Biosci Biotechnol Res Asia 20:449–463
Mehta S (2023) Biodegradable textile polymers: a review of current scenario and future opportunities. Environ Technol Rev 12:441–457
Mitra S, Chakraborty AJ, Tareq AM, Emran TB, Nainu F, Khusro A et al (2022) Impact of heavy metals on the environment and human health: novel therapeutic insights to counter the toxicity. J King Saud Univ Sci 34:101865
Morrow R, Ribul M, Eastmond H, Lanot A, Baurley S (2023) Bio-producing bacterial cellulose filaments through co-designing with biological characteristics. Materials 16:4893
Mudhoo A, Mohee R, Unmar GD, Sharma SK (2011) In: Sharma SK, Mudhoo A (eds) A handbook of applied biopolymer technology: synthesis, degradation and applications. Royal Society of Chemistry, Washington, DC, pp 332–364
Mudoi MP, Sinha S (2022) Natural fiber composites. CRC Press, Boca Raton, FL, pp 279–324
Mukopadhyay S (2010) Technical textile yarns. Elsevier, Amsterdam, pp 534–567
Müller RJ (2005) Biopolymers online: biology chemistry biotechnology applications, vol 10. Wiley, Hoboken, NJ
Mutmainna I, Gareso PL, Suryani S, Tahir D (2024) Microplastics from petroleum-based plastics and their effects: a systematic literature review and science mapping of global bioplastics production. Integr Environ Assess Manag. 20(6):1892–1911
Naikwadi S (2024) In: Sadhna, Kumar R, Greeshma S (eds) Climate action through eco-friendly textiles. Springer, Cham, pp 163–178
Nair AB, Sivasubramanian P, Balakrishnan P, Ajith Kumar KAN, Sreekala MS (2013) Environmental effects, biodegradation, and life cycle analysis of fully biodegradable "green" composites. Polym Compos:515–568
Nayak R, Cleveland D, Tran G, Joseph F (2024) Potential of bacterial cellulose for sustainable fashion and textile applications: a review. J Mater Sci:1–26
Nyholm N (1991) The European system of standardized legal tests for assessing the biodegradability of chemicals. Environ Toxicol Chem Int J. 10:1237–1246
Okolie O, Kumar A, Edwards C, Lawton LA, Oke A, Mcdonald S et al (2023) Bio-based sustainable polymers and materials: from processing to biodegradation. J Compos Sci 7:213
Oladimeji T, Oyedemi M, Emetere M, Agboola O, Adeoye J, Odunlami O (2024) Review on the impact of heavy metals from industrial wastewater effluent and removal technologies. Heliyon 10:e40370
Ortelli S, Costa AL, Torri C, Samorì C, Galletti P, Vineis C et al (2019) Innovative and sustainable production of biopolymers. In: Factories of the future: the Italian flagship initiative. Springer, Cham, pp 131–148
Patel B, Toksha B, Gupta P (2023) In: Ajay P et al (eds) Handbook of sustainable materials: modelling, characterization, and optimization. CRC Press, Boca Raton, FL, pp 19–38
Pedroso-Roussado C (2023) The fashion industry needs microbiology: opportunities and challenges. mSphere 8:e00681–e00622
Peng Y, Chen J, Song AY, Catrysse PB, Hsu P-C, Cai L et al (2018) Nanoporous polyethylene microfibres for large-scale radiative cooling fabric. Nat Sustain 1:105–112
Pinaeva LG, Noskov AS (2024) Biodegradable biopolymers: Real impact to environment pollution. Sci Total Environ 945:174445
Plakantonaki S, Kiskira K, Zacharopoulos N, Chronis I, Coelho F, Togiani A et al (2023) A review of sustainability standards and ecolabeling in the textile industry. Sustainability 15:11589
Pooja N, Chakraborty I, Rahman MH, Mazumder N (2023) An insight on sources and biodegradation of bioplastics: a review. 3 Biotech 13:220
Popescu A (2024) Enhancing environmental standards in textiles: the role of OEKO-TEX® standard 100. Bull Transilv Univ Bras V Econ Sci 17:59–68
Prakasam M, Locs J, Salma-Ancane K, Loca D, Largeteau A, Berzina-Cimdina L (2017) Biodegradable materials and metallic implants—a review. J Funct Biomater. 26;8(4):44. https://doi.org/10.3390/jfb8040044. PMID: 28954399; PMCID: PMC5748551

Pranta AD, Rahaman MT (2024) Extraction of eco-friendly natural dyes and biomordants for textile coloration: a critical review. Nano-Struct Nano-Objects 39:101243
Qayoom K, Manzoor S (2024) Recent sustainable practices for eco-friendly utilisation of sericultural waste products. J Exp Agric Int 46:918–927
Rahaman MT, Pranta AD, Repon MR, Ahmed MS, Islam T (2024) Green production and consumption of textiles and apparel: importance, fabrication, challenges and future prospects. J Open Innov Technol Mark Complex. 10:100280
Rahman SS, Siddiqua S, Cherian C (2022) Sustainable applications of textile waste fiber in the construction and geotechnical industries: a retrospect. Clean Eng Technol. 6:100420
Raj H, Tripathi S, Bauri S, Choudhury AM, Mandal SS, Maiti P (2023) Green composites using naturally occurring fibers: a comprehensive review. Sustain Polym Energy 1:10010
Ramírez-Escamilla HG, Martínez-Rodríguez MC, Padilla-Rivera A, Domínguez-Solís D, Campos-Villegas LE (2024) Advancing toward sustainability: a systematic review of circular economy strategies in the textile industry. Recycling 9:95
Ramkumar SS, Sata U, Hussain M (2008) In: Kendall RJ, Presley SM, Austin GP, Smith PN (eds) Advances in biological and chemical terrorism countermeasures. CRC Press, Boca Raton, FL, pp 203–227
Rana S, Pichandi S, Parveen S, Fangueiro R (2014) Biodegradation studies of textiles and clothing products. In: Roadmap to sustainable textiles and clothing: environmental and social aspects of textiles and clothing supply chain. Springer, Cham, pp 83–123
Rasheed T, Vattathurvalappil SH, Shaukat MM, Theravalappil R, Ali U, Ummer AC et al (2024) Recent updates on biodegradability and recyclability of bioplastics-Towards a new era in sustainability. Sustain Mater Technol 41:e01051
Razzaq S, Shahid S, Nawab Y (2024) Applications and environmental impact of biodegradable polymers in textile industry: a review. Int J Biol Macromol 282:136791
Reineke W (2001) In: Beek B (ed) Biodegradation and persistence. Springer, Cham, pp 1–161
Roadkasamsri V, Teekalee A, Asana U, Noulnim P, Sungvondee A, Li Y (2024) Interdisciplinary approach to design-led innovation for eco-friendly customer-centric clothing entrepreneurs. Int J Prof Bus Rev 9:e04476
Roy Choudhury A (2014) Environmental impacts of the textile industry and its assessment through life cycle assessment. In: Roadmap to sustainable textiles and clothing: environmental and social aspects of textiles and clothing supply chain. Springer, Cham, pp 1–39
Sabina R, Hussain N (2024) In: Sabina R, Hussain N (eds) Dye pollution from textile industry: challenges and opportunities for sustainable development. Springer, Cham, pp 351–404
Sadeghianmaryan A, Ahmadian N, Wheatley S, Sardroud HA, Nasrollah SS, Naseri E et al (2024) Advancements in 3D-printable polysaccharides, proteins, and synthetic polymers for wound dressing and skin scaffolding–A review. Int J Biol Macromol 266:131207
Saharan BS, Dhanda D, Mandal NK, Kumar R, Sharma D, Sadh PK et al (2024) Microbial contributions to sustainable paddy straw utilization for economic gain and environmental conservation. Curr Res Microb Sci. 7:100264
Samir A, Ashour FH, Hakim AA, Bassyouni M (2022) Recent advances in biodegradable polymers for sustainable applications. npj Mater Degrad 6:68
Samuel H, Makong F-D, Ori M (2024) Green chemistry strategies for mitigating microplastic pollution in aquatic environments. Asian J Environ Res 1:73–82
Šantek V, Vujasinović E (2023) Genetically engineered/modified fibres for the 21st century textiles and fashion. Eng Power Bull Croat Acad Eng. 18:20–35
Sarkar M, Upadhyay A, Pandey D, Sarkar C, Saha S (2023) In: Saha S, Sarkar C (eds) Biodegradable polymers and their emerging applications. Springer, Cham, pp 89–114
Shah F, Ranawat B, Patel V, Patidar SK, Thomas RM, Mishra S (2024) Environmental engineering and waste management: recent trends and perspectives. Springer, Cham, pp 297–318
Shaikh S, Yaqoob M, Aggarwal P (2021) An overview of biodegradable packaging in food industry. Curr Res Food Sci. 4:503–520. https://doi.org/10.1016/j.crfs.2021.07.005. PMID: 34401747; PMCID: PMC8349771

Sharma N, Allardyce B, Rajkhowa R, Adholeya A, Agrawal R (2022) A substantial role of agro-textiles in agricultural applications. Front Plant Sci. 13:895740. https://doi.org/10.3389/fpls.2022.895740. PMID: 35800605; PMCID: PMC9253683
Sharma P, Dutta D, Udayan A, Kumar S (2021) Industrial wastewater purification through metal pollution reduction employing microbes and magnetic nanocomposites. J Environ Chem Eng 9:106673
Sharma D, Srivastava S, Kumar S, Sharma PK, Hassani R, Dailah HG et al (2023) Biodegradable electrospun scaffolds as an emerging tool for skin wound regeneration: a comprehensive review. Pharmaceuticals 16:325
Shukla A, Parmar P, Goswami D, Gehlot Y, Vala J, Parmar N et al (2021) Green chemistry for sustainable textiles. Elsevier, Amsterdam, pp 173–189
Silva RRA, Marques C, Arruda T, Teixeira S, De Oliveira T (2023) Biodegradation of polymers: stages, measurement, standards and prospects. Macromolecules 3:371–399
Singh S (2024) Artificial intelligence in the fashion and apparel industry. Tekstilec 67:225–240
Singh J, Singhal N, Singhal S, Sharma M, Agarwal S, Arora S (2017) Environmental implications of rice and wheat stubble burning in north-western states of India. In: Advances in health and environment safety: select proceedings of HSFEA 2016. Springer, Cham, pp 47–55
Singh BJ, Chakraborty A, Sehgal R (2023) A systematic review of industrial wastewater management: evaluating challenges and enablers. J Environ Manag 348:119230
Sivaramakrishnan CN (2019) Life cycle analysis of textiles. https://textilevaluechain.in/in-depth-analysis/articles/research-paper/%C2%AClife-cycle-analysis-of-textiles
Song JH, Murphy R, Narayan R, Davies G (2009) Biodegradable and compostable alternatives to conventional plastics. Philos Trans R Soc B Biol Sci. 364:2127–2139
Sridhar S, Murugesan N, Gopalakrishnan M, Janjoren D, Ganesan S (2024) Removal of microplastic for a sustainable strategy by microbial biodegradation. Sustain Chem Environ 6:100088
Stasiškienė Ž, Barbir J, Draudvilienė L, Chong ZK, Kuchta K, Voronova V et al (2022) Challenges and strategies for bio-based and biodegradable plastic waste management in Europe. Sustainability 14:16476
Sun H, Zhao W, Mao X, Li Y, Wu T, Chen F (2018) High-value biomass from microalgae production platforms: strategies and progress based on carbon metabolism and energy conversion. Biotechnol Biofuels 11:1–23
Thakur N, Munjal K (2023) A study on consumers'perceptions towards eco-friendly fibre and its utilization in apparel industry in India. In: International Conference on Emerging Trends in Design & Arts. p 60–71
Thomas L, Neelima T, Archana T (2024) In: Thomas S, Chirayil CJ, Varghese RT (eds) Handbook of advanced biomass materials for environmental remediation. Sprinerg, Cham, pp 43–73
Tian X (2023) Study of multifunctional one-, two-and three-dimensional fiber-based triboelectric nanogenerators and self-powered sensors PhD thesis, The Hong Kong Polytechnic University
Tokiwa Y, Calabia BP, Ugwu CU, Aiba S (2009) Biodegradability of plastics. Int J Mol Sci. 10:3722–3742
Tournier V, Duquesne S, Guillamot F, Cramail H, Taton D, Marty A et al (2023) Enzymes' power for plastics degradation. Chem Rev 123:5612–5701
Uddin MA, Begum MS, Ashraf M, Azad AK, Adhikary AC, Hossain MS (2023) Water and chemical consumption in the textile processing industry of Bangladesh. PLoS Sustain Transform 2:e0000072
Uddin MK, Novembre L, Greco A, Sannino A (2024) Polyhydroxyalkanoates, A prospective solution in the textile industry-A review. Polym Degrad Stab 219:110619
Umaquinga-Criollo AC, Molina-Martínez CE, Guerrero-Loyola WM, Saltos-Echeverria TK, Jaramillo-Vinueza ED (2021) Innovation in biodegradable textile fibers for the creation of ecological textiles. In: Innovation and research: a driving force for socio-econo-technological development 1st. Springer, Cham, pp 545–554
Varnaitė-Žuravliova S, Baltušnikaitė-Guzaitienė J (2025) Biodegradable polymers and textiles. J Funct Biomater 16:26

Venkatraman P, Sayed U, Parte S, Korgaonkar S (2021) Consumer perception of environmentally friendly antimicrobial textiles: a case study from India. Int J Adv Sci Eng 7:1782–1793

Vieira AC, Guedes RM, Tita V (2013) Considerations for the design of polymeric biodegradable products. J Polym Eng 33:293–302

Vinay M (2022) Sustainable textiles through microbe-produced bioleather. Commun Eng 1:22

Voukkali I, Papamichael I, Loizia P, Economou F, Stylianou M, Naddeo V et al (2024) Fashioning the future: green chemistry and engineering innovations in biofashion. Chem Eng J 497:155039

Wu H, Yao C, Li C, Miao M, Zhong Y, Lu Y, Liu T (2020) Review of application and innovation of geotextiles in geotechnical engineering. Materials (Basel). 13(7):1774. https://doi.org/10.3390/ma13071774. PMID: 32290049; PMCID: PMC7179040

Yang M, Xu H, Hou X, Zhang J, Yang Y (2017) Biodegradable sizing agents from soy protein via controlled hydrolysis and dis-entanglement for remediation of textile effluents. J Environ Manag 188:26–31

Yaseen DA, Scholz M (2019) Textile dye wastewater characteristics and constituents of synthetic effluents: a critical review. Int J Environ Sci Technol. 16:1193–1226

Younes S, Arnold N, Paper M, Keil L, Awad D, Brueck T (2023) In: Abomohra A, Ende S (eds) Value-added products from algae: phycochemical production and applications. Springer, Cham, pp 403–459

Young B, Ingwersen WW, Bergmann M, Hernandez-Betancur JD, Ghosh T, Bell E et al (2022) A system for standardizing and combining us environmental protection agency emissions and waste inventory data. Appl Sci 12:3447

Yousaf AM, Aqsa R (2023) Integrating circular economy, SBTI, digital LCA, and ESG benchmarks for sustainable textile dyeing: a critical review of industrial textile practices. Global NEST J 25:39–51

Zhang H, Lin X, Cao X, Wang Y, Wang J, Zhao Y (2024) Developing natural polymers for skin wound healing. Bioact Mater 33:355–376

Zhu J, Yang Y, Li Y, Xu P, Wang L (2020) Water footprint calculation and assessment of viscose textile. Ind Text 71:33–40

Zuiderveen EA, Kuipers KJ, Caldeira C, Hanssen SV, Van Der Hulst MK, De Jonge MM et al (2023) The potential of emerging bio-based products to reduce environmental impacts. Nat Commun 14:8521

Chapter 10
Microbial Biotechnology for Textile Waste Management

Samman Munir, Sehar Aslam, Usman Ali Ashfaq, Muhammad Qasim, Ahsen Taqveem, Muhammad Khuram Razzaq, and Mohsin Khurshid

Abstract The extensive use of synthetic dyes in textile industry particularly anthraquinone and azo compounds has become a significant environmental concern. These dyes often contain heavy metals which results in increased Chemical Oxygen Demand (COD), Biological Oxygen Demand (BOD), and water toxicity of textile effluents. While conventional physical and chemical treatment methods exist, their high energy consumption and operational costs necessitate alternative solutions. Microbial bioremediation has emerged as a promising strategy for detoxifying and degrading these harmful compounds. Various microorganisms, including specific bacteria, fungi, and algae, can effectively degrade dyes through enzymes such as azoreductase, peroxidase, and laccase. Recent technological advances have enhanced bioremediation efficiency through the integration of microbial biosorbents, nanotechnology, genetic engineering, and specialized bioreactors. This chapter examines the recent biological approaches for dye contamination reduction and addresses the technical challenges in scaling these solutions for industrial applications.

Keywords Synthetic dyes · Azo compounds · Heavy metals · Water toxicity · Textile effluents · Microbial bioremediation · Biodegradation · Biosorbents

S. Munir · S. Aslam · U. A. Ashfaq · M. Qasim
Department of Bioinformatics and Biotechnology, Government College University Faisalabad, Faisalabad, Pakistan
e-mail: ashfaqua@gcuf.edu.pk; qasimawan@gcuf.edu.pk

A. Taqveem
School of Allied Health Sciences, Faculty of Sciences, Multan University of Science and Technology, Multan, Pakistan

M. K. Razzaq
Faculty of Agriculture and Veterinary Sciences, Superior University, Lahore, Pakistan

M. Khurshid (✉)
Institute of Microbiology, Government College University Faisalabad, Faisalabad, Pakistan
e-mail: mohsinkhurshid@gcuf.edu.pk

M. Khurshid et al. (eds.), *Microbial Biotechnology in Textile Innovation*, SDGs and Textiles, https://doi.org/10.1007/978-981-95-0347-6_10

10.1 Introduction

Water is an essential component of human life. Reports indicate that approximately 71% of the surface of earth is covered by water, with 97.5% being seawater and only 2.5% classified as freshwater (Petersen et al. n.d.). Industrial wastewater poses a significant threat to the environment and human health, with major contributions from the printing, rubber, textile, paper, and cosmetics industries (Younis et al. 2021). Among these, the textile industry is a major polluter, generating large volumes of harmful substances at various production stages (Kishor et al. 2021).

Synthetic textile dyes, if not treated properly, pose serious toxicological risks to living organisms (Shah and Rodriguez-Couto 2022). The textile dyeing industry consumes vast amounts of water, which is subsequently discharged as wastewater (Andleeb et al. 2010). It is estimated that between 40 and 65 L of effluent is released for every kilogram of fabric produced (Andleeb et al. 2010). This effluent contains significant quantities of residual dyes, varying in concentration depending on the dye class used (Cervantes and Dos Santos 2011; Al-Tohamy et al. 2022). Azo dyes are the most widely used in the textile sector, accounting for approximately 80% of all commercial dyes globally (Fu and Viraraghavan 2001; Shetty and Krishnakumar 2020; Tripathi et al. 2023a, b, c). The release of effluent containing these dyes in large volumes poses serious health risks, including shortness of breath, skin irritation, cancer, irregular heart rhythms, and dizziness (Khan and Malik 2018; Al-Tohamy et al. 2022).

In developed countries, various physical and chemical techniques are employed to treat dye-contaminated effluents. However, in developing nations, where textile production is largely carried out by small-scale industries, such techniques are rarely implemented. Physical methods for dye removal include natural biosorbents, coagulants, and filtration processes. Biosorbents such as chitosan, clays, rice husk, red mud, silica gel, peat, alumina, zeolite, activated carbon, bagasse pith, and chitin are commonly used to eliminate dyes from textile effluent (Gupta 2009; Prasetyo et al. 2022). However, a major challenge with biosorbents is their removal after treatment, as they retain harmful dyes without breaking them down. Similarly, chemical methods such as irradiation, oxidation, ozonation, electro-oxidation, sonolysis, and Fenton oxidation are restricted due to the high costs associated with ozone, radiation, and electricity consumption (Garg and Tripathi 2017).

In contrast, microbial treatment of textile effluents is considered a cost-effective and eco-friendly alternative. Unlike chemical treatments that require continuous application, microbes need to be introduced only once, as they proliferate while degrading textile dyes. Moreover, biological methods completely eliminate contaminants from wastewater, making them an environmentally sustainable solution. Microorganisms capable of degrading azo dyes produce various enzymes, including tyrosinase, laccase, azoreductase, MG reductase, and peroxidases (Rita de Cássia et al. 2013). Among these, peroxidases, laccases, and azoreductases play a crucial role in azo dye degradation (Joshni and Subramaniam 2011).

Textile wastewater is complex, containing various synthetic dyes, high concentrations of salts, and heavy metals (Alinsafi et al. 2006). Additionally, it often has high temperatures and fluctuating pH levels, which can hinder the ability of microorganisms to efficiently decolorize dyes when applied to untreated sewage. This chapter explores various technologies for treating textile wastewater, including microbial processes, bioreactors, phytoremediation techniques, and nanotechnology. Additionally, it discusses the challenges associated with these methods and how innovative approaches can be effectively implemented to purify dye-contaminated water, promoting a cleaner and greener environment.

10.2 Chemical Characteristics of Textile Effluent

The effluent generated by the clothing manufacturing sector is among the most hazardous of all industrial wastewater, both in terms of volume and composition (Yaseen and Scholz 2019). Various salts, including caustic soda (sodium hydroxide), are used in wet-processing procedures to fix colors, making textile wastewater highly alkaline, with a pH range of 10 to 11. This effluent is characterized by its vivid coloration, elevated temperature, high salt content, and significant oxygen demand for both chemical and biological processes. Additionally, it contains heavy metals such as zinc, chromium, copper, and cadmium, as well as surfactant molecules, all of which contribute to its environmental and human health hazards (Periyasamy 2024).

Textile dyes have been linked to various health issues, including dermatitis, ulceration, skin cancer, nasal septum damage, skin ulcers, bleeding, and nausea (Islam et al. 2011). Furthermore, numerous chemicals, such as leveling agents, dispersants, and fixatives, are commonly used in fabric dyeing. The improper discharge of textile wastewater into water bodies leads to severe water toxicity and environmental degradation (Al-Tohamy et al. 2022). One of its major consequences is significant water discoloration, which reduces light penetration in aquatic ecosystems, thereby hindering photosynthesis. This disruption also enhances heterotrophic processes, increasing both Chemical and Biological Oxygen Demand, which depletes dissolved oxygen levels and negatively impacts aquatic flora and fauna (Berradi et al. 2019; Dutta et al. 2024).

Many fish species are particularly vulnerable to the toxic effects of textile dyes, which have been reported to be carcinogenic, mitotically toxic, and mutagenic. These dyes can cause skin and gastrointestinal irritation and may contribute to kidney and respiratory failure. Several textile dyes, including Congo blue, rhodamine, procion red, and amido black, have been documented for their ecotoxic effects (Berradi et al. 2019; Dutta et al. 2024).

10.3 Methods of Dye Degradation

Given the aforementioned concerns, the detoxification and degradation of textile dyes are crucial. Traditional wastewater treatment methods, including physical and chemical processes, are often expensive, energy-intensive, and environmentally harmful. In contrast, biological treatment techniques have recently gained attention due to their cost-effectiveness, efficiency, and reduced production of secondary sludge (Bala et al. 2022a, b; Senthil Rathi and Senthil Kumar 2022). Moreover, microbial approaches have proven effective in degrading a wide range of organic and inorganic contaminants (Tripathi and Singh 2022; Das et al. 2023). Several studies have highlighted microbial treatment as a highly effective strategy for dye remediation (Pourbabaee et al. 2006). various approaches for the degradation of dyes in textile wastewater are illustrated in Fig. 10.1.

10.3.1 *Enzymatic Method*

Bacterial enzymes such as laccase (LAC), peroxidase (POD), and azoreductase (AZR) are widely used for dye degradation in wastewater treatment. These enzymes offer a more effective alternative to traditional dye removal techniques, as they can

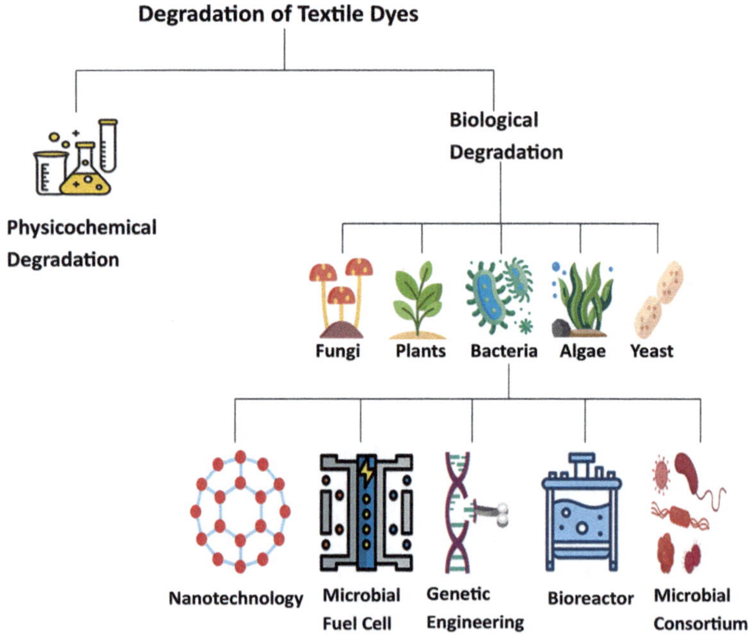

Fig. 10.1 Overview of the methods for the degradation of textile dyes

efficiently break down complex dyes while producing fewer toxic aromatic amines (Fig. 10.2). POD enzymes operate under moderate reaction conditions and exhibit broad substrate specificity, whereas LAC enzymes are particularly effective in degrading phenolic compounds. AZR enzymes, on the other hand, specialize in the breakdown of azo dyes. A deeper understanding of these enzymatic mechanisms can facilitate the development of improved strategies for textile effluent treatment.

Oxidoreductases (such as oxidase, peroxidase, and oxygenase) facilitate electron transfer from electron donors (reductants) to electron acceptors (oxidants) using cofactors like FMN, NADP, and FAD (Sarkar et al. 2017). Unlike LACs, PODs oxidize azo dyes through a similar mechanism but utilize hydrogen peroxide as an electron acceptor. The most commonly used ligninolytic peroxidases for dye degradation include manganese peroxidase (MnP) and lignin peroxidase (LiP), both of which are derived from *Phanerochaete chrysosporium* and have demonstrated a strong reduction potential for dye removal (Pham et al. 2018).

The optimal temperature for oxidoreductase activity typically ranges from 20 °C to 40 °C, though this may vary depending on the enzyme and the microbial source. Combining enzymes such as LAC, oxidoreductase, and AZR can significantly enhance the biodecolorization of synthetic dyes. However, the selection of specific enzymes and their combinations depends on the bacterial species involved and the structural characteristics of the dye. AZR, primarily produced by fungi and bacteria, plays a key role in the biodegradation of synthetic dyes. This reductive enzyme catalyzes the cleavage of azo bonds, enabling endophytic bacteria to break down azo dyes. AZRs are generally constitutively expressed but can also be induced upon exposure to dyes (Chen et al. 2011). They exhibit both broad and specific substrate ranges and are classified as either cytoplasmic or membrane-bound

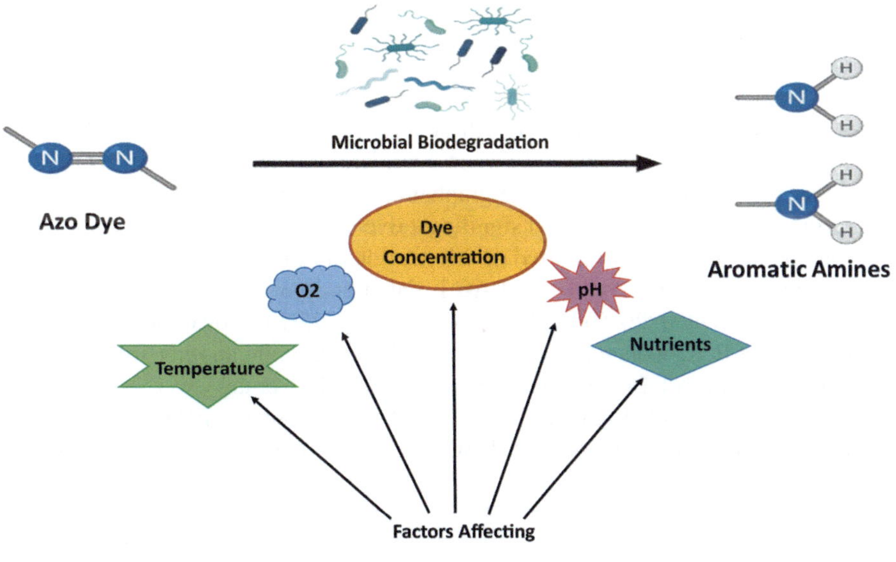

Fig. 10.2 Factors influencing the microbial degradation of azo dyes

(Jalandoni-Buan et al. 2010; Misal et al. 2011). The optimal pH for AZR activity typically ranges from 6 to 8, although certain enzymes may function efficiently under slightly acidic or basic conditions. The ideal temperature for AZR activity usually falls between 25 °C and 40 °C, though this can vary depending on the organism and enzyme. This enzymatic process requires reducing equivalents such as NADH, NADPH, and FADH2. AZRs cleave azo bonds, leading to the formation of aromatic amines (AAs), which are subsequently degraded aerobically into carbon dioxide and water (Chang et al. 2000).

LAC belongs to the oxidase enzyme family and contains multiple copper atoms within its active sites (Sridharan et al. 2021). Due to its broad substrate specificity, LAC can effectively degrade various textile dyes, including triarylmethane dyes, anthraquinone acid dyes, and azo dyes. Unlike azoreductases (AZRs), LACs degrade dyes without generating harmful by-products such as aromatic amines (AAs). Additionally, LACs can further break down AAs, providing an advantage over AZRs. The degradation of azo dyes by LAC begins with the removal of electrons, which are transferred to oxygen via the enzyme. Subsequently, the azo dye molecule undergoes hydrolytic cleavage by a water molecule, resulting in the release of nitrogen and protons. While LACs are constitutively expressed, their production has been observed to increase upon exposure to dye-contaminated water (D'souza et al. 2006).

Certain bacterial LACs degrade azo dyes by generating phenolic compounds rather than breaking the azo bond via a free radical reaction mechanism. The optimal pH for LAC activity typically ranges from 4 to 7, though some LACs remain functional at pH levels as high as 9. The ideal temperature for LAC activity generally falls between 25 and 60 °C, with some enzymes retaining functionality at even higher temperatures. A study demonstrated that LAC in combination with CuI nanoparticles (NPs) led to rapid azo dye degradation, outperforming other dye-removal methods, with in-silico studies predicting strong enzyme-dye interactions (Sridharan et al. 2021). Another study reported that *Bacillus* sp. strain MR-1/2 efficiently degraded textile dyes such as Congo red and Reactive Black 5 (RB5), simultaneously promoting mung bean growth by reducing azo dye toxicity (Shahid et al. 2018). Similarly, another study observed complete dye degradation at pH 7 using peroxidase (POD) immobilized onto graphene oxide (Vineh et al. 2020).

Further investigations revealed significant dye decolorization efficiencies across varying pH levels. A study reported decolorization rates ranging from 20% to 100% for different dyes at pH 5–9 using LAC isolated and purified from *Thermus thermophilus* (Navas et al. 2020). A study documented a 95% removal rate for Reactive Blue 21 using POD isolated from *Brassica oleracea*, while another study demonstrated that Remazol Brilliant Blue R and 2-naphthol orange were successfully decolorized at relatively high temperatures and neutral-to-basic pH levels following treatment with vault-encapsulated LAC enzymes. (Gao et al. 2022; Kalsoom et al. 2023).

10.3.2 Microbial Bioremediation

Various microorganisms, including yeast, bacteria, algae, actinomycetes, and fungi, have been utilized for dye degradation in textile effluents (Zimmermann et al. 1982; Zhou and Zimmermann 1993; Tripathi et al. 2023a, b, c) (Table 10.1).

10.3.2.1 Bacterial Bioremediation

Among microbial groups, bacteria play a crucial role in decolorizing textile dyes. From a biotechnological perspective, bacteria offer significant advantages due to their abundance of degradative enzymes, enabling the breakdown of various dye types (Saratale et al. 2011; Khan et al. 2013). A key benefit of using bacteria is their rapid growth rate and ease of cultivation. Additionally, bacteria can degrade chlorinated and aromatic hydrocarbon-based organic pollutants, utilizing them as energy sources (Yang et al. 2014). Through oxidation, they also convert sulfur-containing textile dyes into sulfuric acid (H_2SO_4) (Nguyen et al. 2016).

Numerous studies have documented that a single bacterial strain in a pure culture can effectively decolorize textile dyes. For example, Louati and colleagues demonstrated complete decolorization of all dyes at pH 8 using a strain of *Pseudomonas aeruginosa* (Gb30) (Louati et al. 2020). Meanwhile, Montañez-Barragán and his team observed that over 90% of dye removal occurred between pH 6 and 11 using *Halomonas* species (Montañez-Barragán et al. 2020). Another study showed that *Brilliant Crocein* dye was completely decolorized by *Providencia rettgeri* bacteria (Shi et al. 2021). Additionally, another group of researchers reported between 80% and 100% dye removal with both immobilized and free *Bacillus cereus* cells at temperatures of 45 °C, 37 °C, and 32 °C (Fareed et al. 2022). Srinivasan and colleagues achieved decolorization of 88.35–96.30% of various azo dyes (Srinivasan et al. 2022). In contrast, Du and colleagues reported 100% removal of crystal violet and malachite green dyes at pH levels of 3–10 and temperatures between 20 and 45 °C, after 12 h of incubation under optimal conditions, using a novel bacterial species, *Serratia* species (Du et al. 2023). Another study found that 98% of gentian violet was decolorized using an MT-1 isolate of naturally occurring *Aeromonas caviae* (Tripathi et al. 2023a, b, c).

Mixed bacterial cultures have been evaluated for their ability to degrade dyes in textile wastewater. Although single-species bacterial cultures can effectively remove dyes from textile effluent, mixed bacterial cultures can sometimes outperform single-species cultures due to their synergistic activities (Joshi et al. 2022). For instance, a study found 90% decolorization of dye at 35 °C through a combination of bacteria, including *Pseudomonas putida*, *Lactobacillus acidophilus*, and *Sphingomonas paucimobilis* (Ayed et al. 2021). Similarly, another investigation

Table 10.1 Degradation of textile dyes using microorganisms

Dye	Organism type	Species name	% Degradation	References
Malachite green	Bacteria	*Streptomyces chrestomyceticus*	96	Vignesh et al. (2020)
Methyl orange		*Aeromonas species*	99	Velusamy et al. (2022)
Rhodamine B		*Brevundimonas diminuta*	90–95	Saravanan et al. (2022)
Reactive black 5		*Sterigmatomyce Halophilus*		Dhanve et al. (2023)
Disperse blue		*Klebsiella pneumoniae*	90	Mustafa et al. (2021)
Malachite green, methylene blue, crystal violet		*Iodidimonas spp.*	>90	Ebihara et al. (2022)
Acid orange, rhodamine		*Bacillus subtilis*	78–98	Teja et al. (2022)
Methyl orange, methyl red		*Acinetobacter calcoaceticus*	90–98	Modi et al. (2022)
Fast red, Amaranth RI		*Gerobacillus thermoleovorans*	99	Gianolini et al. (2020)
Congo red		*Enterococcus Faecalis*	90.7	Wang et al. (2022)
Red HE8B, Remazol black		*Providencia species*	84	Uduma et al. (2023)
Reactive yellow		*Pseudomonas putida*	90	Ayed et al. (2021)
Carmoisine		*Oerskovia paurometabola*	91	Franca et al. (2020)
Malachite green		*Synechococcus elongatus*	99.5	Han et al. (2020)
Reactive black 5		*Bacillus cereus*	88.7	Sheela and Sadasivam (2020)
Reactive red		*Shewanella haliotis*	100	Radhika and Aruna (2022)
Rhodamine B		*Brevundimonas diminuta*	90–95	Saravanan et al. (2022)
Procion red		*Pseudomonas stutzeri*	90	Bera and Tank (2021)
Reactive blue		*Staphylococcus species*	97	Moyo et al. (2022)
Acid black		*Bacillus pseudomycoides*	96	Shamim Hossain et al. (2022)
Anthraquinone		*Pseudomonas species*	80	Kalyani et al. (2011)
Gentian violet, crystal violet		*Serratia spp.*	96.5	Du et al. (2023)
Joyfix red, reactive yellow		*Lysinibacillus sphaericus*	92.71, 96.30	Srinivasan et al. (2022)

Drimarene, Azo dyes	Fungi	*Aspergillus Foetidus*	95	Sumathi and Manju (2000)
Direct violet		*Candida albicans*	73–87	Vitor and Corso (2008)
Crystal violet		*Trichoderma asperellum*	97.05	Shanmugam et al. (2017)
Poly R		*Pseudoglarobasidium acaciicola*	82	Barathikannan et al. (2017)
Direct blue		*Aspergillus terreus*	98.4	Singh and Dwivedi (2020)
Acid orange, acid red		*Trametes villosa*	>90	Ortiz-Monsalve et al. (2019)
Methyl red, methylene blue, methyl orange	Algae	*Sargassum natanes*, *Sargassum fluitanes*	65, 100, 25	López-Miranda et al. (2020)
Direct red		*Chlorella pyrenoidosa*	96	Sinha et al. (2016)
Indigo		*Phormidium autumnale*	60	Dellamatrice et al. (2017)
Methylene blue		*Nostoc muscorum*	96	Sinha et al. (2015)
Malachite green, methylene blue		*Desmodesmus sp*	88.90, 64.4–98.4	Al-Fawwaz and Abdullah (2016)
Acid orange, methylene blue		*Chlorella vulgaris*	83	Srivastava et al. (2021)
C.I. acid red 2	Yeast	*Saccharomyces cerevisiae*	75	Vatandoostarani et al. (2017)

revealed that 93% of *Metanil Yellow* was degraded using a combination of *Marinobacter*, *Clostridium salibacter*, *Marinobacter*, and *Halomonas* (Guo et al. 2020). In another study, Joshi and colleagues demonstrated decolorization rates ranging from 24% to 94% with a combination of six microbial strains, including *Pseudomonas stutzeri* (AK1), *P. stutzeri* (AK3), *P. stutzeri* (AK2), *Bacillus* species (AK4), *P. stutzeri* (AK6), and *P. stutzeri* (AK5) (Joshi et al. 2022). A group of researchers similarly observed 85% decolorization of acid orange after nearly 23 h using a microbial consortium, SPB92, which consisted of four different bacterial species: *P. stutzeri*, *Kocuria rosea*, *B. flexus*, and *B. tequilensis*, along with yeast as an additional source (Bera et al. 2022). On the other hand, Neihsial and colleagues reported a similar level of dye degradation, ranging from 85% to 97%, using a mixture of bacteria from various genera, including *Comamonas*, *Acinetobacter*, *Citrobacter*, *Trichococcus*, *Dysgonomonas*, and *Erwinia* (Neihsial et al. 2022). In a recent study, it was shown that a consortium of three bacterial species, *Brevibacillus borstelensis*, *B. firmus*, and *B. subtilis*, decolorized dye more rapidly at lower dye concentrations; however, their degradation capacity decreased as the dye concentrations increased (Barathi et al. 2022).

10.3.2.2 Actinomycetes

Bacteria, particularly those in the actinomycete group, play a significant role in the degradation of dyes from textile effluents. Numerous studies have demonstrated that actinomycetes can successfully degrade textile dyes. For instance, one study found that *Streptomyces* species could decolorize dyes by approximately 3–10% (Zhou and Zimmermann 1993). In contrast, more recent research revealed that *Streptomyces* species (S27) effectively degraded dyes by up to 99% (Dong et al. 2019). Similarly, another group of researchers reported a decolorization rate of 64–94.7% for various dyes when treated with actinomycetes. They documented the degradation of triarylmethane dyes through *S. bacillaris*, utilizing both biodegradation and biosorption mechanisms (Adenan et al. 2022). A study on marine actinomycetes showed that these microorganisms could degrade up to 88% of acid black 1 dye at a concentration of 5 parts per million (Raja et al. 2016). In another investigation, a research group used four species of *Streptomyces* isolated from different soil samples to examine the degradation of the azo dye T-1824. They found that *Streptomyces* species could remove up to 97% of the dye from a starting concentration of 50 milligrams per liter of T-1824 (Kameche et al. 2022).

10.3.2.3 Fungi

Several fungal strains have been found to be effective in the treatment of textile effluent, specifically for the elimination of dye pollutants. White-rot fungi (WRF), in particular, are known to decolorize a wide range of xenobiotics, including synthetic dyes (Wesenberg et al. 2003). One study reported that *Ganoderma sp. En3*, a

WRF, successfully degraded over 90% of the dye in textile effluent (Zhuo et al. 2011). Another investigation found that 93% of the dye could be decolorized from textile wastewater using the WRF *Irpex lacteus* (Shin 2004). Similarly, *Tinea versicolor*, another WRF, exhibited significant potential for degrading textile wastewater, showing a 60% reduction in dye, while another researcher reported a 92% reduction (Amaral et al. 2004; Blánquez et al. 2008). A recent study found a 98% reduction in color using *Phanerochaete chrysosporium*, while another reported a 78% decrease in the original color of textile effluent after treatment (Assadi et al. 2001; Sedighi et al. 2009). Several studies have also reported that *Aspergillus* strains, including *Aspergillus fumigatus*, *A. lentulus*, and *A. niger*, can decolorize textile dyes (Jin et al. 2007; Kaushik and Malik 2010). Additionally, a study demonstrated that *KCTC 18061P*, a strain of *Citrobacter* species, rapidly degraded dyes from textile wastewater (Jang et al. 2007), while another investigation found that the intensity of the hue in wastewater was significantly diminished after treatment with *Clitocybula dusenii* (Wesenberg et al. 2003). Two other fungal species, *Phanerochaete chrysosporium* and *Curvularia lunata*, have also been shown to effectively degrade dye-containing textile wastewater in a bioreactor (Miranda et al. 2013). Furthermore, several other fungal species, such as *Thelephora* species, *Pleurotus flabellatus*, *Basidiomycetes*, *Pleurotus florida*, *Bjerkandera adusta*, and *Ganoderma lucidum*, have demonstrated potential for removing color from untreated wastewater generated by the textile sector (Diwaniyan et al. 2010; Anastasi et al. 2011).

While fungal species have shown potential for degrading dyes from textile wastewater, there are certain limitations to their use at a larger scale for water treatment. For example, fungal species require relatively acidic environments and long hydraulic retention times to fully degrade dyes (Rajamohan and Rajasimman 2013). The low pH requirement for fungal growth limits their applicability in many wastewater streams containing dyes, as these streams typically have a basic pH (Nordin et al. 2013). Furthermore, microbial consortia containing fungal species are not recommended for effluent treatment, as numerous fungal species have been shown to interfere with the activity and growth of other beneficial microbes (Chen et al. 2003). Additionally, while white-rot fungi (WRF) have been documented for their wide range of applications in textile wastewater treatment, their usage is limited due to their sensitivity to shear stress, need for oxygen, and competition with native microorganisms (Valentín et al. 2007). In comparison to filamentous bacteria and fungi, yeast has not been as extensively investigated or utilized in dye decolorization processes (Ngo and Tischler 2022). However, yeast has been shown to degrade dyes through biosorption (Ngo and Tischler 2022), with the AZR enzyme in yeast playing a key role in the degradation of synthetic dyes (Jafari et al. 2014). For instance, *Galactomyces geotrichum* (MTCC1360), a yeast strain, has been reported to degrade both azo and triarylmethane dyes (Jadhav et al. 2008). Another study demonstrated that this yeast strain could achieve 92% degradation of the azo dye Acid Scarlet GR (Guo et al. 2019). More recently, a study showed 82% degradation of synthetic dyes present in textile effluents using a yeast consortium comprising *Sterigmatomyces halophilus*, *Barnettozyma californica*, and *Yarrowia* species (Ali et al. 2021).

10.3.3 Phycoremediation

Studies indicate that synthetic dyes can serve as a source of energy and carbon for algae, which subsequently degrade azo dyes into aromatic amines (AAs) that are further transformed into organic and inorganic substances (Jamee and Siddique 2019). Several species, including *Phormidium*, *Chlorella*, *Synechocystis*, and *Oscillatoria*, have been employed in the investigation of dye degradation from textile effluent samples (Tripathi et al. 2023a, b, c). These algae contain various functional groups such as hydroxy, carboxy, amide, carbonyl, and phosphoryl within their cell walls, which facilitate the dye degradation process (Mahalakshmi et al. 2015). One study found that *C.I. Acid Red 2* dye was decolorized by 70% to 100% at a pH range of 3.5–9.5 when treated with *Chara vulgaris* (Mahajan and Kaushal 2020). Similarly, another study reported that *Chlorella vulgaris* decolorized between 45% and 80% of the dyes, while a separate investigation observed a remarkable 91% degradation of dye with *Cyanobacterium phormidium* (Dellamatrice et al. 2017; Boulkhessaim et al. 2022). Another study revealed that using the lipid-free and dry biomass of *Arthrospira platensis* resulted in 61.11% and 75.7% degradation of dyes, respectively (Alprol et al. 2021). Furthermore, *Arthrospira platensis* led to a 93% reduction in the color of methylene blue (Mansour et al. 2022). These findings suggest that algae could offer a promising and environmentally friendly solution for the effective degradation of dye-containing textile effluents.

10.3.4 Phytoremediation

Phytoremediation is a plant-based technique used to treat environmental pollutants, offering an affordable solution for removing various types of contaminants (Sharma et al. 2023). This technique involves several processes or mechanisms to degrade contaminants in wastewater. Widely accepted phytoremediation techniques include phytostabilization where compounds produced by plant species immobilize pollutants rather than degrade them. Whereas the phytoextraction is the process by which roots absorb pollutants and transfer them into the plant's shoots for breakdown. The rhizofiltration is a technique where pollutants are absorbed by roots and either stored or metabolized. The phytovolatilization is the transfer of pollutants from roots to leaves, where they evaporate into the air (Viana et al. 2021; Alsafran et al. 2022). Phytoremediation utilizes various plant interactions, including biological, microbial, chemical, and physical processes, in contaminated areas to reduce the harmful effects of pollutants (Wei et al. 2021). A study found that *Salvinia* species effectively degraded between 75% and 82% of various azo dyes (Biju et al. 2023). Similarly, another study reported the complete decolorization of remazol red azo dye and textile wastewater using *Alternanthera philoxeroides* (Rane et al. 2015). A recent investigation demonstrated that *Lemna minor* (duckweed) successfully decolorized 80.56% of methylthioninium chloride dye, while another study revealed 85% degradation of the same dye using *Azolla pinnata*, a floating plant (Imron et al. 2019).

10.4 Recent Techniques for Bioremediation of Dyes

Modern techniques, including nanotechnology and microbial fuel cells (MFCs), have shown significant potential in degrading contaminants from textile industries.

10.4.1 Microbial Fuel Cells

Microbial fuel cells (MFCs), a promising bioelectrochemical technology, can simultaneously generate electricity and remove both inorganic and organic contaminants. MFCs function by oxidizing natural organic matter, acting as catalysts in the process. This oxidation generates protons and electrons at the anode, which are then transferred to the cathode through an external circuit, completing the reaction and producing energy. The system consists of external resistance, applied voltages, electrode materials, co-substrates, and synthetic dyes (Sun et al. 2022).

In MFCs, microbes are commonly employed to generate electricity from biological sources (Das 2022). These systems are categorized as either single-chamber MFCs or two-chamber MFCs, the latter separated by a membrane. Within the anaerobic (anodic) chamber, microorganisms oxidize substrates to release electrons. One study described an electrochemical treatment process in which electron donors and acceptors interact to maximize the degradation of textile dyes in polluted effluents while generating electrical energy (Jain and He 2018). Another investigation documented dye decolorization using an electrochemical system, revealing that microorganisms oxidized acetate as an electron donor (Mu et al. 2009).

Further research demonstrated that MFCs not only degrade dyes but also generate electrical power. The study reported a maximum dye decolorization efficiency ranging from 96% to 100% and examined key factors influencing both decolorization and electricity generation, including anode and cathode materials, catalysts, substrates, cathode membranes, and bacterial strains (Yadav et al. 2022). Similarly, the application of MFCs in treating dyes and other environmental contaminants has been explored (Yaqoob et al. 2020).

10.4.2 Microbial Biosorbents

Biosorption is an energy-independent process that involves the binding of contaminants to biomolecules (Fomina and Gadd 2014). Microbial biosorbents are materials capable of adsorbing or absorbing xenobiotics from soil and water, including metals, drugs, carcinogens, and dyes (Kour et al. 2021; Tripathi et al. 2023a, b, c).

One study found that *Chromochloris zofingiensis*, a dye biosorbent, achieved 97% decolorization of textile dyes (Bayat et al. 2023). Similarly, another investigation demonstrated that *Enteromorpha prolifera*, a species of marine algae, removed

92.2% of Direct Red 23 from water at a pH of 2 through biosorption (Sun et al. 2019). Meanwhile, research on biosorbents derived from weeds revealed an adsorption capacity of 76.34 mg of methylthioninium chloride dye per gram, highlighting their potential for pollutant removal (Silva et al. 2019). Another study reported that 95% of dye was decolorized from textile dye-containing wastewater at a pH of 2.6 using orange seed (OS) powder as an effective adsorbent (Flores Alarcón et al. 2022). Furthermore, recent research has emphasized the effectiveness of microorganisms as biosorbents for removing heavy metals and textile dye contaminants from water (Tripathi et al. 2023a, b, c).

10.4.3 Genetically Engineered Microbes

Bioaugmentation using genetically modified strains has proven to be an efficient approach for removing environmental contaminants (Liu et al. 2019). Genetic engineering enables the modification of microorganisms, including yeast, fungi, and bacteria, through molecular techniques. These genetically engineered microbes (GEMs) can then be employed for environmental cleanup (Liu et al. 2019; Singh and Dwivedi, 2020).

One study investigated dye degradation using wild-type and mutant strains of the laccase enzyme, specifically D550G and Lacep69. The findings revealed that the D550G mutant exhibited a higher decolorization efficiency than the wild-type strain, achieving a 78% degradation rate for the azo dye acid violet (Bu et al. 2020). Similarly, another study demonstrated the rapid degradation of various azo dyes within 24 h by transferring the *azoK* gene, which encodes the AZR enzyme, from *Klebsiella pneumoniae* into *Escherichia coli* (Dixit and Garg 2018). Further research showed that the degradation rate of synthetic dyes could be increased by approximately 10% through the genetic modification of *E. coli* genes coding for AZR, an enzyme naturally present in *Pseudomonas luteola* (Chang et al. 2000).

10.4.4 Bioreactors

The development of highly efficient bioreactors is crucial for the effective treatment of dyes in industrial wastewater. These bioreactors facilitate complex chemical and biological processes, making them indispensable for wastewater management. Various types of bioreactors, including membrane bioreactors (MBRs), packed-bed, fixed-bed, fluidized-bed, rotating biological filters, and air-lift bioreactors, have gained significant attention in recent years (Tiwari et al. 2023).

Among these, MBRs have proven to be a viable and effective method for treating textile wastewater (Elmoutez et al. 2023). A study reported that the use of anaerobic MBRs could achieve a 99% elimination of textile dyes (Manzoor et al. 2022). Similarly, another investigation demonstrated the removal of more than 90% of azo dyes using anaerobic MBRs, highlighting their potential for large-scale application in dye contamination reduction.

In addition to MBRs, rotating bioreactors have shown promising results. A study utilizing white-rot fungi (WRF) in a rotating bioreactor for industrial effluent treatment successfully removed the dye RB5. The use of mixed microbial cultures resulted in over 93% decolorization of dye-contaminated wastewater (Kalnake et al. 2023).

10.4.5 Nanoparticles-Based Bioremediation

Nanotechnology-based bioremediation has emerged as a promising eco-friendly and efficient approach for eliminating dyes from textile effluents in recent years (Yaou Balarabe et al. 2023). The application of nanotechnology spans various industrial sectors, including medicine, environmental science, and agriculture (Bala et al. 2022a, b; Garg et al. 2023).

Nanoparticles (NPs) synthesized using bacteria play a crucial role in the biodegradation of pollutants at the nanoscale. Their unique structural characteristics have garnered significant attention for textile dye remediation. A study reported that eosin Y was effectively removed (approximately 95%) within 15 min of treatment using iron oxide NPs at a concentration of 75 ppm and pH 6 (Yaou Balarabe et al. 2023). Similarly, another investigation demonstrated that immobilized copper oxide NPs achieved nearly 90% dye removal efficiency from textile effluents (Darwesh et al. 2023).

Functionalized NPs further enhance the removal of synthetic dyes. Recent research has highlighted their potential for wastewater treatment, although further studies are needed to explore large-scale applications in dye remediation (Rani and Shanker 2023).

10.5 Future Perspectives

Although biological methods for treating textile wastewater have shown great potential, they still face several challenges, including the diverse and complex nature of synthetic dyes, the presence of inhibitory compounds, and the toxic properties of certain dyes. Further research is needed to assess the efficiency of bioreactors and to understand how variations in wastewater composition impact microbial adaptation. One major challenge is the degradation of dyes with intricate molecular structures and high molecular weights, which require specialized methods and advanced research to enhance their breakdown. Additionally, the decolorization of different synthetic dyes may necessitate multiple microbial strains, as some dyes require specific microbial species or tailored microbial consortia. Identifying the most effective microbial combinations remains an essential area of study. Another concern is the formation of toxic byproducts during dye degradation, which may have harmful environmental impacts and must be carefully evaluated. Furthermore, maintaining optimal enzyme activity for effective degradation requires careful monitoring of key parameters and supplementation with necessary cofactors. Identifying and optimizing these conditions is crucial to improving dye remediation efficiency. Addressing these challenges through continued research and technological advancements will enhance the sustainability and effectiveness of biological approaches for textile wastewater treatment (Patel et al. 2022).

Further studies are required to identify microorganisms capable of degrading and decomposing dye combinations, as real textile wastewater contains a diverse range of colors. It is essential to assess the actual ability of microbes to remove color from such wastewater in bioreactor setups constructed at industrial facilities that collect water directly from fabric dyeing units. Additionally, the process of immobilizing enzymes on solid supports needs further real-world experiments to evaluate the dye degradation rate. Ongoing research is also focused on identifying mixed microbial populations that can withstand stressors, such as metal ions and salts, commonly found in textile effluent. The metabolic pathways responsible for the biodegradation of textile effluent remain unclear. Therefore, identifying these pathways and understanding the underlying processes, by targeting metabolites and genes involved, is crucial. Moreover, gaining insights into the genetics of salt and metal tolerance could aid in identifying more potent microbial species that can efficiently decolorize various azo dyes from untreated textile wastewater containing mixed pollutants. More in-depth studies are necessary to better understand the mechanisms of dye decolorization, the genetic makeup of microorganisms involved in dye degradation, and the practical applications of new techniques for effective dye degradation.

Dye decolorization can yield useful by-products, such as energy sources and food, which can be exchanged as goods. However, the complete decolorization of dyes remains a significant challenge for scientists. Omics techniques, including meta-proteomics, meta-transcriptomics, metabolomics, and metagenomics, have revolutionized bioremediation by providing a deeper understanding of the processes and essential enzymes involved in the degradation of contaminants (Sharma et al. 2022). These technologies enhance the effectiveness of bioremediation processes by advancing our knowledge of the activities and dynamics of microbial communities in disturbed environmental conditions. In summary, biodecolorization offers numerous advantages over conventional methods, including the production of less waste, and being non-toxic, economical, and eco-friendly. Nevertheless, further research is required to improve the usability and effectiveness of biodegradation methods.

10.6 Conclusion

Textile wastewater containing dyes must be adequately treated before being discharged into the natural environment. The release of hazardous dyes into water bodies can have significant repercussions, such as disrupting sunlight absorption, affecting the metabolic processes of photosynthetic organisms, reducing the oxygen content in water, and disturbing the functioning of plants and animals. If these dyes come into contact with human skin or are consumed, they can pose serious risks to both humans and animals. Biotechnological applications have been widely employed for dye decolorization, primarily because biological methods are more effective and environmentally friendly. However, current bioremediation techniques are mostly conducted on a small laboratory scale, and a major technological challenge lies in integrating these methods for large-scale industrial applications. Several factors need to be considered when implementing bioremediation, including the type of waste, the toxicity of by-products, the intended use of the treated water, and the associated costs. In this chapter, we have reviewed the biodecolorization of synthetic textile dyes based on available scientific literature. To achieve optimal dye decolorization using specific microbial species, it is crucial to adjust all relevant conditions. Moreover, with the development of nanotechnology and advancements in omics research, the efficiency of bacterial enzymes in degrading hazardous textile dyes has significantly improved.

References

Adenan NH, Lim YY, Ting ASY (2022) Removal of triphenylmethane dyes by Streptomyces bacillaris: a study on decolorization, enzymatic reactions and toxicity of treated dye solutions. J Environ Manag 318:115520

Al-Fawwaz AT, Abdullah M (2016) Decolorization of methylene blue and malachite green by immobilized Desmodesmus sp. isolated from North Jordan. Int J Environ Sci Dev 7:95

Ali SS, Al-Tohamy R, Koutra E, El-Naggar AH, Kornaros M, Sun J (2021) Valorizing lignin-like dyes and textile dyeing wastewater by a newly constructed lipid-producing and lignin modifying oleaginous yeast consortium valued for biodiesel and bioremediation. J Hazard Mater 403:123575

Alinsafi A, Da Motta M, Le Bonté S, Pons M, Benhammou A (2006) Effect of variability on the treatment of textile dyeing wastewater by activated sludge. Dyes Pigments 69:31–39

Alprol AE, Heneash AMM, Ashour M, Abualnaja KM, Alhashmialameer D, Mansour AT et al (2021) Potential applications of arthrospira platensis lipid-free biomass in bioremediation of organic dye from industrial textile effluents and its influence on Marine rotifer (Brachionus plicatilis). Materials 14:4446

Alsafran M, Usman K, Ahmed B, Rizwan M, Saleem MH, Al Jabri H (2022) Understanding the phytoremediation mechanisms of potentially toxic elements: a proteomic overview of recent advances. Front Plant Sci 13:881242

Al-Tohamy R, Ali SS, Li F, Okasha KM, Mahmoud YA, Elsamahy T et al (2022) A critical review on the treatment of dye-containing wastewater: ecotoxicological and health concerns of textile dyes and possible remediation approaches for environmental safety. Ecotoxicol Environ Saf 231:113160

Amaral P, Fernandes D, Tavares A, Xavier A, Cammarota M, Coutinho J et al (2004) Decolorization of dyes from textile wastewater by Trametes versicolor. Environ Technol 25:1313–1320

Anastasi A, Parato B, Spina F, Tigini V, Prigione V, Varese GC (2011) Decolourisation and detoxification in the fungal treatment of textile wastewaters from dyeing processes. New Biotechnol 29:38–45

Andleeb S, Atiq N, Ali MI, Razi-Ul-Hussnain R, Shafique M, Ahmad B et al (2010) Biological treatment of textile effluent in stirred tank bioreactor. Int J Agric Biol 12:256–260

Assadi MM, Rostami K, Shahvali M, Azin M (2001) Decolorization of textile wastewater by Phanerochaete chrysosporium. Desalination 141:331–336

Ayed L, Ladhari N, Achour S, Chaieb K (2021) Decolorization of reactive yellow 174 dye in real textile wastewater by active consortium: experimental factorial design for bioremediation process optimization. J Text Inst 112:1449–1459

Bala S, Garg D, Thirumalesh BV, Sharma M, Sridhar K, Inbaraj BS et al (2022a) Recent strategies for bioremediation of emerging pollutants: a review for a green and sustainable environment. Toxics 10:484

Bala S, Sharma M, Dashora K, Siddiqui S, Diwan D, Tripathi M (2022b) Nanomaterials based sustainable bioenergy production systems: current trends and future prospects. Nanofabrication 7:314–324

Barathi S, Aruljothi KN, Karthik C, Padikasan IA, Ashokkumar V (2022) Biofilm mediated decolorization and degradation of reactive red 170 dye by the bacterial consortium isolated from the dyeing industry wastewater sediments. Chemosphere 286:131914

Barathikannan K, Ramasamy KP, Manohar CS, Meena RM (2017) Diversity and decolorization potential of fungi isolated from the coral reef regions off Kavaratti, India. Indian J Geo-Mar Sci 46:497–503

Bayat M, Salehi E, Mahdieh M (2023) Chromochloris zofingiensis microalgae as a potential dye adsorbent: adsorption thermo-kinetic, isothermal, and process optimization. Algal Res 71:103043

Bera SP, Tank SK (2021) Microbial degradation of procion red by Pseudomonas stutzeri. Sci Rep 11:3075

Bera SP, Shah MP, Godhaniya M (2022) Microbial remediation of textile dye acid orange by a novel bacterial Consortium SPB92. Front Environ Sci 10:1–8

Berradi M, Hsissou R, Khudhair M, Assouag M, Cherkaoui O, El Bachiri A et al (2019) Textile finishing dyes and their impact on aquatic environs. Heliyon 5:e02711

Biju LM, Veena Gayathri K, Senthil Kumar P, Kavitha R, Rajagopal R, Rangasamy G (2023) Application of Salvinia sps. in remediation of reactive mixed azo dyes and Cr (VI) - Its pathway elucidation. Environ Res 216:114635

Blánquez P, Sarrà M, Vicent T (2008) Development of a continuous process to adapt the textile wastewater treatment by fungi to industrial conditions. Process Biochem 43:1–7

Boulkhessaim S, Gacem A, Khan SH, Amari A, Yadav VK, Harharah HN et al (2022) Emerging trends in the remediation of persistent organic pollutants using nanomaterials and related processes: a review. Nanomaterials 12:2148

Bu T, Yang R, Zhang Y, Cai Y, Tang Z, Li C et al (2020) Improving decolorization of dyes by laccase from Bacillus licheniformis by random and site-directed mutagenesis. PeerJ 8:e10267

Cervantes FJ, Dos Santos AB (2011) Reduction of azo dyes by anaerobic bacteria: microbiological and biochemical aspects. Rev Environ Sci Biotechnol 10:125–137

Chang J-S, Kuo T-S, Chao Y-P, Ho J-Y, Lin P-J (2000) Azo dye decolorization with a mutant Escherichia coli strain. Biotechnol Lett 22:807–812

Chen K-C, Wu J-Y, Liou D-J, Hwang S-CJ (2003) Decolorization of the textile dyes by newly isolated bacterial strains. J Biotechnol 101:57–68

Chen G, Huang MH, Chen L, Chen DH (2011) A batch decolorization and kinetic study of Reactive Black 5 by a bacterial strain Enterobacter sp. GY-1. Int Biodeterior Biodegradation 65:790–796

Darwesh OM, Li H, Matter IA (2023) Nano-bioremediation of textile industry wastewater using immobilized CuO-NPs myco-synthesized by a novel Cu-resistant Fusarium oxysporum OSF18. Environ Sci Pollut Res Int 30:16694–16706

Das A (2022) Dye degradation using microbial fuel cell: a critical review. Int J Res Appl Sci Eng Technol 10:1745–1754

Das S, Cherwoo L, Singh R (2023) Decoding dye degradation: Microbial remediation of textile industry effluents. Biotechnol Notes 4:64–76

Dellamatrice PM, Silva-Stenico ME, Moraes LA, Fiore MF, Monteiro RT (2017) Degradation of textile dyes by cyanobacteria. Braz J Microbiol 48:25–31

Dhanve RS, Naidu CU, Jadhav JP (2023) In: Govindwar SP, Kurade MB, Jeon B-H, Pandey A (eds) Current developments in bioengineering and biotechnology. Elsevier, Amsteram, pp 429–455
Diwaniyan S, Kharb D, Raghukumar C, Kuhad RC (2010) Decolorization of synthetic dyes and textile effluents by basidiomycetous fungi. Water Air Soil Pollut 210:409–419
Dixit S, Garg S (2018) Biodegradation of environmentally hazardous azo dyes and aromatic amines using Klebsiella pneumoniae. J Environ Eng 144:04018035
Dong H, Guo T, Zhang W, Ying H, Wang P, Wang Y et al (2019) Biochemical characterization of a novel azoreductase from Streptomyces sp.: application in eco-friendly decolorization of azo dye wastewater. Int J Biol Macromol 140:1037–1046
D'souza DT, Tiwari R, Sah AK, Raghukumar C (2006) Enhanced production of laccase by a marine fungus during treatment of colored effluents and synthetic dyes. Enzym Microb Technol 38:504–511
Du L, Wu H, Li G, Wei Y, Wang F, Xu L et al (2023) Efficient degradation and decolorization of triphenylmethane dyes by Serratia sp. WKD under extreme environmental conditions and the mechanism. Int Biodeterior. Biodegradation 179:105565
Dutta S, Adhikary S, Bhattacharya S, Roy D, Chatterjee S, Chakraborty A et al (2024) Contamination of textile dyes in aquatic environment: adverse impacts on aquatic ecosystem and human health, and its management using bioremediation. J Environ Manag 353:120103
Ebihara K, Yoshikawa J, Horiguchi H, Amachi S (2022) Decolorization of cationic dyes under alkaline conditions by Iodidimonas sp. Q-1 multicopper oxidase. J Biosci Bioeng 133:323–328
Elmoutez S, Abushaban A, Necibi MC, Sillanpää M, Liu J, Dhiba D et al (2023) Design and operational aspects of anaerobic membrane bioreactor for efficient wastewater treatment and biogas production. Environ Chall 10:100671
Fareed A, Zaffar H, Bilal M, Hussain J, Jackson C, Naqvi TA (2022) Decolorization of azo dyes by a novel aerobic bacterial strain Bacillus cereus strain ROC. PLoS One 17:e0269559
Flores Alarcón MAD, Revilla Pacheco C, Garcia Bustos K, Tejada Meza K, Terán-Hilares F, Pacheco Tanaka DA et al (2022) Efficient dye removal from real textile wastewater using orange seed powder as suitable bio-adsorbent and membrane technology. Water 14:4104
Fomina M, Gadd GM (2014) Biosorption: current perspectives on concept, definition and application. Bioresour Technol 160:3–14
Franca RDG, Vieira A, Carvalho G, Oehmen A, Pinheiro HM, Barreto Crespo MT et al (2020) Oerskovia paurometabola can efficiently decolorize azo dye Acid Red 14 and remove its recalcitrant metabolite. Ecotoxicol Environ Saf 191:110007
Fu Y, Viraraghavan T (2001) Fungal decolorization of dye wastewaters: a review. Bioresour Technol 79:251–262
Gao Y, Wang M, Shah K, Singh Kalra S, Rome LH, Mahendra S (2022) Decolorization and detoxification of synthetic dye compounds by laccase immobilized in vault nanoparticles. Bioresour Technol 351:127040
Garg SK, Tripathi M (2017) Microbial strategies for discoloration and detoxification of azo dyes from textile effluents. Res J Microbiol 12:1–19
Garg D, Sridhar K, Stephen Inbaraj B, Chawla P, Tripathi M, Sharma M (2023) Nano-biofertilizer formulations for agriculture: a systematic review on recent advances and prospective applications. Bioengineering 10:1010
Gianolini JE, Britos CN, Mulreedy CB, Trelles JA (2020) Hyperstabilization of a thermophile bacterial laccase and its application for industrial dyes degradation. 3 Biotech 10:288
Guo G, Tian F, Zhao Y, Tang M, Liu W, Liu C et al (2019) Aerobic decolorization and detoxification of Acid Scarlet GR by a newly isolated salt-tolerant yeast strain Galactomyces geotrichum GG. Int Biodeterior Biodegradation 145:104818
Guo G, Hao J, Tian F, Liu C, Ding K, Xu J et al (2020) Decolorization and detoxification of azo dye by halo-alkaliphilic bacterial consortium: systematic investigations of performance, pathway and metagenome. Ecotoxicol Environ Saf 204:111073
Gupta V (2009) Application of low-cost adsorbents for dye removal–a review. J Environ Manag 90:2313–2342

Han S, Han W, Chen J, Sun Y, Dai M, Zhao G (2020) Bioremediation of malachite green by cyanobacterium Synechococcus elongatus PCC 7942 engineered with a triphenylmethane reductase gene. Appl Microbiol Biotechnol 104:3193–3204

Imron MF, Kurniawan SB, Soegianto A, Wahyudianto FE (2019) Phytoremediation of methylene blue using duckweed (Lemna minor). Heliyon 5:e02206

Islam MM, Mahmud K, Faruk O, Billah M (2011) Textile dyeing industries in Bangladesh for sustainable development. Int J Environ Sci Dev 2:428

Jadhav SU, Kalme SD, Govindwar SP (2008) Biodegradation of methyl red by galactomyces geotrichum MTCC 1360. Int Biodeterior Biodegradation 62:135–142

Jafari N, Soudi MR, Kasra-Kermanshahi R (2014) Biodegradation perspectives of azo dyes by yeasts. Microbiology 83:484–497

Jain A, He Z (2018) Cathode-enhanced wastewater treatment in bioelectrochemical systems. npj Clean Water 1:23

Jalandoni-Buan AC, Decena-Soliven ALA, Cao EP, Barraquio VL, Barraquio WL (2010) Characterization and identification of Congo red decolorizing bacteria from monocultures and consortia. Philipp J Sci 139:71–78

Jamee R, Siddique R (2019) Biodegradation of synthetic dyes of textile effluent by microorganisms: an environmentally and economically sustainable approach. Eur J Microbiol Immunol 9:114–118

Jang MS, Jung BG, Sung NC, Lee YC (2007) Decolorization of textile plant effluent by Citrobacter sp. strain KCTC 18061P. J Gen Appl Microbiol 53:339–343

Jin XC, Liu GQ, Xu ZH, Tao WY (2007) Decolorization of a dye industry effluent by Aspergillus fumigatus XC6. Appl Microbiol Biotechnol 74:239–243

Joshi A, Hinsu A, Kothari R (2022) Evaluating the efficacy of bacterial consortium for decolorization of diazo dye mixture. Arch Microbiol 204:515

Joshni TC, Subramaniam K (2011) Enzymatic degradation of azo dyes-a review. Int J Environ Sci 1:1250–1260

Kalnake RP, Raval R, Murthy DVR, Vanzara PB, Raval K (2023) Enhanced degradation of azo dye using mixed cultures of white-rot fungi in a modified rotating packed disc bioreactor and reuse of treated water. Bioresour Technol Rep 22:101449

Kalsoom U, Bhatti HN, Aftab K, Amin F, Jesionowski T, Bilal M (2023) Biocatalytic potential of Brassica oleracea L. var. botrytis leaves peroxidase for efficient degradation of textile dyes in aqueous medium. Bioprocess Biosyst Eng 46:453–465

Kalyani DC, Phugare SS, Shedbalkar UU, Jadhav JP (2011) Purification and characterization of a bacterial peroxidase from the isolated strain Pseudomonas sp. SUK1 and its application for textile dye decolorization. Ann Microbiol 61:483–491

Kameche K, Amrani S, Mouzaoui S, Aït-Amar H (2022) Biodegradation of diazo dye Evans blue by four strains of Streptomyces isolated from soils of Algeria. Biocatal Agric Biotechnol 46:102529

Kaushik P, Malik A (2010) Effect of nutritional conditions on dye removal from textile effluent by Aspergillus lentulus. World J Microbiol Biotechnol 26:1957–1964

Khan S, Malik A (2018) Toxicity evaluation of textile effluents and role of native soil bacterium in biodegradation of a textile dye. Environ Sci Pollut Res Int 25:4446–4458

Khan R, Bhawana P, Fulekar MH (2013) Microbial decolorization and degradation of synthetic dyes: a review. Rev Environ Sci Biotechnol 12:75–97

Kishor R, Purchase D, Saratale GD, Saratale RG, Ferreira LFR, Bilal M et al (2021) Ecotoxicological and health concerns of persistent coloring pollutants of textile industry wastewater and treatment approaches for environmental safety. J Environ Chem Eng 9:105012

Kour D, Kaur T, Devi R, Yadav A, Singh M, Joshi D et al (2021) Beneficial microbiomes for bioremediation of diverse contaminated environments for environmental sustainability: present status and future challenges. Environ Sci Pollut Res Int 28:24917–24939

Liu L, Bilal M, Duan X, Iqbal HMN (2019) Mitigation of environmental pollution by genetically engineered bacteria - Current challenges and future perspectives. Sci Total Environ 667:444–454

López-Miranda JL, Silva R, Molina GA, Esparza R, Hernandez-Martinez A, Hernández-Carteño J et al (2020) Evaluation of a dynamic bioremediation system for the removal of metal ions and toxic dyes using Sargassum Spp. J Mar Sci Eng 8:899
Louati I, Elloumi-Mseddi J, Cheikhrouhou W, Hadrich B, Nasri M, Aifa S et al (2020) Simultaneous cleanup of Reactive Black 5 and cadmium by a desert soil bacterium. Ecotoxicol Environ Saf 190:110103
Mahajan P, Kaushal J (2020) Phytoremediation of azo dye methyl red by macroalgae Chara vulgaris L.: kinetic and equilibrium studies. Environ Sci Pollut Res Int 27:26406–26418
Mahalakshmi S, Lakshmi D, Menaga U (2015) Biodegradation of different concentration of dye (Congo red dye) by using green and blue green algae. Int J Environ Res 9:735–744
Mansour AT, Alprol AE, Abualnaja KM, El-Beltagi HS, Ramadan KMA, Ashour M (2022) The using of nanoparticles of microalgae in remediation of toxic dye from industrial wastewater: kinetic and isotherm studies. Materials 15:3922
Manzoor K, Khan SJ, Yasmeen M, Jamal Y, Arshad M (2022) Assessment of anaerobic membrane distillation bioreactor hybrid system at mesophilic and thermophilic temperatures treating textile wastewater. J Water Process Eng 46:102603
Miranda RDCMD, Gomes EDB, Pereira N, Marin-Morales MA, Machado KMG, Gusmão NBD (2013) Biotreatment of textile effluent in static bioreactor by Curvularia lunata URM 6179 and Phanerochaete chrysosporium URM 6181. Bioresour Technol 142:361–367
Misal SA, Lingojwar DP, Shinde RM, Gawai KR (2011) Purification and characterization of azoreductase from alkaliphilic strain Bacillus badius. Process Biochem 46:1264–1269
Modi A, Singh S, Patki J, Padmadas N (2022) Screening and identification of azo dye decolorizers from mangrove rhizospheric soil. Environ Sci Pollut Res 29:83496–83511
Montañez-Barragán B, Sanz-Martín JL, Gutiérrez-Macías P, Morato-Cerro A, Rodríguez-Vázquez R, Barragán-Huerta BE (2020) Azo dyes decolorization under high alkalinity and salinity conditions by Halomonas sp. in batch and packed bed reactor. Extremophiles 24:239–247
Moyo S, Makhanya BP, Zwane PE (2022) Use of bacterial isolates in the treatment of textile dye wastewater: a review. Heliyon 8:e09632
Mu Y, Rabaey K, Rozendal RA, Yuan Z, Keller J (2009) Decolorization of azo dyes in bioelectrochemical systems. Environ Sci Technol 43:5137–5143
Mustafa G, Tariq Zahid M, Ali S, Zaghum Abbas S, Rafatullah M (2021) Biodegradation and discoloration of disperse blue-284 textile dye by Klebsiella pneumoniae GM-04 bacterial isolate. J King Saud Univ Sci 33:101442
Navas LE, Carballo R, Levin L, Berretta MF (2020) Fast decolorization of azo dyes in alkaline solutions by a thermostable metal-tolerant bacterial laccase and proposed degradation pathways. Extremophiles 24:705–719
Neihsial R, Singha NA, Singh AK (2022) Taxonomic diversity and predictive metabolic functions of a heavy metal tolerant multiple azo dye degrading bacterial consortium from textile effluents. Int Biodeterior Biodegradation 171:105421
Ngo ACR, Tischler D (2022) Microbial degradation of azo dyes: approaches and prospects for a hazard-free conversion by microorganisms. Int J Environ Res Public Health 19:4740
Nguyen TA, Fu CC, Juang RS (2016) Biosorption and biodegradation of a sulfur dye in high-strength dyeing wastewater by Acidithiobacillus thiooxidans. J Environ Manag 182:265–271
Nordin N, Amir SFM, Riyanto OMR (2013) Textile industries wastewater treatment by electrochemical oxidation technique using metal plate. Int J Electrochem Sci 8:11403–11415
Ortiz-Monsalve S, Valente P, Poll E, Jaramillo-García V, JaP H, Gutterres M (2019) Biodecolourization and biodetoxification of dye-containing wastewaters from leather dyeing by the native fungal strain Trametes villosa SCS-10. Biochem Eng J 141:19–28
Patel H, Yadav VK, Yadav KK, Choudhary N, Kalasariya H, Alam MM et al (2022) A recent and systemic approach towards microbial biodegradation of dyes from textile industries. Water 14:3163
Periyasamy AP (2024) Recent advances in the remediation of textile-dye-containing wastewater: prioritizing human health and sustainable wastewater treatment. Sustainability 16:495

Petersen L, Heynen M, Pellicciotti F (n.d.) International encyclopedia of geography. Wiley, Hobojen, NJ, pp 1–12

Pham LTM, Seo H, Kim K-J, Kim YH (2018) In silico-designed lignin peroxidase from Phanerochaete chrysosporium shows enhanced acid stability for depolymerization of lignin. Biotechnol Biofuels 11:1–13

Pourbabaee AA, Malekzadeh F, Sarbolouki MN, Najafi F (2006) Aerobic decolorization and detoxification of a disperse dye in textile effluent by a new isolate of Bacillus sp. Biotechnol Bioeng 93:631–635

Prasetyo H, MNAM N, MHD O, Jaafar J, Yoshioka T, Li Z et al (2022) Technologies for treating wastewater from textile industry: a review. Mater Today Proc 65:3066–3072

Radhika B, Aruna K (2022) Elucidating a pathway for degradation of azo dye reactive red 120 BY bacterial consortium. J Appl Biol Sci 16:396–417

Raja MMM, Raja A, Salique SM, Gajalakshmi P (2016) Studies on effect of marine actinomycetes on amido black (azo dye) decolorization. J Chem Pharm Res 8:640–644

Rajamohan N, Rajasimman M (2013) Kinetic modeling of dye effluent biodegradation by Pseudomonas stutzeri. Eng Technol Appl Sci Res 3:387–390

Rane NR, Chandanshive VV, Watharkar AD, Khandare RV, Patil TS, Pawar PK et al (2015) Phytoremediation of sulfonated remazol red dye and textile effluents by Alternanthera philoxeroides: an anatomical, enzymatic and pilot scale study. Water Res 83:271–281

Rani M, Shanker U (2023) In: Shanker U, Hussain CM, Rani M (eds) Handbook of green and sustainable nanotechnology: Fundamentals, developments and applications. Springer, Cham, pp 1267–1298

Rita De Cássia M, De Barros GE, Pereira N Jr, Marin-Morales MA, Machado KMG, De Gusmão NB (2013) Biotreatment of textile effluent in static bioreactor by Curvularia lunata URM 6179 and Phanerochaete chrysosporium URM 6181. Bioresour Technol 142:361–367

Saratale RG, Saratale GD, Chang JS, Govindwar SP (2011) Bacterial decolorization and degradation of azo dyes: a review. J Taiwan Inst Chem Eng 42:138–157

Saravanan S, Carolin CF, Kumar PS, Chitra B, Rangasamy G (2022) Biodegradation of textile dye Rhodamine-B by Brevundimonas diminuta and screening of their breakdown metabolites. Chemosphere 308:136266

Sarkar S, Banerjee A, Halder U, Biswas R, Bandopadhyay R (2017) Degradation of synthetic azo dyes of textile industry: a sustainable approach using microbial enzymes. Water Conserv Sci Eng 2:121–131

Sedighi M, Karimi A, Vahabzadeh F (2009) Involvement of ligninolytic enzymes of Phanerochaete chrysosporium in treating the textile effluent containing Astrazon Red FBL in a packed-bed bioreactor. J Hazard Mater 169:88–93

Senthil Rathi B, Senthil Kumar P (2022) Sustainable approach on the biodegradation of azo dyes: a short review. Curr Opin Green Sustain Chem 33:100578

Shah MP, Rodriguez-Couto S (2022) Development in wastewater treatment research and processes: Microbial degradation of xenobiotics through bacterial and fungal approach. Elsevier, Amsterdam

Shahid M, Mahmood F, Hussain S, Shahzad T, Haider MZ, Noman M et al (2018) Enzymatic detoxification of azo dyes by a multifarious Bacillus sp strain MR-1/2-bearing plant growth-promoting characteristics. 3 Biotech 8:425

Shamim Hossain M, Kumar Paul G, Mahmud S, Abu Saleh M, Salah Uddin M, Kumar Dutta A et al (2022) Mixed dye degradation by Bacillus pseudomycoides and Acinetobacter haemolyticus isolated from industrial effluents: a combined affirmation with wetlab and in silico studies. Arab J Chem 15:104078

Shanmugam S, Ulaganathan P, Sivasubramanian S, Esakkimuthu S, Krishnaswamy S, Subramaniam S (2017) Trichoderma asperellum laccase mediated crystal violet degradation–optimization of experimental conditions and characterization. J Environ Chem Eng 5:222–231

Sharma P, Singh SP, Iqbal HMN, Tong YW (2022) Omics approaches in bioremediation of environmental contaminants: an integrated approach for environmental safety and sustainability. Environ Res 211:113102

Sharma JK, Kumar N, Singh NP, Santal AR (2023) Phytoremediation technologies and their mechanism for removal of heavy metal from contaminated soil: an approach for a sustainable environment. Front Plant Sci 14:1076876
Sheela T, Sadasivam SK (2020) Dye degradation potential and its degradative enzymes synthesis of Bacillus cereus SKB12 isolated from a textile industrial effluent. J Appl Biol Biotechnol 8:042–046
Shetty K, Krishnakumar G (2020) Algal and cyanobacterial biomass as potential dye biodecolorizing material: a review. Biotechnol Lett 42:2467–2488
Shi Y, Yang Z, Xing L, Zhou J, Ren J, Ming L et al (2021) Ethanol as an efficient cosubstrate for the biodegradation of azo dyes by Providencia rettgeri: mechanistic analysis based on kinetics, pathways and genomics. Bioresour Technol 319:124117
Shin K-S (2004) The role of enzymes produced by white-rot fungus Irpex lacteus in the decolorization of the textile industry effluent. J Microbiol 42:37–41
Silva F, Nascimento L, Brito M, Da Silva K, Paschoal W Jr, Fujiyama R (2019) Biosorption of methylene blue dye using natural biosorbents made from weeds. Materials 12:2486
Singh G, Dwivedi S (2020) Decolorization and degradation of Direct Blue-1 (Azo dye) by newly isolated fungus Aspergillus terreus GS28, from sludge of carpet industry. Environ Technol Innov 18:100751
Sinha S, Nigam S, Singh R (2015) Potential of Nostoc muscorum for the decolorisation of textiles dye RGB-red. Int J Pharm Bio Sci 6:1092–1100
Sinha S, Singh R, Chaurasia AK, Nigam S (2016) Self-sustainable Chlorella pyrenoidosa strain NCIM 2738 based photobioreactor for removal of Direct Red-31 dye along with other industrial pollutants to improve the water-quality. J Hazard Mater 306:386–394
Sridharan R, Krishnaswamy VG, Archana K, Rajagopal R, Thirumal Kumar D, George Priya Doss C (2021) Integrated approach on azo dyes degradation using laccase enzyme and Cul nanoparticle. SN Appl Sci 3:1–12
Srinivasan S, Bankole PO, Sadasivam SK (2022) Biodecolorization and degradation of textile azo dyes using Lysinibacillus sphaericus MTCC 9523. Front Environ Sci 10:990855
Srivastava AK, Singh RK, Singh D (2021) In: Kumar A, Singh VK, Singh P, Mishra VK (eds) Microbe mediated remediation of environmental contaminants. Cambridge, Woodhead Publishing, pp 241–253
Sumathi S, Manju BS (2000) Uptake of reactive textile dyes by Aspergillus foetidus. Enzym Microb Technol 27:347–355
Sun W, Sun W, Wang Y (2019) Biosorption of Direct Fast Scarlet 4BS from aqueous solution using the green-tide-causing marine algae Enteromorpha prolifera. Spectrochim Acta A Mol Biomol Spectrosc 223:117347
Sun L, Mo Y, Zhang L (2022) A mini review on bio-electrochemical systems for the treatment of azo dye wastewater: state-of-the-art and future prospects. Chemosphere 294:133801
Teja M, Kumar DM, Bindiya P, Sudhakar G (2022) New innovations in chemistry and biochemistry, vol. 7. Book Publisher International (a part of SCIENCEDOMAIN International
Tiwari H, Sonwani RK, Singh RS (2023) Bioremediation of dyes: a brief review of bioreactor performance. Environ Technol Rev 12:83–128
Tripathi M, Singh DN (2022) Bioremediation: challenges and advancements. Bentham Science Publishers, Sharjah
Tripathi M, Pathak N, Chaudhary VK, Singh P, Singh PK, Thirumalesh BV et al (2023a) Microbial decolorization of crystal violet dye by a native multi-metal tolerant aeromonas caviae MT-1 isolate from dye-contaminated soil: optimization and phytotoxicity study. Toxicol Int 30:83–93
Tripathi M, Singh S, Pathak S, Kasaudhan J, Mishra A, Bala S et al (2023b) Recent strategies for the remediation of textile dyes from wastewater: a systematic review. Toxics 11:940
Tripathi M, Singh P, Singh R, Bala S, Pathak N, Singh S et al (2023c) Microbial biosorbent for remediation of dyes and heavy metals pollution: a green strategy for sustainable environment. Front Microbiol 14:1168954
Uduma RC, Oguzie KL, Chijioke CF, Ogbulie TE, Oguzie EE (2023) Bioelectrochemical technologies for simultaneous treatment of dye wastewater and electricity generation: a review. Int J Sci Environ Technol 20:10415–10434

Valentín L, Lu-Chau TA, López C, Feijoo G, Moreira MT, Lema JM (2007) Biodegradation of dibenzothiophene, fluoranthene, pyrene and chrysene in a soil slurry reactor by the white-rot fungus Bjerkandera sp. BOS55. Process Biochem 42:641–648

Vatandoostarani S, Bagheri Lotfabad T, Heidarinasab A, Yaghmaei S (2017) Degradation of azo dye methyl red by Saccharomyces cerevisiae ATCC 9763. Int Biodeterior Biodegradation 125:62–72

Velusamy K, Periyasamy S, Kumar PS, Femina Carolin C, Jayaraj T, Gokulakrishnan M et al (2022) Transformation of aqueous methyl orange to green metabolites using bacterial strains isolated from textile industry effluent. Environ Technol Innov 25:102126

Viana DG, Egreja Filho FB, Pires FR, Soares MB, Ferreira AD, Bonomo R et al (2021) In situ barium phytoremediation in flooded soil using Typha domingensis under different planting densities. Ecotoxicol Environ Saf 210:111890

Vignesh A, Manigundan K, Santhoshkumar J, Shanmugasundaram T, Gopikrishnan V, Radhakrishnan M et al (2020) Microbial degradation, spectral analysis and toxicological assessment of malachite green by Streptomyces chrestomyceticus S20. Bioprocess Biosyst Eng 43:1457–1468

Vineh MB, Saboury AA, Poostchi AA, Ghasemi A (2020) Biodegradation of phenol and dyes with horseradish peroxidase covalently immobilized on functionalized RGO-SiO(2) nanocomposite. Int J Biol Macromol 164:4403–4414

Vitor V, Corso CR (2008) Decolorization of textile dye by Candida albicans isolated from industrial effluents. J Ind Microbiol Biotechnol 35:1353–1357

Wang R, Li H, Liu Y, Chen J, Peng F, Jiang Z et al (2022) Efficient removal of azo dyes by Enterococcus faecalis R1107 and its application in simulated textile effluent treatment. Ecotoxicol Environ Saf 238:113577

Wei Z, Van Le Q, Peng W, Yang Y, Yang H, Gu H et al (2021) A review on phytoremediation of contaminants in air, water and soil. J Hazard Mater 403:123658

Wesenberg D, Kyriakides I, Agathos SN (2003) White-rot fungi and their enzymes for the treatment of industrial dye effluents. Biotechnol Adv 22:161–187

Yadav A, Kumar P, Rawat D, Garg S, Mukherjee P, Farooqi F et al (2022) Microbial fuel cells for mineralization and decolorization of azo dyes: Recent advances in design and materials. Sci Total Environ 826:154038

Yang HY, Jia RB, Chen B, Li L (2014) Degradation of recalcitrant aliphatic and aromatic hydrocarbons by a dioxin-degrader Rhodococcus sp. strain p52. Environ Sci Pollut Res Int 21:11086–11093

Yaou Balarabe B, Illiassou Oumarou MN, Koroney AS, Adjama I, Ibrahim Baraze AR (2023) Photo-oxidation of organic dye by Fe2O3 nanoparticles: catalyst, electron acceptor, and polyurethane membrane (PU-Fe2O3) effects. J Nanotechnol 2023:1292762

Yaqoob AA, Khatoon A, Mohd Setapar SH, Umar K, Parveen T, Mohamad Ibrahim MN et al (2020) Outlook on the role of microbial fuel cells in remediation of environmental pollutants with electricity generation. Catalysts 10:819

Yaseen DA, Scholz M (2019) Textile dye wastewater characteristics and constituents of synthetic effluents: a critical review. Int J Sci Environ Technol 16:1193–1226

Younis SA, Serp P, Nassar HN (2021) Photocatalytic and biocidal activities of ZnTiO(2) oxynitride heterojunction with MOF-5 and g-C(3)N(4): a case study for textile wastewater treatment under direct sunlight. J Hazard Mater 410:124562

Zhou W, Zimmermann W (1993) Decolorization of industrial effluents containing reactive dyes by actinomycetes. FEMS Microbiol Lett 107:157–161

Zhuo R, Ma L, Fan F, Gong Y, Wan X, Jiang M et al (2011) Decolorization of different dyes by a newly isolated white-rot fungi strain Ganoderma sp. En3 and cloning and functional analysis of its laccase gene. J Hazard Mater 192:855–873

Zimmermann T, Kulla HG, Leisinger T (1982) Properties of purified Orange II azoreductase, the enzyme initiating azo dye degradation by Pseudomonas KF46. Eur J Biochem 129:197–203

GPSR Compliance

The European Union's (EU) General Product Safety Regulation (GPSR) is a set of rules that requires consumer products to be safe and our obligations to ensure this.

If you have any concerns about our products, you can contact us on ProductSafety@springernature.com

In case Publisher is established outside the EU, the EU authorized representative is:

Springer Nature Customer Service Center GmbH
Europaplatz 3
69115 Heidelberg, Germany

Batch number: 10755332

Printed by Printforce, the Netherlands